General Relativity and Gravitation

General Relativity and Gravitation

Proceedings of the 11th International Conference on General Relativity and Gravitation, Stockholm, July 6-12, 1986

Edited by
M.A.H. MacCallum, Queen Mary College, London

The right of the University of Cambridge to print and sell all manner of books was granted by Henry VIII in 1534. The University has printed and published continuously since 1584.

CAMBRIDGE UNIVERSITY PRESS

Cambridge

New York New Rochelle Melbourne Sydney

Published by the Press Syndicate of the University of Cambridge
The Pitt Building, Trumpington Street, Cambridge CB2 1RP
32 East 57th Street, New York, NY 10022, USA
10, Stamford Road, Oakleigh, Melbourne 3166, Australia

First Published 1987

Printed in Great Britain at the University Press, Cambridge

Library of Congress cataloging in publication data:

International Conference on General Relativity and Gravitation
(11th: 1986: Stockholm, Sweden) General relativity and gravitation.
Bibliography: p.
1. General Relativity (Physics) – Congresses. 2. Gravitation – Congresses.
3. Astrophysics – Congresses.
I. MacCallum, M.A.H. II. Title
QCC173.6.I573 1986 530.1'1 87-9351

British Library cataloguing in publication data:

International Conference on General Relativity and Gravitation
(11th: 1986: Stockholm) General relativity and gravitation: proceedings of the 11th International Conference on General Relativity and Gravitation, Stockholm, 6 - 12 July 1986.
1. General Relativity (Physics) 2. Gravitation
I. Title II. MacCallum, M.A.H.
531'.14 QC173.6

ISBN 0 521 33296 6 hard covers

CONTENTS

EDITOR'S PREFACE

One well-known relativist commented during GR11 that it was the best-organised meeting he had ever attended (and he had been to many). The credit for this must go to the very hard-working local committee, and to the professional staff of the RESO Congress Service (led by Mona Moback and Marianne Hanson) who handled all the practical aspects. The local committee was in fact rather small - just three members, all from the Institute of Theoretical Physics of the University of Stockholm: Bertel Laurent, the chairman and chief begetter of the whole meeting, Ulf Lindstrom, and the very capable organisational and scientific secretary, Kjell Rosquist. I know that they would also wish to thank the President of the Society on GRG, Dennis Sciama, and its Secretary, Alan Held, for their advice and help.

I am also aware that a large number of the staff and students of the Institute of Theoretical Physics put in an immense amount of voluntary (and in some cases involuntary?) work, both beforehand and during the meeting. The technical support they provided in sessions was exemplary: any hiatus of more than a few seconds brought out one of the hurrying figures in a GR11 T-shirt. Unfortunately, I do not have anything like a complete list of their names to record here, but Bertel and Kjell in their foreword to the Abstracts mentioned Michael Bradley, Johan Grundberg, Anders Karlhede, Hans Rang and Jan Aman in particular.

My own task, and that of my fellow members of the Scientific organising committee, was a much easier one, made easier still by the excellent local organisation. We had only to decide on the form of the programme and invite the speakers and chairmen, who were then faced with the real effort.

The meeting consisted of plenary lectures, given in the mornings, and symposia based on contributed papers, which were held mainly in the afternoon. There were well over 500 abstracts submitted for the meeting. All these, except those which failed to meet the time deadlines or space limitations, and those which the organisers felt were clearly of unacceptable quality, were reproduced in the two volumes of Abstracts distributed to all participants (*Abstracts of contributed papers, 11th International Conference on General Relativity and Gravitation,* ed. B. Laurent and K. Rosquist, 1986, of which copies may still be available from Stockholm).

All contributors were also invited to display a poster, and the excellent facilities of the Folkets Hus enabled these to be on show for the whole of the meeting, providing a focal point for those who wished to pursue informal discussion of particular topics.

The number of papers, as had been anticipated, was clearly too great for all to be given any reasonable amount of time for verbal presentation. Therefore the chairmen of the symposia were faced with the difficult task of selecting among the contributors those who were to actually speak. This was thankless, as one can only make enemies and not win friends in such a situation. Equally, the contents of all these papers could not possibly be communicated to the whole meeting or reproduced in this book, and so the symposia chairmen were charged with two further demanding tasks: to give a brief review of the major points at the closing session, and to produce a report for this book. We were extremely fortunate that the people involved made an excellent job of these onerous duties, and it is a pleasure for me, on behalf of the Scientific Committee, to thank them for all their work.

We were also, of course, grateful to the plenary lecturers, who gave an exciting series of talks on varied topics, the diversity of which still betrayed an underlying unity of purpose in the field. I find it interesting how strong the links are even between the apparently most widely-separated branches of the subject, and I hope this comes out to some extent in the contributions in this volume.

Reproduced here are the texts of the plenary lectures, together with the reports of the symposia chairmen. There are a couple of exceptions: with the agreement of its chairman, David Boulware of the University of Washington in Seattle, the report of the symposium on Quantum Gravity and Quantum Cosmology has been incorporated into the written version of Chris Isham's plenary talk on the same area, and no report of the symposium on Relativistic Astrophysics was received in time for publication. I have put the plenary lectures first, in alphabetical order of the speaker's name, and the symposia reports second, in alphabetical order of symposium title (which is almost the same as the numerical order of the codes assigned to the symposia in the Abstracts volume), thus avoiding any attempt to draw demarcation lines between sub-areas of the subject of the meeting.

As already mentioned, Folkets Hus, with its very large circulation area (eminently useful in the coffee breaks), and many smaller rooms for committee meetings or informal discussions (as well as book exhibitions, a computer show, and a cafeteria) proved an admirable site, especially as it was also within convenient reach of many hotels and well placed for those who wanted to stroll round the more scenic parts of Stockholm during their breaks. Participants also enjoyed a reception at the Stockholm City Hall, and a tour of the open air museum, Skansen, and a dinner there, as well as a boat trip through the Stockholm archipelago to Vaxholm, where we picnicked at the old fort (the organisers managing to schedule the torrential rain just while we were on the return boat trip, not in the open).

A meeting of this size (about 600 participants), with speakers from all over the world, is expensive to mount. We are therefore happy to acknowledge the generous support of the financial sponsors whose names appear on a subsequent page, and to the Honorary Committee and Gunnar Hoppe who helped us to obtain financial and other support.

We are also grateful to the Royal Swedish Academy of Sciences, who acted as our main scientific sponsors (in collaboration, of course, with the International Society on General Relativity and Gravitation) and to its President, Professor Sven Johansson, who opened the meeting and welcomed the visitors.

Thanks are also due to the chairmen of the plenary sessions, who, apart from myself and Bertel Laurent, were Peter Bergmann (Syracuse University), Isaac Khalatnikov (Landau Institute, Moscow) and Dennis Sciama (SISSA, Trieste), the outgoing President of GRG.

I owe a personal debt to the members of the Scientific Organising Committee (listed on the next page) whose constant flow of pertinent and imaginative suggestions enabled us to put together a most enjoyable conference programme, to the editors at Cambridge University Press, especially Rufus Neal, Simon Mitton and David Tranah, and to my friends in Stockholm, particularly Inge Frick, Ian Cohen, Jan Aman, Kjell Rosquist, Anders Karlhede, and Bertel Laurent, for their hospitality and help during the meeting and its preceding organisation.

GR11 Scientific Organising Committee

Malcolm A.H. MacCallum (Chairman), Queen Mary College, University of London
Allen J. Anderson, University of Uppsala
Abhay Ashtekar, Syracuse University
Bruno Bertotti, University of Pavia
Vladimir B. Braginskii, Moscow State University
Yvonne Choquet-Bruhat, University of Paris VI
Thibault Damour, Paris Observatory, Meudon
George F.R. Ellis, University of Capetown
John Ellis, CERN
Leonid P. Grishchuk, Sternberg Astronomical Institute, Moscow
Alan Held, University of Bern (Secretary, GRG)
Christopher J. Isham, Imperial College, University of London
Bertel Laurent, University of Stockholm
Andrei D. Linde, Lebedev Institute, Moscow
Ulf Lindstrom, University of Stockholm
Martin J. Rees, University of Cambridge
Dennis W. Sciama, SISSA, Trieste (President, GRG)
Larry Smarr, University of Illinois
Andrzej Trautman, Academy of Sciences, Warsaw
C.V. Vishveshwara, Raman Research Institute, Bangalore
J. Winicour, University of Pittsburgh

Honorary Committee

Professor Kerstin Fredga, Stockholm Observatory
Professor Staffan Helmfrid, Rector of the University of Stockholm
Professor Bengt Nagel, Royal Institute of Technology, Stockholm
Professor Olof Rydbeck, Onsala Space Observatory
Professor Kai Siegbahn, University of Uppsala

GR11 Financial Sponsors

Main Financial Sponsor
The Marcus Wallenberg Foundation for International Cooperation in Science

Other Financial Sponsors
The Swedish Natural Science Research Council
The Nobel Foundation
The City of Stockholm
The International Society on General Relativity and Gravitation
The Swedish Department of Education
The Marianne and Marcus Wallenberg Foundation
International Union of Pure and Applied Physics (IUPAP)
Nordisk Institut fur Teoretisk Atomfysik (NORDITA)
The University of Stockholm
The Scandinavian Airlines System (SAS)
The Swedish Institute

CONTRIBUTORS TO THIS BOOK

Marek A. Abramowicz, Scuola Internazionale Superiore de Studi Avanzati (SISSA), Strada Costiera 11, 34014 Trieste, ITALY.

James M. Bardeen, Department of Physics, University of Washington, Seattle, WA 98195, U.S.A.

Dieter Brill, Department of Physics and Astronomy, University of Maryland, College Park, MD 20742, U.S.A.

Lars Brink, Institute of Theoretical Physics, Chalmers University of Technology, Fack, S-40220 Goteborg 5, SWEDEN, and Institute for Theoretical Physics, University of California, Santa Barbara, CA 93106, U.S.A.

Yvonne Choquet-Bruhat, Mécanique Relativiste, Institut de Mécanique, Université de Paris VI, Tour 66, 4 Place Jussieu, 75252 Paris Cedex 05, FRANCE.

Christopher J.S. Clarke, Faculty of Mathematical Studies, University of Southampton, Southampton SO9 5NH. U.K.

Michael J. Duff, Theory Division, CERN, CH-1211, Geneva 23, SWITZERLAND.

Jürgen Ehlers, Max-Planck-Institut für Physik und Astrophysik, Institut für Astrophysik, Karl Schwarzschild Strasse 1, D-8046 Garching bei München, F.R.G.

George F.R. Ellis, Department of Applied Mathematics, University of Capetown, Rondebosch, 7700 C.P., SOUTH AFRICA, and School of Mathematical Sciences, Queen Mary College, University of London, Mile End Road, London E1 4NS, U.K.

William M. Fairbank, Department of Physics, Stanford University, Stanford, CA 94305-2196, U.S.A.

Inge Frick, Institute of Theoretical Physics, University of Stockholm, Vanadisvagen 9, S-113 46 Stockholm, SWEDEN.

Hubert F.M. Goenner, Institute of Theoretical Physics, University of Göttingen, Bunsenstrasse 9, D-3400 Göttingen, F.R.G.

Joshua N. Goldberg, Department of Physics, Syracuse University, Syracuse, NY 13244-1130, U.S.A.

Michael B. Green, Department of Physics, Queen Mary College, University of London, Mile End Road, London E1 4NS, U.K.

Leonid P. Grishchuk, Sternberg Astronomical Institute, 119899 Moscow V-234, U.S.S.R.

Petr Hajicek, Institut für Theoretische Physik, University of Bern, Sidlerstrasse 5, CH-3012 Bern, SWITZERLAND.

Christopher J. Isham, Department of Theoretical Physics, The Blackett Laboratory, Imperial College, University of London, Prince Consort Road, London, SW7 2BZ, U.K.

Dietrich Kramer, Sektion Physik, Friedrich-Schiller-Universität Jena, Max Wien Platz 1, DDR-6900 Jena, G.D.R..

Pawel O. Mazur, Department of Physics, Syracuse University, Syracuse, NY 13244-1130, U.S.A.

Takashi Nakamura, Department of Physics, Kyoto University, Kyoto 606, JAPAN.

Dmitri V. Nanopoulos, Department of Physics, University of Wisconsin-Madison, Madison, WI 53706, U.S.A.

Ho Jung Paik, Department of Physics and Astronomy, University of Maryland, College Park, MD 20742, U.S.A.

Roger Penrose, Mathematical Institute, 24-29 St. Giles, Oxford OX1 3BD, U.K.

Tsvi Piran, Racah Institute for Physics, The Hebrew University, Jerusalem, ISRAEL, and Institute of Advanced Study, Princeton, NJ 08544, U.S.A.

Robert D. Reasenberg, Smithsonian Astrophysical Observatory, Center for Astrophysics, 60 Garden Street, Room B217, Cambridge, MA 02138, U.S.A.

Bernard F. Schutz, Department of Applied Mathematics and Astronomy, University College Cardiff, P.O. Box 78, Cardiff CF1 1XL, U.K.

John J. Stachel, Einstein Project, Boston University, 745 Commonwealth Avenue, Boston, MA 02215, U.S.A.

Joseph H. Taylor, Joseph Henry Laboratories and Physics Department, Jadwin Hall, Princeton University, P.O. Box 708, Princeton, NJ 08544, U.S.A.

K. P. Tod, Mathematical Institute, 24-29 St. Giles, Oxford OX1 3BD, U.K.

Michael S. Turner, Fermilab Astrophysics, Fermi National Accelerator Laboratory, Box 500, Batavia, IL 60510, U.S.A., and Departments of Physics and Astronomy and Astrophysics, Enrico Fermi Institute, University of Chicago, Chicago, IL 60637, U.S.A.

Shing-Tung Yau, Department of Mathematics, University of California at San Diego, La Jolla, CA 92093, U.S.A.

ACCRETION DISKS AROUND BLACK HOLES

M.A. Abramowicz
Scuola Internazionale Superiore di Studi Avanzati
34014 Trieste, Strada Costiera 11, Italy

Abstract. Accretion disks around black holes differ qualitatively, especially when stability is considered, from those around other compact objects (neutron stars, white dwarfs) even when the accretion rate is very small. The so called "standard" accretion disk model is never adequate for the black hole accretion.

INTRODUCTION

One of the most important predictions of the General Theory of Relativity is the existence of black holes. Astronomical black holes cannot be observed directly, but their presence may be inferred from observations of high energy processes connected with the accretion of matter onto the black holes. I shall argue in this lecture that, however shocking this may sound, the most standard model of accretion is totally inadequate for describing accretion onto black holes. The standard model of Shakura and Sunyaev (1973) describes very successfully accretion onto white dwarfs and neutron stars, using Newtonian hydrodynamics. It makes several assumptions which, although quite correct in the case of accretion onto white dwarfs and neutron stars, are wrong in the black hole case. The standard model was generalized by Novikov and Thorne (1973) to the Kerr spacetime. However, these authors made exactly the same physical assumptions as Shakura and Sunyaev and only changed the Newtonian gravity to the Kerr case. Their model is qualitatively the same as the Newtonian model of Shakura and Sunyaev, despite the relativistic formalism used. It played at one time a very important role in bringing the attention of astrophysicists interested in accretion to the relativistic formalism. It should be replaced, however, by a truly general relativistic model, which incorporates important relativistic effects reviewed in this lecture.

ACCRETION DISKS IN ASTRONOMICAL OBJECTS

Accretion disks are quite common in close binary systems consisting of a compact object (primary) and a late-type star (secondary) closely orbiting each other. The gravitational attractions of the primary and secondary stars are equal at the "Lagrangian point" which lies very close to the surface of the secondary. At this point matter is not gravitationally bound and mass transfer from the secondary to the compact primary takes place. The gas streaming from the secondary has, however, too much angular momentum to be accreted directly onto the compact star. Instead it forms a rapidly rotating accretion disk around it. The disk operates through the action of viscous stresses. Gas orbiting a compact central body gradually loses its angular momentum to the gas further out. As the balance of the centrifugal force against gravity weakens in gravity's favour, the inner gas slowly spirals further inwards. Viscous stresses not only drive accretion by transporting mass inwards and angular momentum outwards, but they also convert the gravitational energy of matter into heat. The heat diffuses towards the top and bottom surfaces and is radiated away by the disk's photosphere. Radiation from the hot innermost region can be absorbed by much cooler outer parts, heating them up, forming a tenuous corona and blowing off a wind. A large scale ordered magnetic field may influence the accretion flow by preventing it from reaching the surface of the primary except in the regions where the magnetic pressure is low. A small scale chaotic magnetic field may be responsible for the turbulence and viscosity.

Accretion disks are subject to many instabilities: *dynamical*, connected with the angular momentum redistribution, *thermal*, governed by the details of the balance between heating up by viscous stresses and cooling down by radiation (or other processes) and *viscous* instabilities arising when the diffusion connected with the viscosity formally has a negative coefficient, so an initially small overdense perturbation will gain more and more mass and become unstable. Other instabilities are due to plasma processes (e.g. pair production), self gravity, magnetic field, etc. Because of their complex structure and instabilities real accretion disks are non stationary, non symmetric, very noisy and lumpy. A realistic model of an accretion disk must therefore contain a large number of parameters, not amenable to computation by even the most advanced numerical codes and the biggest computers available today. Most of the present theoretical models of accretion disks assume instead a rather smooth, axially symmetric distribution of physical characteristics of the disks, and use only three "control" parameters:

The mass of the central object M. For white dwarfs and neutron stars this mass is $M \approx 1M_{\odot}$. For the black holes which are believed to be present in some binary systems in our Galaxy or in the Large Magellanic Cloud (Cygnus X-1, LMC X-1, LMC X-3, A0620-00, SS433) one has $3M_{\odot} < M < 10\ M_{\odot}$. The quasars and other active galactic nuclei may be connected with supermassive, $M \approx 10^{8}M_{\odot}$, accreting black holes in galactic centers.

The accretion rate $\dot{M}$. This is the rate at which the mass is supplied to the disk from outside. In the case of close binaries the source of the matter is the secondary star. In the case of accretion disks in active galactic nuclei the mass supply can be due to stellar disruption and collisions in the dense stellar population at the galactic center. The accretion rate can be conveniently scaled by the Eddington accretion rate,

$$\dot{M}_E = L_E/C^2 = 1.4 = 10^{17}(M/M_{\odot})\ [\text{g/sec}]. \qquad (1)$$

Here L_E is the Eddington luminosity, equal to the power emitted by a spherical object which is in equilibrium between the pull of its gravity and the push of the internal radiation pressure. There are no general limits for $\dot{M}$, but when $\dot{M} \ll \dot{M}_E$ the accretion is called subcritical.

The viscosity parameter α. In the absence of a microphysical theory for viscosity a phenomenological formula for the kinematical viscosity coefficient ν is adopted

$$\nu = \alpha v_S H. \tag{2}$$

Here α = const is the viscosity parameter (assumed ad hoc), v_S is the sound velocity and H is the vertical thickness of the accretion flow. General arguments give $\alpha \leqslant 1$.

Table 1: The three control parameters for astrophysical accretion disks

Central accreting object	Central mass	Accretion rate	Viscosity	Recent review papers
	$M/M_\odot$	$\dot{M}/\dot{M}_E$	α	
White dwarf (*)	~1	$\sim 10^{-6}$	~0.1	Smak (1984) Bath and Pringle (1985)
Neutron star (*)	~1	$\sim 10^{-1}$	~0.1(?)	Hale et al (1983)
Black hole: galactic	~10	$\sim 10^{-1}$(?)	(?)	Ilovaiski (1986)
Black hole: extragalactic	$\sim 10^{8}$	<<1 thin disk	~0.1(?)	
		<<1 ion tori	$\sim 10^{-2}$(?)	Rees (1984)
		$\gtrsim$1 radiation tori	<<1	
Black hole: (*) SS433	~10	>>1	~0.1(?)	Sharp et al (1985) Margon (1984) Katz (1986)

Table 1 shows typical values for the three control parameters M, $\dot{M}$, and α for directly observed (*) or hypothetical accretion disks around compact objects.

The theory of accretion disks around white dwarfs is quite successful in explaining the detailed observational data. In this case the accretion rate is very small, $\dot{M} \ll \dot{M}_E$, and the viscosity parameter is quite high, $\alpha \approx 0.1$. For the hypothetical black hole accretion, however, one needs (at least in some cases) the opposite regime of the control parameters: high accretion rate and small viscosity.

The "standard" accretion disk model assumes $\dot{M} \ll \dot{M}_E$. It is based on a very approximate, semi-analytic approach developed by Prendergast and Burbidge, Shakura, Pringle and Rees and finally shaped by Shakura and Sunyaev (1973). There is an excellent review of it by Pringle (1981). I shall present the standard model here using the pseudo-Newtonian approach of Paczynski and Wiita (1980) in which one describes the black holes as Newtonian objects with the potential

$$\Phi = -GM/(r - r_G), \quad r_G \equiv 2GM/c^2 \tag{3a}$$

and uses non relativistic hydrodynamics. The circular orbits of free particles have the "Keplerian" angular velocity and angular momentum

$$\Omega = \Omega_k \equiv (GM/r^3)^{1/2} (1 - r_G/r)^{-1}, \tag{3b}$$

$$\ell = \ell_k \equiv (GMr)^{1/2} (1 - r_G/r)^{-1}, \tag{3c}$$

where r denotes the radial coordinate. A circular orbit is marginally stable for $r = r_{MS} = 3r_G$ and marginally bound for $r = r_{MB} = 2r_G$, just like in the Schwarzschild metric. The pseudo-Newtonian approach gives results which differ from the correct general relativistic treatment (for the Schwarzschild spacetime) typically by only a few percent.

SMALL ACCRETION RATES: THE STANDARD MODEL

The standard accretion disk model is based on rather restrictive assumptions on its time dependence, geometry, dynamics, kinematics and dissipation. The vertical half-thickness $H = H(R)$, the angular velocity Ω and the surface density of the flow appear in some of these assumptions.

$$\Sigma(R) = \int_{-H(R)}^{+H(R)} \rho(R,Z)\,dZ \tag{4}$$

Here ρ is the density of matter and R, Z, Φ are cylindrical coordinates.

Time dependence. The disk is stationary, none of its properties depending on time.

Geometry. The flow is axially symmetric and plane symmetric: its properties do not depend on Φ and they depend on Z only through Z^2. The vertical half-thickness is very small

$$H \ll R \tag{5}$$

Kinematics The azimuthal rotation agrees with the Keplerian (i.e. circular geodesic) motion

$$v_\Phi = \Omega R = v_k = (GM/R)^{1/2}(1 - R_G/R)^{-1} \tag{6}$$

The last equality is valid for the Paczyński pseudo-Newtonian potential. Only the v_Φ component of velocity contributes to the shear

$$\sigma = \tfrac{1}{2}R(d\Omega/dR) =$$

$$- \tfrac{1}{4}(GM/R^5)^{1/2}R(1 - 3R_G/R)(1 - R_G/R)^{-2} \tag{7}$$

Again, the last equality is valid for the Paczyński potential. The horizontal component of velocity is very subsonic, $v_R \equiv v \ll v_S \equiv (\partial P/\partial \rho)_S^{1/2}$. The vertical component of velocity is negligible, $v_Z \ll v_R \ll v_\Phi$.

Dynamics. The horizontal pressure gradient is negligible. At the inner edge of the accretion flow, $R = R_{in}$, the viscous torque vanishes. The accretion rate is much smaller than the Eddington one $\dot{M} \ll \dot{M}_E$. There are no large scale ordered electromagnetic fields.

Dissipation. The horizontal heat flux is negligible. The Z-averaged kinematic viscosity

$$\bar{\nu}(R) = \frac{1}{\Sigma(R)} \int_{-H(R)}^{+H(R)} \rho(R, Z)\nu(R, Z)dZ \tag{8}$$

is given by eq.(2): $\bar{\nu} = \alpha v_s H$. The opacity is dominated by electron scattering

$$\kappa = \kappa_{es} = 0.4[g/cm^2] \tag{9}$$

The action of viscous stresses is so important for the theory of accretion disks that our discussion must start from this point. Consider a particular cylindrical surface $R = R_o$, crossing the accretion flow. The rates of mass and angular momentum flow across this surface are $\dot{M}(R_o)$ and $\dot{J}(R_o)$:

$$\dot{J}(R_o) = \dot{M}(R_o)\, \ell(R_o) - g(R_o), \tag{10}$$

where ℓ is the specific angular momentum and $g(R_o)$ the viscous torque acting through $R = R_o$:

$$g(R_o) = 2\pi \int R_o^3 \left[\frac{d\Omega}{dR} \right]_{R_o} \rho(R_o, Z)\ \nu(R_o, Z)dZ$$
$$= 2\pi R_o^3(d\Omega/dR)_{R_o}\Sigma(R_o)\bar{\nu}(R_o)\ . \tag{11}$$

When $(d\Omega/dR) < 0$, as in the case of Keplerian motion, the viscous torque transports angular momentum outwards $(\dot{J} > 0)$. At the inner edge of the disk, $R = R_{in}$, the torque is assumed to vanish, and therefore

$\dot{J}(R_{in}) = \dot{M}(R_{in})\ell(R_{in})$. In the stationary case one must have $\dot{M}(R) = \dot{M}(R_{in})$ = const and $\dot{J}(R) = \dot{J}(R_{in})$ = const, otherwise there would be matter and angular momentum deposition somewhere. Therefore $\dot{M}[\ell(R) - \ell(R_{in})] = 2\pi R^3(d\Omega/dR)\Sigma\nu$. For the Keplerian motion in the Paczyński potential this can be written as

Table 2: The list of the standard accretion disk structure quantities

a) Structure quantities

M	= Mass of the central object	Q^+	= viscous heat production rate
$\dot{M}$	= accretion rate	Q^-	= vertical radiation heat loss
R	= distance from the axis	q	= non standard heat flux
α	= viscosity coefficient	F	= radiation flux from the surface
α_0	= viscosity constant	R_{in}	= inner edge
ρ	= density	H	= half thickness
P	= pressure	f	= radial function
T	= temperature	$\mathcal{D}$	= radial function
v_s	= sound velocity	$\mathcal{C}$	= radial function
Σ	= surface density	v	= accretion velocity
ν	= kinematic viscosity	ℓ	= specific angular momentum
κ	= opacity coefficient	ℓ_k	= Keplerian angular momentum
γ	= heat transport coefficient	v_k	= Keplerian velocity

b) Physical constants and useful scaling factors

a	$= 7.6 \times 10^{-15}$ erg cm^{-3} deg^{-4}	radiation constant
G	$= 6.7 = 10^{-8}$ dyn cm^{2}g^{-2}	gravitational constant
c	$= 3.0 \times 10^{10}$ cm sec^{-1}	velocity of light
$\mathcal{R}$	$= 8.3 \times 10^7$ erg deg^{-1} mole^{-1}	gas constant
$\bar{\mu}$	$= 0.6$ g mole^{-1}	mean molecular weight (cosmic abundances)
κ_{es}	$= 0.4$cm^{2}g^{-1}	electron scattering opacity
$M_\odot$	$= 2.0 \times 10^{33}$g	solar mass
R_G	$= 2.9 \times 10^5 (M/M_\odot)$cm	gravitational radius
L_E	$= 1.3 \times 10^{38} (M/M_\odot)$erg sec^{-1}	Eddington luminosity
$\dot{M}_E$	$= 1.4 \times 10^{17} (M/M_\odot)$ g sec^{-1}	Eddington accretion rate

$$3\pi\nu\Sigma = \dot{M} f \mathcal{D} \mathcal{C}^{-1} \tag{12}$$

$$f \equiv 1 - \ell_k(R_{in})/\ell_k(R), \quad \mathcal{D} \equiv 1 - R_G/R,$$

$$\mathcal{C} \equiv 1 - R_G/3R. \tag{13}$$

For the purely Newtonian potential $-GM/r$ one has $f = 1 - (R_{in}/R)^{1/2}$, $D = 1$, $C = 1$.

The viscous stresses transform the kinetic energy to heat at the rate

$$\dot{E} = 2\pi \iint R^3 (d\Omega/dR)^2 \, \nu \rho dR dZ \tag{14}$$

The viscous heat production rate per unit surface area is therefore given by

$$Q^+(R) \equiv \frac{1}{2\pi R}\frac{d\dot{E}}{dR} = \left[R\frac{d\Omega}{dR}\right]^2 \Sigma\bar{\nu} = \frac{3GM\dot{M} f C}{4\pi R^3 D^3} . \tag{15}$$

The last equality is valid for the Paczyński potential. In thermal equilibrium the radiation losses from the surface are given by $Q^-(R) = 2F$ and the other (e.g. convection, advection) possible heat losses q must exactly compensate the heat production rate: $Q^+ = Q^- + q$. The radiative flux F in the case of optically thick flow can be approximated by $F = (\alpha c T^4)/(3\kappa\rho H)$ with T being the temperature at the equatorial plane $Z = 0$.

Table 3: The structure equations for the standard disk model

[1]	$\Sigma = 2H\rho$	definition of the surface density
[2]	$(H/R) = (v_s/v_k)$	vertical hydrostatic equilibrium
[3]	$v_s^2 = (P/\rho)$	sound velocity
[4]	$\dot{M} = 2\pi R\Sigma V$	mass conservation
[5]	$P = (\mathcal{R}/\bar{\mu})\rho T + (\alpha/3)T^4$	equation of state
[6]	$Q^+ = (3GM\dot{M} fC)/(4\pi R^3 D^3)$	viscous heat production rate
[7]	$3\pi\nu\Sigma = \dot{M} f D C^{-1}$	angular momentum conservation
[8]	$F = (acT^4)/(3\kappa\rho H)$	radiative flux (optically thick case)
[9]	$Q^- = 2F$	radiative heat loss
[10]	$Q^+ = Q^- + q$	heat balance (energy conservation)
[11]	$q = (1 - \gamma)\, Q^+$	non standard heat flux
[12]	$[\ell - \ell(R_{in})]/\ell = f$	radial function
[13]	$D = 1 - R_G/R$	radial function
[14]	$C = 1 - R_G/3R$	radial function
[15]	$\ell_k^2 = GMR^3/(R-R_G)^2$	Keplerian angular momentum
[16]	$v_k^2 = GMR/(R-R_G)^2$	Keplerian velocity
[17]	$\ell = \ell_k$	angular momentum in the disk
[18]	$R_{in} = 6GM/c^2 = 3R_G$	inner edge
[19]	$\nu = \alpha v_s H$	viscosity law
[20]	$\alpha = \alpha(\dots) \overset{*}{=} \alpha_0 = \text{const}$	viscosity coefficient
[21]	$\kappa = \kappa(\dots) \overset{*}{=} \kappa_{es} = \text{const}$	opacity cefficient
[22]	$\gamma = \gamma(\dots) \overset{*}{=} 1 = \text{const}$	convection or advection law

Table 4: The standard solution. Gas pressure dominates.

Quantity	X_o	A(M)	B($\dot{M}$)	C(α)	D(R)	E(f)	F(D)	G(C)
Σ	4.80×10^{4} g cm^{-2}	1/5	3/5	-4/5	-3/5	1	3/5	-1
H/R	1.82×10^{-2}	-1/10	1/5	-1/10	1/20	1/5	7/10	0
ρ	4.82 g cm^{-3}	-7/10	2/5	-7/10	-33/10	2/5	-1/10	-1
P	1.24×10^{17} dyn cm^{-2}	-9/10	4/5	-9/10	-51/20	4/5	-7/10	-1
T	3.29×10^{8} ^{o}k	-1/5	2/5	-1/5	-9/10	2/5	-3/5	0
v/c	5.35×10^{-5}	-1/5	2/5	4/5	-2/5	-3/5	-3/5	1
F	9.18×10^{25} erg sec^{-1}cm^{-2}	-1	1	0	-3	1	-3	1

Table 5: The standard solution. Radiation pressure dominates.

Quantity	X_o	A(M)	B($\dot{M}$)	C(α)	D(R)	E(f)	F(D)	G(C)
Σ	4.08 g cm^{-2}	0	-1	-1	3/2	-1	4	-3
H/R	0.718	0	1	0	-1	1	-1	1
ρ	9.04×10^{-6} g cm^{-3}	-1	0	-1	3/2	0	-2	1
P	2.45×10^{15} dyn cm^{-2}	-1	0	-1	-3/2	0	-2	1
T	3.14×10^{7} ^{o}K	-1/4	0	-1/4	-3/4	0	-1/2	¼
v/c	0.64	0	2	1	-5/2	1	-4	3
F	9.18×10^{25} erg sec^{-1}cm^{-2}	-1	1	0	-3	1	-3	1

The other standard disk structure equations, listed in Table 3, are self-explanatory. Table 2 gives the list of all the accretion disk structure quantities.

Most of the equations in Table 3 are derived by approximate integrations, taking Z-averaged quantities. For example, the exact relation (4) is replaced by [1] in the Table, the exact hydrostatic equilibrium condition in the vertical direction, $(\partial P/\partial Z) = \rho(\partial\Phi/\partial Z)$ is replaced by [2], etc. The equation $\alpha = \alpha(..)$ indicates that the quantity α depends on <u>some</u> of the disk parameters, and $\alpha \overset{*}{=} \alpha_0 = \mathrm{const}$ states that in the <u>standard</u> disk model α is constant.

All the standard disk structure equations are algebraic ones. In their logarithmic version they are all linear, except the equation of state [5]. However, when either the gas pressure $P_{gas} = (\mathcal{R}/\bar{\mu})\rho T$ or radiation pressure $P_{rad} = (a/3)T^4$ dominates the total pressure $P = P_{gas} + P_{rad}$, all the equations are linear in their logarithmic version. In this case the system of 22 equations for 26 quantities can be explicitly solved, i.e. every structure quantity X can be expressed in terms of 26 - 22 = 4 other structure quantities: for example, the three control parameters M, $\dot{M}$, α_0 and the cylindrical coordinate R. Because the equations are logarithmically linear, the solution can be put in the explicit, analytical form:

$$X = X_0(M/M_\odot)^A(\dot{M}/\dot{M}_E)^B\alpha_0{}^C(R/R_G)^D\mathcal{f}^E\mathcal{D}^F\mathcal{C}^G, \tag{16}$$

with X_0, A,..., G being constant This was done by Shakura and Sunyaev (1973) in the Newtonian potential (with $\mathcal{D} = 1 = \mathcal{C}$) and by Novikov and Thorne (1973) in the Kerr spacetime (in this case the radial functions $\mathcal{f}$, $\mathcal{D}$, $\mathcal{C}$,... are different). Tables 4 and 5 adopted from an unpublished work of Lasota, Xu Cong and myself, show the solution in the Paczyński potential.

Tables 4 and 5 describe either the radial distribution of the physical properties in one particular disk characterized by a particular choice of the control parameters M, $\dot{M}$, α_0:

$$X = X(R) \text{ for } M = \mathrm{const},\ \dot{M} = \mathrm{const},\ \alpha_0 = \mathrm{const} \tag{17}$$

or sequences of disk models at a fixed radius R but with different values of the control parameters. Astrophysically the most important and the most realistic case is

$$X = X(\dot{M}) \text{ for } M = \mathrm{const},\ \alpha_0 = \mathrm{const},\ R = \mathrm{const}\ . \tag{18}$$

In both cases the solutions described in the Tables are valid, strictly speaking, only in the asymptotic cases $P_{gas} \gg P_{rad}$ and $P_{rad} \gg P_{gas}$. The solution for $P_{gas} \approx P_{rad}$ can be easily obtained numerically, however.

<u>A particular model for a given $\dot{M}$. The spectrum.</u>

The important fact to notice is that independently of M, $\dot{M}$, α_0 the temperature maximum is reached at about $R \approx 2R_{in}$, because $dQ^+/dR = 0$ there. In the case of black hole accretion with $\dot{M} \ll \dot{M}_E$ one has (as we shall prove later) $R_{in} \approx R_{MS} = 3R_G$. The maximal temperature is

$$T_{max} \approx 10^7[^oK](M/M_\odot)^{-1/4}(\dot{M}/\dot{M}_E)^{1/4} \tag{19}$$

The thermal radiation connected with this temperature has maximal frequency

$$\omega_{max} \approx kT_{max}/h \approx 1[keV](M/M_\odot)^{-1/4}(\dot{M}/\dot{M}_E)^{1/4}. \tag{20}$$

The optical photons have frequencies of about 10^{-3} keV, the X-ray photons 1keV – 10^2 keV and γ -ray photons higher than 10^2 keV. Therefore, thermal photons from optical to X-ray can be produced by accretion disks around galactic black holes ($M \approx 10M_\odot$). Accretion onto supermassive black holes ($M \approx 10^8 M_\odot$) can produce optical and ultraviolet photons. The temperature at the outer parts of the disk ($R/R_{in} \to \infty$) is given by

$$T_{out} \approx 5 \times 10^7[^oK](M/M_\odot)^{-1/4}(\dot{M}/\dot{M}_E)^{1/4}(R/R_{in})^{-3/4}. \tag{21}$$

At a given radius one can assume the Planck formula for the thermal spectrum $S_\omega(\omega,T)$ with the temperature corresponding to this radius. For small frequencies, connected with T_{out}, one has $S_\omega \sim \omega^2$ and for the high frequencies connected with T_{max} one has $S_\omega \sim \exp(-\omega)$. In between one must integrate the contributions from the different parts of the disk to get the spectrum. The result can be approximated by $S_\omega \sim \omega^P$ with $0 \lesssim P \lesssim 1/3$. In the real disks the shape of the spectrum is influenced by non-thermal processes and by the fine details of the radiative transfer.

Sequences of disk models with varying accretion rate The $\Sigma(\dot{M})$ relation. By putting $X \equiv \Sigma$ in equation (18) one defines a particular sequence of the standard disk models in the Σ – $\dot{M}$ plane. Asymptotically for $P_{gas} \gg P_{rad}$ one has for this sequence, $d\ell n\Sigma/d\ell n\dot{M} = 3/5$ and for $P_{rad} \gg P_{gas}$ $d\ell n\Sigma/d\ell n\dot{M} = -1$, so the curve $\Sigma = \Sigma(\dot{M})$ has somewhere for $P_{gas} \approx P_{rad}$ a turning point, $d\ell n\Sigma/d\ell n\dot{M} = 0$. Although the exact shape of the $\Sigma = \Sigma(\dot{M})$ curve can be found only numerically, it is not difficult to show that:

$$d\ell n\Sigma/d\ell n\dot{M} = 5(\beta - 2/5)/(2 + 3\beta), \tag{22}$$

where $\beta \equiv P_{gas}/P$. It is clear that the turning point for the sequence always coincides with $\beta = 2/5$ and that for $\beta > 2/5$ the slope of the $\Sigma(\dot{M})$ curve is positive, while for $\beta < 2/5$ it is negative. The accretion rate at the turning point

$$(\dot{M}/\dot{M}_E)_{turn} =$$

$$3.85 \times 10^{-3}(R/R_G)^{21/16}(M/M_\odot)^{-1/8}\alpha_0^{-1/8}\mathcal{F}^{-1}\mathcal{D}^{17/8}\mathcal{C}^{-5/4} \tag{23}$$

depends only very weakly (note the small powers -1/8) on the control parameters M and α_0. Formula (22) is valid for any gravitational potential, but (23) only for the pseudo-Newtonian potential (3a). They have both been derived by Lasota, Xu Cong and myself in an unpublished paper. We have also discussed how the turning point of the $\Sigma(\dot{M})$ curve indicates a transition from the stable to unstable disk models. The discussion starts from well known, classical, results obtained by Pringle, Shakura, Sunyaev, Pacholczyk, Rees, Piran, Eardley and Lightman and goes as follows:

The characteristic time for the thermal processes t_{th} can be estimated as the time of radiating away the thermal energy stored in a given location of the disk. The viscous time t_{vis} can be determined from the small drift velocity, connected with the viscous dissipation of angular momentum: $t_{th} = (2\pi/\Omega\alpha)$, $t_{vis} = t_{th}(H/R)^{-2}$. Because $H \ll R$, one has for the standard disk models

$$t_{th} \ll t_{vis} \tag{24}$$

i.e. that thermal processes go much faster than the viscous ones. It would be convenient to discuss the criteria for thermal and viscous stability in terms of the phenomenological coefficients introduced by Piran (1978):

$$\begin{aligned} K &\equiv (\partial \ln Q^-/\partial \ln H)_\Sigma, \quad L \equiv (\partial \ln Q^-/\partial \ln \Sigma)_H, \\ M &\equiv (\partial \ln Q^+/\partial \ln H)_\Sigma, \quad N \equiv (\partial \ln Q^+/\partial \ln \Sigma)_H. \end{aligned} \tag{25}$$

<u>Thermal instabilities</u>. In thermal equilibrium $Q^+ = Q^-$. Suppose that a perturbation is introduced and that $Q^+ > Q^-$. The distribution of Σ can be changed only by viscous stresses. Because $t_{vis} \gg t_{th}$, one can assume that $\Sigma \approx$ const during the thermal perturbation. With $Q^+ > Q^-$ the disk is overheated and will expand, i.e. H will increase. If the equilibrium heat production rate Q^+ grows faster with H than the cooling rate Q^- in a particular place in the sequence, the disk will not be able to find an appropriate equilibrium - there will be a runaway thermal instability. The case with $Q^+ < Q^-$ initially is obviously analogous. Therefore, the criterion for the thermal stability is

$$\{(\partial \ln Q^+/\partial \ln H)_\Sigma < (\partial \ln Q^-/\partial \ln H)_\Sigma\} \Rightarrow \{K - M > 0\}. \tag{26a}$$

<u>Viscous instability.</u> Because $t_{vis} \gg t_{th}$, one can assume that $Q^+ = Q^-$ during the viscous perturbation. Suppose that an equilibrium surface density is Σ_0 and that the perturbed one is $\Sigma > \Sigma_0$. If along the equilibrium sequence the accretion rate decreases with the increasing surface density, it would be impossible for the disk to find an appropriate equilibrium: the surface density will keep increasing causing an increase of the mass supply rate $\dot{M}_0 - \dot{M}$ and therefore a further increase of the surface density. Thus, the criterion for viscous stability is

$$\{(\partial \ln \Sigma/\partial \ln \dot{M})_{Q^+ = Q^-} > 0\} \Rightarrow \{(NK - ML)/(K - M) > 0\} \tag{26b}$$

A negative slope of the $\Sigma(\dot{M})$ curve indicates a viscous instability. Note, that for the standard disk model

$$K = 4(1 + \beta)/(4 - 3\beta), \quad L = -\beta/(4 - 3\beta), \quad M = 2, \quad N = 1, \tag{27}$$

so the quantity $NK - ML = (4 + 6\beta)/(4 - 3\beta)$ is always positive and therefore the condition for both thermal and viscous stability is

$$\begin{aligned} &\{\text{stability}\} \Leftrightarrow \{K - M > 0\} \Leftrightarrow \{\beta < 2/5\} \\ &\Leftrightarrow \{d\ln \Sigma/d\ln \dot{M} > 0\}. \end{aligned} \tag{28}$$

This proves formally our point.

The above stability criterion involves theoretical parameters. A criterion more attached to observations can be obtained in the following way: The efficiency of a stationary accretion disk is defined by $\eta = L/\dot{M}C^2$, where L is the total emitted power of radiation (luminosity). It can be rigorously shown that the efficiency is equal to the binding energy of fluid elements at $R = R_{in}$:

$$\eta = -e_k(R_{in}) = (GM/2R)(1 - 2R_G/R)(1 - R_G/R)^{-1} = 1/16 \qquad (29)$$

The last equality, valid for the Paczyński potential, assumes that there is no torque at the marginally stable circular orbit at $R_{in} = 3R_G = R_{MS}$. Most of the radiated power originates, as we have seen, at $R \approx 2R_{in} \approx 5R_G$. By expressing $\dot{M}$ in terms of L, with efficiency given by (29), one can use formula (23) to write the stability criterion for the most luminous part of the disk in the form $L < 3.12 \times 10^{36} (M/M_\odot)^{7/8}$ erg sec^{-1}. However, the mass of the black hole can be accurately deduced from observations only for black holes in galactic binary systems. In the case of individual accreting black holes an indirect method must be used. One argues, for example, that the shortest observed time scale of variability t_{min} corresponds to the shortest orbital period, $t_{min} = 2\pi/\Omega_k(R_{MS}) = 3.03 \times 10^{-4}(M/M_\odot)$sec. This allows one to write the stability condition in terms of the observed quantities.

$$\{\text{stability}\} \Leftrightarrow \log L < \log L_{turn} = \begin{cases} (7/8)\log(M/M_\odot) + 36.49 \\ (7/8)\log t_{min} + 39.57 \end{cases} \qquad (30)$$

An unpublished review of the observational data made by Ramella, Szuszkiewicz and myself, shows that there are many objects with $L > L_{turn}$ which should be unstable on viscous and thermal timescales, but they are not. One of the reasons for the discrepancy between the standard disk model instability theory and observations may be the fundamental, but ad hoc, assumption about the nature of viscosity, $\nu = \alpha v_S H$. This possibility was discussed e.g. by Piran (1978). In an unpublished work of Lasota, Xu Cong and myself we have examined this idea by assuming (ad hoc) a very strong dependence of viscosity coefficient on the accretion rate: $\alpha = \alpha_0 \exp(-K\dot{M}/\dot{M}_E)$, with α_0 = const, K = const. For $R = 5R_G$, say, and $K \gg 1$ the instabilities discussed above are either not present or they are present only in a finite range of accretion rates, $\dot{m}_1 < \dot{M}/\dot{M}_E < \dot{m}_2$. The existence of the two stable solutions for a given Σ, one with $\dot{M}/\dot{M}_E < \dot{m}_1$ and the other one with $\dot{M}/\dot{M}_E > \dot{m}_2$ may introduce a limit cycle behaviour of the accretion flow.

LIMITS OF VALIDITY OF THE STANDARD MODEL

The assumption about the viscosity is not the only ad hoc one made in the standard model. In Table 6 I present the set of equations being used by Frank and myself in our, not yet completed, work on non-stationary accretion disks with $\dot{M} \ll \dot{M}_E$. The left-hand sides of the corresponding equations from the standard model are always equal to zero. I show the time derivatives which should be added to the left-hand side if the disk is not stationary. These equations were first derived by

Muchotrzeb and Paczyński (1982) who also integrated them for the two sets of the control parameters: $M = 10M_\odot$, $\alpha = 0.001$, $\dot{M}/\dot{M}_E = 1.6$ or 16.

Table 6: The accurate accretion disk equations ($\dot{M} \ll \dot{M}_E$).

[I]	$0 = -\frac{1}{R}\frac{\partial}{\partial R}(R\Sigma v)$	mass conservation	$\left[\frac{\partial\Sigma}{\partial t}\right]$
[II]	$0 = -\frac{1}{R}\frac{\partial}{\partial R}(R\Sigma v\ell) + \frac{1}{2\pi R}\frac{\partial g}{\partial R}$	angular momentum conservation	$\left[\frac{\partial\Sigma\ell}{\partial t}\right]$
[III]	$0 = \frac{v_s^2}{H} - \frac{v_k^2 H}{R^2}$	dynamical vertical equilibrium	$\left[\frac{\partial^2 H}{\partial t^2}\right]$
[IV]	$v\frac{\partial v}{\partial R} + \frac{1}{\rho}\frac{\partial P}{\partial R} = \frac{\ell^2 - \ell_K^2}{R^3}$	dynamical horizontal equilibrium	$\left[\frac{\partial V}{\partial t}\right]$
[V]	$v\Sigma T\frac{\partial S}{\partial R} + 2\frac{H}{R}\frac{\partial}{\partial R}(R\,\chi\,\frac{\partial T}{\partial R}) = \Sigma\nu R^2\left[\frac{d\Omega}{dR}\right]^2 - 2F$	energy conservation	$\left[\Sigma T\frac{\partial S}{\partial t}\right]$
[VI]	$0 = \Sigma - 2H\rho$	surface density	
[VII]	$0 = g - 2\pi R\nu\Sigma R^2\frac{d\Omega}{dR}$	viscous torque	
[VIII	$0 = \ell - \Omega R^2$	angular momentum	
[IX]	$0 = \nu - \alpha v_s H$	viscosity	
[X]	$0 = v_s^2 - (P/\rho)$	sound velocity	
[XI]	$0 = P - (\mathcal{R}/\bar{\mu})\rho T - \frac{a}{3}T^4$	equation of state	
[XII]	$0 = F - \frac{acT^4}{3\kappa\rho H}$	radiative flux	

Additional definitions and equations

$S(\rho,T)$ = entropy; $\kappa(\rho,T)$ = opacity; $\chi(\rho,T)$ = conductivity
α = const = viscosity parameter, ℓ_k = Keplerian angular momentum (3c) $v_k = \ell_k/R$ = Keplerian velocity,
$-\ell_k^2/R^3$ = gravitational force

Note, that only two equations: the equation for the dynamical horizontal equilibrium [IV] and the equation for the energy conservation [V] are explicitly different from the corresponding ones in the standard model. An implicit difference is that one does not assume now that $g(R_{in}) = 0$ for a given priori radius $R = R_{in}$. These differences may seem not to be very important, but in fact they completely change the properties of the solution. I shall now discuss them.

The effective potential and the Roche lobe overflow mechanism. A particle with non-zero angular momentum ℓ cannot be accreted by a Newtonian center with the potential $-GM/r$, because the attractive gravity force grows like GM/r^2 and the repulsive centrifugal one like ℓ/r^3 when r goes to zero. There is an infinite effective potential barrier connected with angular momentum. However, in the case of a black hole, and in the case of the pseudo-Newtonian potential, the potential barrier connected with the angular momentum is always finite and for $\ell < \ell_{MS} \equiv \ell_k(r_{MS})$ there is no barrier at all. When the energy of the particle is greater than the top of the effective potential barrier, the particle can be accreted with no dissipation of the angular momentum. What is true for particles, remains true for fluids: the effective potential for the radial motion

$$W(\ell,R,Z) = \ell^2/2R^2 - GM/(r - r_G) \tag{31}$$

has only a finite barrier $W = W_0$ located at $Z = 0$, $R = R_0$. The equipotential surface which goes through $Z = 0$, $R = R_0$ crosses itself there. By analogy with the situation in close binary stars this selfcrossing equipotential is called "the Roche lobe" and the crossing place is called "the cusp". The potential barrier holds the fluid at $R \gg R_0$ except the part of it with energy greater than the top of the barrier, $E(R_0) > W_0$. The fluid with $E > W_0$ overflows the Roche lobe and flows towards the hole with the accretion rate (Abramowicz, 1985):

$$\dot{M}_0 = \frac{\pi\Gamma^2(n + 3/2)v_k(R_0)\Sigma(R_0)}{(2n + 1)^{1/2}\,\Gamma(n + 1)\,\Gamma(n + 2)}\,H(R_0) \tag{32}$$

where Γ is the Euler gamma function and n the polytropic index. The accretion rate is governed by dynamical and not viscous processes: in fact it is the horizontal pressure gradient which pushes the fluid through the cusp. Because the term with the horizontal pressure gradient $(1/\rho)(\partial P/\partial R)$ - cf. equation [IV] in Table 6 - is absent in the standard theory, the Roche lobe overflow mechanism cannot be properly accounted for in terms of the standard model. This mechanism provides a self-regulatory device between the accretion rate close to the hole [which is determined by the thickness, cf. eq. (32)] and the competition of cooling and heating processes (which determines the thickness). Note also that R_0, the location of the top of the potential barrier is determined by the angular momentum distribution. It is not Keplerian when $(1/\rho)\,(\partial P/\partial R) \neq 0$.

Instead of computing proper values of R_0, W_0, Σ_0 and H_0 by considering the horizontal pressure gradient and the Roche lobe overflow mechanism, the standard model implicitly assumes $R_0 = 3R_G$, $W_0 = 0$, $H_0 = 0$, $\Sigma_0 = \infty$.

Transonic flow. Very far from the black hole the velocity of the accretion flow tends to zero and the sound velocity is finite, $v \ll v_s$, while on the horizon the accretion velocity tends to that of light, i.e. $v > v_s$. Thus, one must have $v = v_s$ somewhere, i.e. the flow must be transonic for some radius $R = R_s$. At this radius (at least for the adiabatic, stationary accretion flow):

$$(dv/dR) = N(R,C_i)/(v^2 - v_s^2), \tag{33}$$

where C_i $(i = 1,2,\ldots)$ are the integration constants. Because for $R = R_s$ one has $v^2 = v_s^2$, it must be also that $N(R_s, C_i) = 0$. This introduces one more unknown constant R_s and two conditions. Thus, the integration constants C_i cannot be independent. A regular, $dv/dR \neq \infty$, transonic solution exists only for a particular choice of C_i. The eigenvalue nature of the problem is quite obvious. In the standard model this important property is completely lost. The only place where the velocity enters the equations is eq. [4] in Table 2. It could be dropped and the rest of the equations solved: the standard model does not contain at all the accretion velocity.

Abramowicz and Zurek (1981) have pointed out that because of the general relativistic effects there are two disconnected, physically different, families of transonic accretion flows with $\ell \neq 0$: the first with $R_0 \approx R_s$ and the second with $R_s \gg R_0$. The stationary accretion disks belong to the first family. It was suggested that some real accretion flows oscillate between these two states (see Fukue, 1986 for the references).

The horizontal heat flux. The Roche lobe overflow provides a very efficient mechanism of advective cooling by the horizontal heat flux. The heat is removed at a speed close to the speed of sound, because $R_0 \approx R_s$. This cooling mechanism works for any accretion rates, including very small ones $\dot{M} \ll \dot{M}_E$. It is very efficient when its characteristic time scale is not longer than the thermal one. This is true when $\dot{M} \approx \dot{M}_E$; for much smaller accretion rates the cooling mechanism is relevant only in the immediate vicinity of the cusp. Close to the cusp (or for $\dot{M} \gg \dot{M}_E$.) the angular momentum distribution changes from being almost Keplerian to being almost constant. To move a matter element from place to place in a fluid with constant angular momentum no viscous torque is necessary. The total amount of heat dQ carried in the horizontal direction by advection from a volume element $dW = H d\mathcal{S} = 2\pi H R dR$ in a unit time is $dQ = (dv_s^2/dR)\rho V dW$. Therefore, the heat flux per unit surface area, $q = dQ/d\mathcal{S}$, equals:

$$\begin{aligned} q_{ad} &= (dv_s^2/dR)Hv \approx (H/R)v_s^2\, v\rho \approx \\ &\approx (2/3)\mathcal{f}^{-1}(H/R)^2 Q^+. \end{aligned} \tag{34}$$

Equations of the standard theory (Table 3) have been used. The horizontal heat flux carried by radiation is

$$q_{th} \approx (H/R)Q^-. \tag{35}$$

Note, that for the equilibrium structure the horizontal flux of radiation q_{th} is more important than the advection flux $(H/R \ll 1)$. However, for the stability the advection flux is more important. The factor $\mathcal{f}^{-1}$

dominates when $\Omega \approx$ const. The key element is therefore the angular momentum distribution which determines the factor f. From (34) it follows that the advective flux can stabilize the accretion flow because the stability condition $K - M > 0$ is formally fulfilled.

CONCLUSIONS

I have reviewed three important general relativistic effects which dominate the accretion flow close to the black hole even for very small accretion rates:

i) the Roche lobe overflow mechanism
ii) the two regimes of the transonic accretion
iii) the relativistic advective cooling

These effects are absent in the standard accretion disk model. Until they are well understood, especially in the context of non-stationary accretion flows and stability, no one can be sure that our theories of black hole accretion are in touch with reality.

References

Abramowicz M.A., 1985, Publ. Astron. Soc. Japan 37, 72.

Abramowicz M.A. and Zurek W.H., 1981, Astrophys. J. 246, 314.

Bath G.T. and Pringle J.E., 1985, in Interacting binary stars, edited by J.E. Pringle & R.A. Wade (Cambridge University Press).

Fukue J., 1986, Publ. Astron. Soc. Japan. (in press).

Hale V.D., McClintock B. and J.E., 1983, Ann. Rev. Astron, Astrophys. 21, 13.

Ilovaiski S.A., 1986, lecture at Villa Olmo Conference.

Katz J., 1986, Comm. Astrophys, 11, 201.

Margon B.,1984, Ann, Rev. Astron. Astrophys. 22, 507.

Muchotrzeb B. and Paczyński B., 1982, Acta Astron, 32, 1.

Novikov I.D. and Thorne K.S., 1973, in Black Holes, edited by B. and C. Dewitt (Gordon and Breach, New York).

Paczyński B. and Wiita P.J., 1980, Astron. Astrophys., 88, 23.

Piran T., 1978, Astrophys. J., 221, 652.

Pringle J.E., 1981, Ann Rev. Astron. Astrophys. 19, 137.

Rees M.J., 1984, Ann. Rev. Astron. Astrophys. 22, 471.

Shakura N.I. and Sunyaev R.A., 1973, Astron. Astrophys. 24, 337.

Sharp N.A., Calvani M. and Turolla R., 1984, Comments Astrophys 10, 53.

Smak J.I., 1984, Pub. Astron. Soc. Pacific, 96, 5.

SUPERGRAVITY, KALUZA-KLEIN AND SUPERSTRINGS

M.J. Duff
Theory Division, CERN, 1211 Geneva 23, Switzerland

Abstract. We review the last three years of developments in supergravity and Kaluza-Klein theories and describe their merger with string theory. We also discuss recent applications of Kaluza-Klein techniques to strings on group manifolds.

I GENERAL RELATIVITY

I would like to begin with a little quiz. Many physicists believe, as did Einstein himself, that general relativity (GR) though beautiful, is not the final theory. Rather it is a stepping-stone to a more fundamental theory which would encompass not only gravity but all the other fundamental forces; a theory which would of necessity be quantum. Over the years, many suggestions have been made as to how GR might be modified or generalized in order to bring about this unification. So the question is:

Q. Which of the following do we need?

1) Kaluza-Klein?
2) Supergravity?
3) Higher derivatives?
4) Conformal invariance?
5) Non-symmetric metric?
6) Twistors?

If recent developments are to be believed, the answer is:

A. All of them!

because all these ideas appear in one guise or another in the theory of "superstrings", a theory which holds promise of the first consistent quantum theory of gravitation. I will have time to cover only the first two of these ideas, namely Kaluza-Klein and supergravity. Indeed, the conference organizers asked me to review the progress in these two subjects since the GR 10 conference three years ago. However, the truth is that in this short period both Kaluza-Klein and supergravity have in effect ceased to exist as separate disciplines! Of course, this does not mean that all that work on merging supersymmetry with gravity was futile nor that the idea of extra space-time dimensions is wrong. What it means is that Kaluza-Klein and supergravity, together with the string ideas of the 1960s, have been subsumed by a yet grander theory: the theory of superstrings, see Fig. 1.

Fig. 1

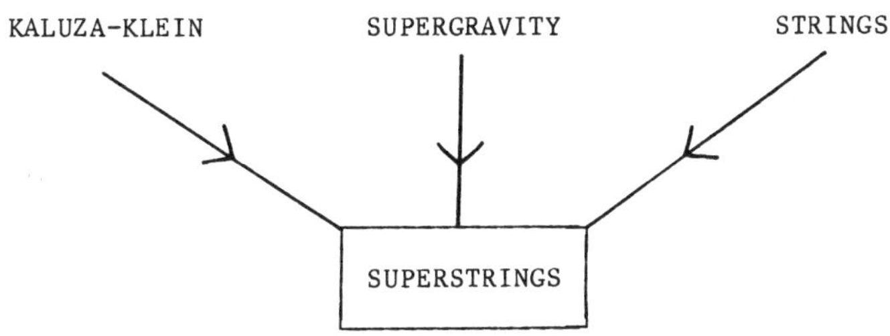

The direction of the arrows is to indicate that these three strands of theoretical thought have now merged. From a strictly logical point of view, however, one could draw the first two in the opposite direction, since it now turns out that a consistent quantum theory of strings implies both higher space-time dimensions and an underlying Bose-Fermi symmetry.

The hope is that superstrings will solve not only the problems of general relativity but also of particle physics, so let us briefly remind ourselves of these.

2 PARTICLE PHYSICS

Our experimental colleagues tell us that the standard SU(3)×SU(2)×U(1) model with three families of quarks and leptons is still in good shape, although there is some uncertainty about the top quark. Moreover, the Higgs bosons necessary for symmetry breaking have yet to be observed.

There are as yet no signs of proton decay, no signs of monopoles, no signs of axions and no signs of supersymmetry. In short, there is no empirical evidence at all that we should go beyond the standard model. So why bother? Although the standard [SU(3)×SU(2)×U(1)] model is probably the truth, and although grand unified theories [SU(5), SO(10), E_6, ...] are possibly the truth, none can claim to be the whole truth. There are still too many unanswered questions. For example:

* Why chiral fermions?
* How many families?
* What dictates the Higgs sector?
* Why SU(3)×SU(2)×U(1)?

By "chiral" we mean that we put the left-handed quarks and leptons into doublets of the weak-isospin and the right-handed into singlets. It is true that with these assignments we avoid the otherwise disastrous chiral anomalies in triangle graphs provided we group the quarks and leptons into families, but this logic does not tell us how many families there are; and whereas the gauge principle beautifully pins down the gauge boson sector and its coupling to fermions, the scalar Higgs sector remains quite arbitrary. Nor does the gauge principle alone tell us which gauge group we should start from.

If we now worry about gravity, the mystery deepens:

* How to incorporate gravity?
* The ultra-violet problem?
* Why is the cosmological constant $\Lambda \simeq 0$?

It seems unlikely that the fundamental theory will be obtained merely by tacking gravity, as an afterthought, on to one's favourite GUT theory. Moreover, the success in formulating the theory of strong, weak and electromagnetic forces relied crucially on the criterion of renormalizability, i.e., the ability to absorb into the masses and coupling constants those infinities which arise from ultra-violet divergent integrals in radiative corrections. Yet Einstein's theory of general relativity, the cornerstone of classical gravity, is at the quantum level non-renormalizable! (just because Newton's constant G has dimensions). Thirdly, there is the problem of the cosmological constant Λ. Originally Einstein suggested that the field equations of gravity should be obtained by equating geometry (the Einstein tensor $G_{\mu\nu}$) to matter (the energy-momentum tensor $T_{\mu\nu}$), i.e.,

$$G_{\mu\nu} = 8\pi G T_{\mu\nu} \tag{1}$$

where ($\mu,\nu = 1,2,3,4$). But he later realized that the principle of general covariance did not rule out a term proportional to the metric tensor $g_{\mu\nu}$, i.e.,

$$G_{\mu\nu} + \Lambda g_{\mu\nu} = 8\pi G T_{\mu\nu}. \tag{2}$$

Upper bounds on the cosmological constant Λ can be put by studying the rate of expansion of the Universe, and the result is one of the most accurate of experimental statements

$$\Lambda < 10^{-120}\ M_P^2 \qquad M_P = \text{Planck mass} \sim 10^{19}\ \text{GeV} \tag{3}$$

Yet theoretically the most natural value seems to be $\Lambda \sim M_P^2$ or perhaps $\Lambda \sim M_W^2$ where M_W = W-boson mass $\sim 10^2$ GeV. No one knows why $\Lambda \sim 0$.

There are even more fundamental questions one might ask: questions so fundamental that their answer is usually taken for granted but which have recently been brought into focus by superstrings, namely:

* Are ultra-violet divergences inevitable?
* Is field theory right?
* Why four apparent space-time dimensions?

Ultra-violet divergences have been with us for so long now that we forget the horror with which they were first greeted. Indeed, Dirac maintained to the end that they were only artifacts of our feeble approximations to Nature and that the fundamental theory should be finite. Field theory has been very successful at low energies but should we expect the field concept to persist beyond the Planck scale? And finally why does space-time appear to have four dimensions, as opposed to some other number?

We want to see how far Kaluza-Klein theories and superstrings have gone in answering these questions, but first let us cast our minds back to 1919 and Kaluza's original paper on the subject.

3 THE KALUZA-KLEIN IDEA

By 1919, Maxwell's theory of electromagnetism had been around for some time and Einstein had recently formulated his general theory of relativity. Of course, very little was known about the strong and weak nuclear forces at this time so it was natural, in searching for a unified field theory, to try and combine gravity with electromagnetism. This Kaluza (1921) was able to do by the ingenious device of postulating an extra, fifth dimension for space-time. Kaluza considered Einstein's theory of gravity alone but with a five-dimensional line element

$$d\hat{s}^2 = \hat{g}_{MN}(z)dx^M dx^N \tag{4}$$

where the indices M, N now run over 1 to 5 and x^M are the co-ordinates of a five-dimensional space-time. Thus

$$x^M = (x^\mu, y) \tag{5}$$

where x^μ are the usual co-ordinates of space-time and $y = x^5$ is the extra fifth co-ordinate. We put hats on $d\hat{s}$ and $\hat{g}_{MN}$ to distinguish them from the corresponding four-dimensional quantities. Kaluza then suggested the following four-dimensional interpretation

$\hat{g}_{\mu\nu} = g_{\mu\nu}(x)$	TENSOR	SPIN-2	GRAVITON	
$\hat{g}_{\mu 5} = A_\mu(x)$	VECTOR	SPIN-1	PHOTON	(6)
$\hat{g}_{55} = \phi(x)$	SCALAR	SPIN-0	DILATON	

i.e., the $\mu\nu$ components have the four-dimensional interpretation of a tensor which we identify with the usual gravitational field; the $\mu 5$ components have the four-dimensional interpretation of a vector which (and this is the crucial step) we identify with Maxwell's electromagnetic vector potential; and the 55 components have the four-dimensional interpretation of a scalar which is a new (and in 1919, rather embarrassing) field. All this was before the invention of quantum field theory but nowadays we would make the identification of these fields with particles, namely a spin 2 graviton, a spin 1 photon, and a spin 0 dilaton (so called because, in Klein's interpretation to be discussed below, the field ϕ measures the size of the fifth dimension.)

Of course, it is not enough to call $g_{\mu 5}$ by the name photon, one must also check that it obeys the correct equations and here we see the Kaluza-Klein miracle at work: Einstein's pure gravity equations in $d = 5$ yielded not only the correct gravity equations for $g_{\mu\nu}(x)$ in $d = 4$ but also the correct Maxwell's equations for $A_\mu(x)$ and, as an extra bonus, the Klein-Gordon equation for $\phi(x)$. In other words, Maxwell's theory of electromagnetism was seen to be a consequence of Einstein's theory of gravity, given that one was willing to buy this extra fifth dimension. In particular, conservation of electric charge (gauge invariance) was a consequence of conservation of energy momentum (general co-ordinate invariance).

Attractive though Kaluza's idea was, it contained a rather unsatisfactory ad hoc assumption. Although the indices were allowed to run over 1 to 5, the metric $\hat{g}_{MN}$ depended on only the four space-time co-ordinates x^μ. For no very good reason, the dependence on the fifth co-ordinate y had been suppressed. An attempt to remedy this defect was made by Oskar Klein (1926a,b) who suggested that the fifth dimension should have a different topology from the other four. In other words, whereas

$$-\infty < x^\mu < \infty \tag{7}$$

the fifth co-ordinate is periodic

$$0 \leqslant my \leqslant 2\pi \tag{8}$$

and the fifth dimension has been compactified to a circle of radius m^{-1}, i.e., space-time has topology $R^4 \times S^1$.

It is difficult to conceptualize a Universe with $R^4 \times S^1$ topology, but a simpler model with $R^1 \times S^1$ is provided by the hose-pipe analogy. Consider a long, thin hose-pipe as in Fig. 2. At large distances it looks like a one-dimensional line (topology R^1), but closer inspection reveals that at each point on the line there is a little circle (topology S^1). So it was that Klein suggested a little circle (S^1) at each point of four-dimensional space-time (R^4).

Fig. 2: The hose-pipe analogy. At large distances the hose-pipe looks like R; closer inspection reveals it to be $R \times S^1$.

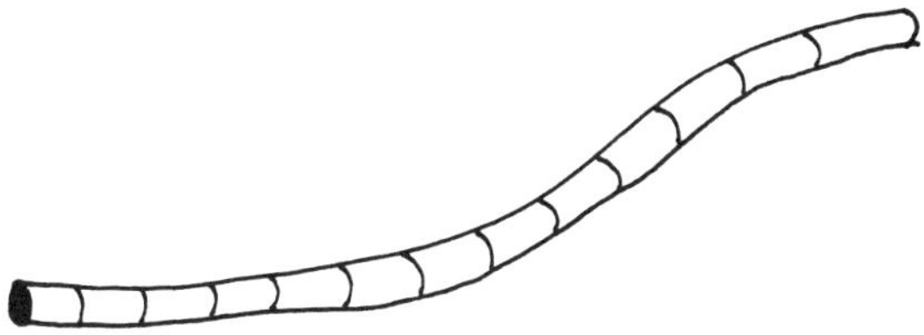

The point of all this is that we can now retain the dependence on the fifth co-ordinate without giving up the four-dimensional interpretation. All we have to do is perform a Fourier expansion

$$\hat{g}_{MN}(x,y) = \Sigma_{n=-\infty}^{\infty} \hat{g}_{MN}{}^{(n)}(x) e^{inmy}. \qquad (9)$$

The $n = 0$ modes in this expansion are just the massless particles that Klein had already discussed, namely the graviton, the photon and the dilaton of Eq. (6). But now we also have an infinite tower of $n = 0$ modes. Substituting this expression (9) into the five-dimensional Einstein equations reveals that these extra modes, which satisfy

$$\hat{g}_{MN}{}^{(n)} = \hat{g}_{MN}{}^{(-n)*}, \qquad (10)$$

admit the four-dimensional interpretation of charged, massive particles with masses

$$m_n = nm \qquad (11)$$

and electric charges

$$q_n = n\kappa m \qquad (12)$$

where

$$\kappa = \sqrt{G}. \qquad (13)$$

Note that the masses are quantized in units of m, the inverse radius of the circle, and that the charges are quantized in units of a fundamental charge

$$e = \kappa m. \tag{14}$$

Although this five-dimensional Kaluza-Klein theory is only a simple model, it illustrates three important points that survive in more realistic models:

(1) The electric charges of the elementary particles are quantized in units of a fundamental charge (a well-known, but hitherto unexplained, empirical fact).

(2) The gravitational constant $\sqrt{G}$ and the electric charge e are proportional to one another, with the constant of proportionality given by the size of the extra compact dimensions, as in Eq. (14).

(3) This means that with e of order unity, we are forced to take the radius of the extra dimensions to be very small; the Planck length

$$m^{-1} \sim 10^{-33} \text{ cm} \tag{15}$$

which is satisfactorily consistent with the fact that these extra space-time dimensions are not apparent in our everyday lives. But this in turn means that we are forced to take the fundamental mass m to be Planck size, i.e.,

$$m \sim 10^{19} \text{ GeV}. \tag{16}$$

Hence all the massive Kaluza-Klein particles have enormous masses way beyond the capability of any current or foreseeable accelerators. If Kaluza-Klein theories are to explain the real world, we must look for the particles we actually observe amongst the massless sector of the theory. These particles must then acquire their small masses (small compared to the Planck scale, $M_W << M_P$) through some symmetry breaking effect, probably at the quantum level. In fact, explaining where this other small scale in Nature comes from remains one of the biggest unsolved problems.

It is ironic that Einstein, who was the referee of Klein's original 1919 paper, did not at first like the idea of a fifth dimension but eventually warmed to it, accepting the paper for publication in 1921. Indeed it was Einstein and Bergmann (1938) who subsequently emphasized

Klein's idea that charge quantization followed from the compactness of the extra dimensions.

Nowadays, of course, we are more ambitious in our unification schemes. We want to unify not only the gravitational and electromagnetic interactions but the strong and weak interactions too. In the language of group theory, we want to replace the Abelian U(1) symmetry of electromagnetism by a non-Abelian group G which contains the SU(3)×SU(2)×U(1) of the standard model. In the language of Kaluza-Klein, this means that we want to replace the one-dimensional circle S^1 whose isometry group is U(1) by some k-dimensional compact space whose isometry group is G. Now we have 4+k co-ordinates

$$x^M = (x^\mu, y^m) \tag{17}$$

where m = 1,...,k. When we say that a compact space has isometry group G we mean that the group of transformations $y^m \to y'^m$ which leaves the metric of the space $\overset{0}{g}_{mn}$ invariant

$$\overset{0}{g}_{mn}(y') = \overset{0}{g}_{mn}(y) \tag{18}$$

is the group G. For example, since the isometry group of a k-dimensional sphere S^k is just the rotation group in k+1 dimensions, this would yield an SO(k+1) Yang-Mills theory.

In the older Kaluza-Klein literature, the compact topology of the extra k dimensions was something imposed on the theory in a somewhat ad hoc fashion. Today we demand that the compactification be "spontaneous", where we use the word in the same sense as in "spontaneous symmetry-breaking". In other words, the compactification should come about dynamically as a consequence of the field equations. The geometrical quantity $\overset{0}{g}_{mn}(y)$, the metric on M_k, is then just the vacuum expectation value of the dynamical variable $\hat{g}_{mn}(x,y)$

$$\langle \hat{g}_{mn}(x,y) \rangle = \overset{0}{g}_{mn}(y) \tag{19}$$

There are by now many examples in the literature of this phenomenon of spontaneous compactification on spaces with non-Abelian isometry group and once again we find that the Yang-Mills coupling constant g is proportional to the gravitational constant κ with the constant of proportionality given by m, the inverse "size" of the compact space

$$g^2 = \kappa^2 m^2 . \tag{20}$$

Now we have the light particles

$g_{\mu\nu}(x)$	TENSOR	SPIN-2	GRAVITON
$g_{\mu n}(x)$	VECTORS	SPIN-1	YANG-MILLS GAUGE BOSONS
$g_{mn}(x)$	SCALARS	SPIN-0	HIGGS BOSONS

together, once again, with an infinite tower of very heavy ($>M_P$) unobservable states. Note incidentally that the scalar Higgs bosons now have an origin every bit as geometrical as the vector bosons. This is important because Kaluza-Klein makes very definite predictions about the Higgs sector that are absent in conventional GUTs.

We conclude this section by summarizing the Kaluza-Klein philosophy as follows: what we perceive to be <u>internal</u> symmetries in d = 4 (electric charge, colour, charge conjugation, etc.) are really <u>space-time</u> symmetries in d = 4+k (general covariance, parity, etc.). For a comprehensive review see Duff et al. (1986a).

4 SUPERSYMMETRY

Progress in the theory of elementary particles has been marked by the discovery of new symmetry principles which group together apparently different particles into the same multiplet, thereby permitting a unified description. Up until the early 70s, however, it was an article of faith that we could never hope to group together particles of different spin. All that changed with the arrival of supersymmetry. Simple (N = 1) supersymmetry relates particles differing by one half unit in helicity, i.e., it transforms boson fields B into fermion fields F and vice-versa. Symbolically, suppressing indices,

$$\delta B = \bar{\varepsilon} F, \qquad \delta F = \partial B \varepsilon \tag{21}$$

Here ε^{α} is an anticommuting fermionic parameter where α is a spinor index which runs over 1 to 4 when we are in four dimensions. Supersymmetry is a space-time symmetry because if we perform two successive transformations

$$(\delta_1 \delta_2 + \delta_2 \delta_1) B = a^{\mu} \partial_{\mu} B \tag{22}$$

where

$$a^{\mu} = \bar{\varepsilon}_2 \gamma^{\mu} \varepsilon_1 \tag{23}$$

and $\gamma^{\mu}{}_{\alpha\beta}$ is the Dirac matrix. But (22) is just the transformation law for a translation through a distance a^{μ}. Thus

$$\text{SUSY} = \sqrt{\text{TRANSLATION}}. \tag{24}$$

If we now allow local supersymmetry so that ε is allowed to depend on the space-time co-ordinates $\varepsilon = \varepsilon(x)$, then $a^{\mu} = a^{\mu}(x)$. But an x-dependent translation is really a general co-ordinate transformation. Thus, had Einstein not already discovered general relativity, supersymmetry would have forced us to invent it! Local supersymmetry ("supergravity") implies the spin 2 gravitational field $g_{\mu\nu}(x)$ and provides it with a spin 3/2 fermionic partner, the gravitino ψ_{μ}^{α}.

There are good reasons for believing that spin 2 is the maximum spin permitted for massless particles and this means there is a maximum gap of eight half-units of helicity between the highest and lowest helicities

$$\begin{array}{ccccccccc} & 1 & 2 & 3 & 4 & 5 & 6 & 7 & 8 \\ -2 \rightarrow & -3/2 \rightarrow & -1 \rightarrow & -1/2 \rightarrow & 0 \rightarrow & 1/2 \rightarrow & 1 \rightarrow & 3/2 \rightarrow & 2. \end{array}$$

This in turn means we can tolerate $N \leqslant 8$ supersymmetries. For example $N = 8$ supersymmetry unifies one spin 2, eight spin 3/2, 28 spin 1, 56 spin $\frac{1}{2}$ and 70 spin 0 into one single supermultiplet. Many people have been attracted by the idea that this single $N = 8$ supermultiplet might describe all the elementary particles.

In the present Kaluza-Klein context, however, the most remarkable feature of supersymmetry is that it places an upper limit on the dimension of space-time! In fact, it says (Nahm, 1979)

$$d \leqslant 11 \tag{25}$$

The reason is that in d dimensions the Dirac spinor index A on the parameter ε^A ranges from 1 to $2^{(d-1)/2}$ if d is odd or $2^{d/2}$ if d is even. In particular, the maximum dimension permitting a 32-component spinor is 11. Now the reason why 32 is the maximum is that, when reinterpreted in terms of ordinary four components in four dimensions, it permits a maximum of eight

$$\varepsilon^A \rightarrow \varepsilon^{\alpha I} \tag{26}$$

where $A = 1,\ldots,32$, $\alpha = 1,\ldots,4$ and $I = 1,\ldots,8$, i.e.,

$$(N = 1 \text{ in } d = 11) \Longleftrightarrow (N = 8 \text{ in } d = 4) \tag{27}$$

and, as we have already seen, we must have $N \leqslant 8$ in $d = 4$. Moreover, the supergravity equations look much simpler and more natural when written in the higher dimension. This obviously suggests a link between supersymmetry and Kaluza-Klein, provided the extra ($\leqslant 7$) dimensions are somehow compactified.

Let us first examine the pros and cons of $N = 1$ supergravity in the maximum dimension, $d = 11$. It describes the interaction of gravity g_{MN}, which in $d = 11$ has 44 degrees of freedom, with a gravitino ψ_M which has 128. Since supersymmetry demands equal numbers of bosons and fermions, however, there must be another bosonic field with 84 degrees of freedom, which turns out to be a three-index antisymmetric tensor gauge field

A_{MNP} (Cremmer et al. 1978). Supersymmetry forbids the presence of any other fields and the resulting equations of motion are unique. Secondly, it fulfils the extreme Kaluza-Klein philosophy in that all its symmetries are space-time symmetries. Thirdly, not only does spontaneous compactification occur but the dimension of space-time (four) is now output rather than input. This is because the field strength F_{MNPQ} of the three-index gauge potential A_{MNP} has four indices and the d = 11 equations of motion admit solutions of the form (Freund & Rubin, 1980)

$$F_{1234} = \text{constant} \tag{28}$$

with all other components vanishing, thus singling out a four-seven split

$$M_{11} = M_4 \times M_7. \tag{29}$$

The resulting four-dimensional physics is determined by the geometry of M_7. Of particular interest is the case $M_7 = S^7$, the seven-dimensional sphere which provides us with a geometrical origin for the Higgs effect and breaks N = 8 supersymmetry down to N = 1. The mathematical reason for N = 1 in d = 4 is that the holonomy group $\mathcal{H}(M_7) = G_2$, (Awada et al. 1983). [Similarly N = 1 if $\mathcal{H}(M_6) = SU(3)$. See Section 5.] This was especially important because despite the beauty of N = 8 supersymmetry all the attempts to relate supersymmetry to phenomenology required N = 1. Here, it seemed, was the missing link.

Unfortunately, the euphoria was short-lived because other phenomenological problems remained just as intractable as before. First, despite the geometrical elegance of S^7 and the breaking to N = 1, the corresponding gauge group SO(8) was not big enough to contain SU(3) × SU(2) × U(1). There were other choices of M_7 which did contain SU(3) × SU(2) × U(1) but these did not admit N = 1.

Secondly, the four-dimensional fermions were not chiral with respect to the gauge group. Indeed Witten (1985) showed that two criteria had to be fulfilled before Kaluza-Klein theories could yield chiral fermions: the fermions must be chiral already in the higher dimension, and this chirality must be preserved in going to four dimensions by some sort of topological twist in the compactification. Since left- and right-handedness for fermions only makes sense in even dimensions, d = 11 was in big trouble. [Attempts to exploit the "hidden" SU(8) symmetry of d = 11 supergravity, to which these objections do not apply, were never very successful largely because they relied on too many unproved assumptions about bound states, even though SU(8) does contain SU(3) × SU(2) × U(1).] Thirdly, the same mechanism of Eq. (28) which was responsible for the compactification, also yielded an enormous cosmological constant $\Lambda \sim M_P^2$. Indeed, this is a problem which all Kaluza-Klein theories have to face up to in one way or another. And finally there was no compelling evidence that d = 11 supergravity, or any other supergravity field theory, could solve the long-outstanding problem of ultra-violet divergences.

The attention of the Kaluza-Klein community therefore turned to supergravity in the next permissible dimensions, d = 10, whose pros and cons we now outline. There had for many years in fact been something magic about d = 10, because ten was the critical dimension for superstrings, which we shall shortly discuss in the next section. Secondly, the first of Witten's objections to chirality did not apply and a possible way of overcoming his second objection was to start not with pure N = 1 supergravity in d = 10 (with fields $g_{\mu\nu}$, ψ_μ, $B_{\mu\nu}$, λ, ϕ) but with supergravity coupled to N = 1 Yang-Mills fields (A_μ^i, χ^i) where i runs over the dimensionality of the gauge group G (Chamseddine 1981; Bergshoeff et al. 1982; Chapline & Manton 1983). The prospects for phenomenology were also better because one was free to choose a group G containing SU(3)×SU(2)×U(1).

There also seemed to be severe drawbacks, however. First, as soon as we introduce chiral fermions we have to worry about chiral anomalies. Just as the presence of anomalous triangle diagrams threatens to spoil the consistency of the standard model in four dimensions, so in ten dimensions there are potentially lethal hexagon diagrams; see Fig. 3. In fact, the situation is much worse in d = 10 supergravity because not only would the anomalies destroy Yang-Mills gauge invariance through a non-conserved current, they would also destroy Einstein's general covariance through a non-conserved energy-momentum tensor. Secondly, although the chirality problem could in principle be solved with the right

Fig. 3: The potentially anomalous hexagon diagrams in d = 10. Gravitons are denoted ---, Yang-Mills bosons and fermions —.

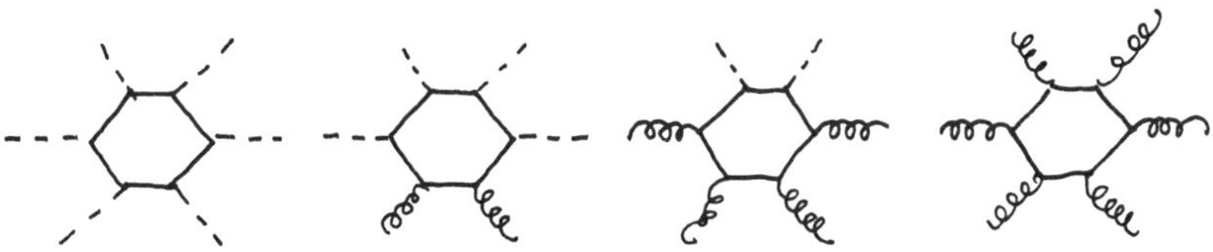

compactification, it was proved that no such compactification existed! And last, but not least, we have apparently lost the original idea of Kaluza-Klein of getting Yang-Mills fields from gravity because the Yang-Mills fields have been put in by hand already in the d = 10 equations.

This was the sorry state of affairs which existed up until the summer of '84. Then came ...

5 THE SEPTEMBER 84 REVOLUTION

Revolution I: Anomaly Cancellations

In a remarkable paper, Green and Schwarz (1984) were able to show that the potentially disastrous gravitational, mixed, and Yang-Mills anomalies exactly cancel provided one chooses as the gauge group

$$G = SO(32) \text{ or } G = E_8 \times E_8$$

Both groups have rank 16 and dimension 496. This cancellation of hexagon anomalies in d = 10 is reminiscent of the triangle anomaly cancellation in d = 4 achieved by putting the quarks and leptons into families, but there is one important difference. Here, and for the first time ever, the Yang-Mills gauge groups have been pinned down! By supersymmetry, moreover, the fermions have been pinned down as well. Not only that, but E_8 contains E_6; E_6 contains SO(10); SO(10) contains SU(5), and SU(5) contains SU(3) × SU(2) × U(1).

One technical, but very important, aspect of the Green-Schwarz anomaly cancellation mechanism was that the conventional N = 1 supergravity coupling to N = 1 Yang-Mills fields had to be modified. The Yang-Mills and gravitational field strengths F and R are related to the corresponding potentials A and ω by

$$F = dA + A \wedge A \qquad R = d\omega + \omega \wedge \omega \tag{30}$$

and obey the usual Bianchi identities. However, the field strength H of the tensor field B did not obey the naive equations H = dB, dH = 0, but instead

$$dH = \kappa^2/g^2 \, \mathrm{Tr}(R \wedge R - F \wedge F) \tag{31}$$

The $F \wedge F$ terms had been introduced earlier (Chamseddine 1981; Bergshoeff et al. 1982; Chapline & Manton 1983) on the grounds of supersymmetry, but the $R \wedge R$ terms were new. These modifications have a deep topological significance and are called "Chern-Simons" terms after the mathematicians Chern and Simons.

Revolution II: Calabi-Yau Compactification

Candelas et al. (1985) showed that these extra $R \wedge R$ Chern-Simons terms implied some corresponding $(\text{Riemann})^2$ modifications to the Lagrangian

$$\mathcal{L} = \sqrt{-g}\, e^{-2\phi}[-R - 4(\partial\phi)^2 + \frac{1}{12}H^2 + \frac{\alpha'}{2}(-F^i_{\mu\nu}F^{\mu\nu i} + R_{\mu\nu\rho\sigma}R^{\mu\nu\rho\sigma})] + O(\alpha'^2) \qquad (32)$$

and together these meant that the problem of compactification from ten to four dimensions could now be overcome, and they were able to find solutions of the form $M_{10} = M_4 \times K$ where K is a particular kind of compact six-dimensional space called a Calabi-Yau manifold, i.e., a Kähler manifold with vanishing first Chern class. Since the Yang-Mills fields were already present in d = 10, it was not necessary that the extra six dimensions have continuous symmetries as in conventional Kaluza-Klein and, indeed, Calabi-Yau manifolds have none. Moreover, these solutions were obtained by the trick of setting the gravitational spin connections in the extra dimensions ω equal to the Yang-Mills potentials A this ensuring a cancellation on the right-hand side of (31). [The idea of solving the coupled Einstein-Yang-Mills equations in Euclidean signature by setting the vector potential equal to the spin connection, and hence relating to the Yang-Mills Dirac index to the gravitational Euler number χ was discussed in a different context by Charap & Duff (1977).] This is just the sort of topological twist necessary for chiral fermions in d = 4 and one obtains the extra bonus of an unbroken N = 1 supersymmetry. Unlike generic compactifications, moreover, the resulting M_4 has zero cosmological constant.

Having made the choice G = SO(32) or $E_8 \times E_8$, the resulting d = 4 physics is then fixed by the geometrical and topological properties of the Calabi-Yau manifold. [It was originally thought that the string equations required the manifold to be Ricci flat, but this is now known not to be the case, owing to $\alpha'^3(\text{Riemann})^4$ terms in Eq. (32) not cancelled by α'^3F^4 terms in the equations of motion (Grisaru et al. 1986, Freeman et al. 1986, Freeman & Pope 1986).] In fact $E_8{\times}E_8$ turns out to be much the more promising and its phenomenology is discussed in Section 6.

Revolution III: The Heterotic String

Before discussing the third wave of this 84 revolution we must first introduce the string. See also the lecture by Mike Green at this conference.

6 STRINGS

Normally one thinks of fundamental objects as being points in space which sweep out one-dimensional world-lines in space-time, as in Fig. 4a. But suppose instead we consider the fundamental objects to be one-dimensional strings in space which sweep out two-dimensional world-sheets in space-time, as in Figs. 4b and 4c. Strings can either be "open" so that their ends move freely in space-time, or "closed" so that the ends are joined to form a closed-loop.

Figure 4: a) Point particle sweeps out world-line in space-time with trajectory $x^M(\tau)$.
b) Open string sweeps out world-sheet in space-time with trajectory $x^M(\sigma,\tau)$.
c) Closed string sweeps out world-sheet in space-time with trajectory $x^M(\sigma,t)$.

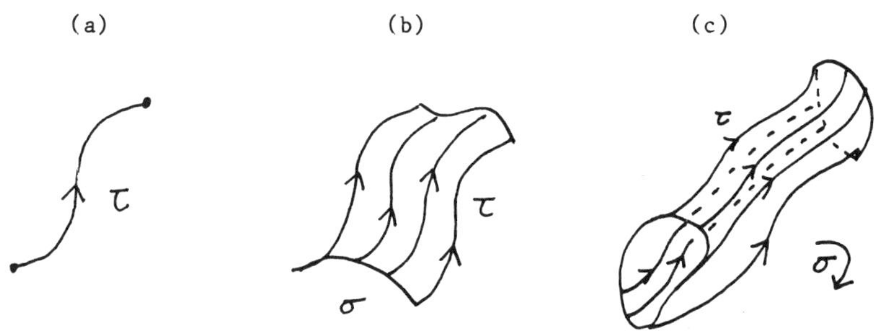

These strings were originally invented in the 1960's by Nambu (1970) to explain the Veneziano (1968) spectrum of hadrons. The string has vibrational modes and each of these modes represented a particle in the spectrum. So a string describes an infinite number of elementary particles: a finite number of massless particles and an infinite tower of massive particles. (This is reminiscent of Kaluza-Klein except that the string spectrum contains massive states of arbitrarily high spin, whereas a consistent field theory of the usual kind is limited to maximum spin 2.) The fundamental scale of string theory is given by the string tension

$$T = \frac{1}{2\pi\alpha'} \tag{33}$$

where α' is the slope of the Regge trajectory and has dimensions of $(\text{mass})^{-2}$. In these early attempts to relate strings to hadrons, one assumed that $1/\sqrt{\alpha'} \sim 1$ GeV. One bizarre aspect, however, was that the equations of the string at the quantum level could be made consistent only in the critical dimension

$$d = 26 \tag{34}$$

It was not until the 1970's, however, that Scherk and Schwarz (1974) abandoned the hadron interpretation and instead realized that the massless modes of the closed string included gravity! In this picture, we take $1/\sqrt{\alpha'} \sim 10^{19}$ GeV, i.e., the Planck scale. In a similar way, the massless modes of the open string included Yang-Mills gauge bosons. There were problems, however. First these strings were purely "bosonic"; fermions were still lacking. Secondly they exhibited an unphysical tachyonic (negative mass^2) mode. Thirdly, they suffered from divergences.

To include fermions, and following earlier work of Neveu and Schwarz (1971) and Ramond (1971), Green and Schwarz (1981) constructed the "superstring" which displayed full space-time supersymmetry which eliminated the tachyon problem. The equations of the superstring at the quantum level could be made consistent only in the critical dimension

$$d = 10 \tag{35}$$

Consistent Green-Schwarz superstrings are now known to come in three types (I, IIA and IIB) as shown in Table I, and they differ according to the different types of spinors which exist in d = 10. A spinor can be Majorana (real) with 32 components; Weyl (right- or left-handed) with 32 components; or both with 16 components. The Type I superstring seems to be most promising phenomenologically because of the SO(32) Yang-Mills fields, but all three are anomaly-free and free of divergences, at least to the one-loop order in perturbation theory. In fact, it has been conjectured that superstrings are finite to all orders. If so, this would be truly remarkable and provide us with a solution to one of the main problems of quantizing gravity discussed in Section 2.

Table I: Consistent Superstrings

TYPE	SPINOR	STRING	MASSLESS STATES
I[SO(32)]	Weyl + Majorana	open + closed	N = 1 SO(32) Yang-Mills +N = 1 supergravity
IIA	Majorana	closed	N = 2 non-chiral supergravity
IIB	Weyl	closed	N = 2 chiral supergravity
Heterotic [SO(32)]	Weyl + Majorana	closed	N = 1 SO(32) Yang-Mills +N = 1 supergravity
Heterotic [$E_8 \times E_8$]	Weyl + Majorana	closed	N = 1 $E_8 \times E_8$ Yang-Mills +N = 1 supergravity

The important point is that the massless sector of the Type I string is just the N = 1 supergravity with SO(32) Yang-Mills of Section 5 and that the Green-Schwarz R×R Chern-Simons terms necessary for anomaly cancellations and the Candelas et al. $(\text{Riemann})^2$ terms necessary for the chiral Calabi-Yau compactification are not accidental, but a direct consequence of the underlying superstring.

Another interesting aspect of the Type I superstring in d = 10 is the relation between the Yang-Mills coupling constant g (dimension mass^{-3}), the gravitational coupling constant κ (dimension mass^{-4}), and the slope parameter α' (dimension mass^{-2}), namely

$$g^2 = \kappa\alpha' \tag{36}$$

At this stage, however, there was still a mystery. Although anomaly cancellations in the SO(32) field theory could be explained by the underlying Type I superstring, where does the $E_8 \times E_8$ come from? The Green-Schwarz superstrings did not allow for this group.

This brings us to Revolution III: the discovery of the "Heterotic" (hybrid) string by Gross et al. (1985), so called because it combined features of the d = 26 bosonic string and the d = 10 Type II superstring. The result is a new kind of d = 10 superstring whose properties are also summarized in Table I. The crucial point is that the heterotic string allows for $E_8 \times E_8$ as well as SO(32) and $E_8 \times E_8$ seems better from the experimental point of view because the $A = \omega$ trick of Section 5 yields realistic chiral fermions in d = 4 in contrast to the SO(32) case. An intriguing difference from the Type I string is the relation between g, κ and α'. Instead of (36), we have

$$\alpha' g^2 = \kappa^2 \qquad (37)$$

[The astute reader may notice a striking similarity between the heterotic string Eq. (37) and the Kaluza-Klein Eq. (20). As we shall see in Section 10 this is no coincidence!]

The effective low-energy theory which emerges after the Calabi-Yau compactification of the $E_8 \times E_8$ heterotic string has the following features:

* There is an unbroken N = 1 supersymmetry [since the holonomy group $\mathscr{H}$ = SU(3)].
* One of the E_8 groups is broken to E_6. [This is again a consequence of $\mathscr{H}$ = SU(3).]
* The other E_8 remains unbroken. These gauge fields and their gluino partners are called the "hidden sector" because they interact with the E_6 sector only gravitationally.
* The fermions belong to chiral families of E_6 and appear in the "right" 27-dimensional representation. (The 27 of E_6 contains all the known quarks and leptons plus some exotics.)
* The number of families, N_F, is determined by $N_F = |\chi|/2$ where χ is the Euler number, a topological invariant of the compactified space. Examples of Calabi-Yau manifolds are known with N_F = 1,2,3,4 and much higher.
* The value of $\sin^2\theta_W$ (θ_W = Weinberg angle) at the compactification scale is the same as in conventional GUTs but the Yukawa couplings are different. This could be an advantage.
* If the Calabi-Yau manifold is not simply connected, the E_6 can be broken further to a rank-6 group like SU(3)×SU(2)×U(1)^3 or a rank-5 group like SU(3)×SU(2)×U(1)^2 or SU(3)×SU(2)×U(1)×SU(2). [Note that this avoids the conventional GUT schemes of SO(10) or SU(5).]

* The right Higgs scalars for electroweak breaking are present.
* The cosmological constant Λ is zero, at least classically.

There are problems, though. For example:

* What theoretical reason have we for picking the $E_8 \times E_8$ heterotic string and rejecting the other four superstring candidates?
* Why single out d = 4 for space-time (c.f., d = 11 supergravity in Section 4).
* Why pick Calabi-Yau manifolds rather than, say, T^6?
* Which Calabi-Yau manifold is the right one? (current estimates are that there are over 10,000 examples!). Some have $N_F \geqslant 4$, for which we lose asymptotic freedom; some give instantaneous proton decay; some yield no satisfactory E_6 breaking. No known example solves all problems.
* How do we break supersymmetry? It has been suggested, though by no means proved, that this comes about by gluino condensates in the hidden sector and that this, in turn, drives the electroweak breaking.
* Does Λ stay zero after we take into account supersymmetry breaking and radiative corrections?

As far as concrete predictions are concerned, there is not very much beyond conventional GUTs at the moment.

* Perhaps extra Z^0's or $W^{\pm}$'s from the extra U(1), $U(1)^2$ or SU(2).
* Magnetic monopoles with charges $2\pi/e \times k$ (k = integer) and corresponding fractional charges e/k. These are unconfined colour singlets but very heavy $\sim M_P$.

My personal feeling is that the right compactification is more subtle than Calabi-Yau and that, contrary to the scheme of Candelas et al., may involve non-vanishing H field in the extra dimensions. In any case, it is early days yet and one should not judge the string on the strength of this preliminary evidence.

There is another major objection one might raise at this stage. Since we are ultimately interested in four dimensions, should we not first compactify the ten-dimensional <u>string</u> to d = 4 and only then consider the low-energy field theory limit, rather than the other way around? A new and very promising development in this direction is string compactification on "orbifolds", which are slight generalizations of tori (Dixon et al. 1985). The string propagates on a background $M_4 \times K$ where K is a torus, points of which have been identified by some discrete isometries of the torus. This results in a space with singularities if these isometries have fixed points. In this way one may obtain a four-dimensional theory with many of the phenomenological features of Calabi-Yau manifolds (indeed, in certain cases, one can obtain a Calabi-Yau manifold by "blowing up" and resolving the singular points) without all the complexity (No-one has yet succeeded in writing down an explicit metric for Calabi-Yau.) Remarkably, the string does not seem to care about the singularities and everything is still well defined. The problem remaining, and this is now emerging as the problem of string theory, is that of vacuum degeneracy. For although the higher-dimensional theory has almost no parameters, all predictive power seems to be lost by the apparent multitude of different compactifications. In particular, Narain (1986) has shown that there are compactified string theories in four dimensions with any rank-22 simply-laced gauge group. See Section 9. Some of these groups, like SO(44) for example, have nothing to do with $E_8 \times E_8$ and do not even contain $E_8 \times E_8$ as a subgroup. This calls into question the current wisdom, based on the 10-dimensional field theory limit, that the standard model SU(3)×SU(2)×U(1) should be embedded in the rank 16 group $E_8 \times E_8$.

Finally, of course, the question arises: "If strings are the ultimate theory, why are there five consistent superstring theories?" Ideally, we would like the ultimate theory to be unique. The answer to this question is still unknown, but an intriguing suggestion has recently been made by Casher et al. (1985) based on an earlier conjecture of Freund (1984): the fundamental theory is the old d = 26 closed bosonic string which when compactified on a 16-dimensional torus (the product of 16 circles $S^1 \times S^1 \ldots S^1$) yields the Type IIA, Type IIB, heterotic SO(32) and heterotic $E_8 \times E_8$ superstring in d = 10; see Table II. Independently, at the same time Duff et al. (1985) attempted to derive the properties of the d = 10 heterotic string by compactifying the bosonic string on the group manifold; see Section 7. We also conjectured (Duff 1986) that the new non-supersymmetric but tachyon-free O(16)×O(16) string (Dixon & Harvey 1986, Alvarez-Gaumé et al. 1986) might have its origin in the bosonic string. (See Cristofano et al. 1986 for a realization of this.)

Table II: The bosonic string strikes back

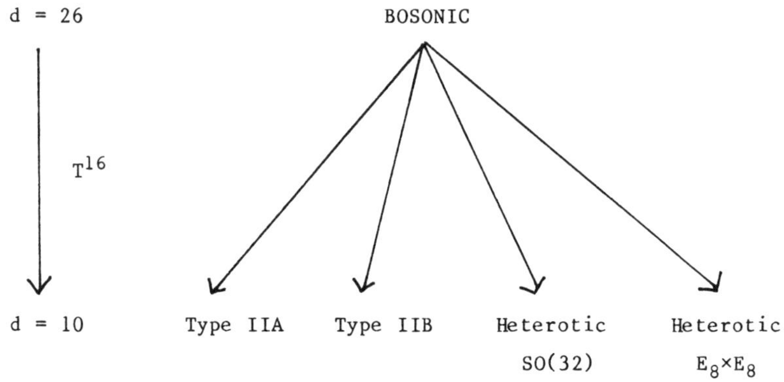

At first sight, these suggestions seem crazy. We have already enumerated the drawbacks of the bosonic string: (a) it describes only bosons (how on earth can we get the fermionic sector of the superstring from a purely bosonic string?); (b) it has a tachyon (the superstrings do not); (c) it suffers from divergences even at one-loop (the superstrings are finite at one-loop, and possibly to all orders). The answer to objection (a) is that the fermions are the "solitons" of the boson theory. (In their broadest definition, solitons are classical lumps of finite energy which exhibit particle-like behaviour.) Fermions as the solitons of an originally bosonic theory are a familiar concept in field theories of two space-time dimensions (where spin is not defined). For example, as shown some time ago by Coleman (1975), the sine-Gordon theory (which consists only of bosons) is equivalent to the Thirring model (which consists only of fermions). The sine-Gordon bosons are the bound states of the Thirring model; the Thirring model fermions are the solitons of the sine-Gordon theory. Of course, Fermi-Bose equivalence in d = 26 or d = 10 is a huge extrapolation from d = 2. The link may well be provided by noting that the string is itself a two-dimensional system moving in a 10 or 26-dimensional space-time. The answer to objection (b) is that the four superstrings and the bosonic string may simply be regarded as different vacuum states of the same theory and that the tachyon of the bosonic string is simply an indication that this particular vacuum is an unstable one: it is somehow energetically more favourable for the system to become supersymmetric. The answer to objection (c) is that, in common with the superstrings, the bosonic string is ultra-violet finite and that its divergences come only from the infra-red, which in turn is just a vacuum-dependent phenomenon which disappears in the true supersymmetric vacuum. Note that by starting with the closed bosonic string in d = 26 one arrives at the four closed superstrings in d = 10. The Type I superstring does not fit neatly into this picture and thus seems to be the black sheep of the family.

It should be emphasized that not everybody accepts that the objections above have been satisfactorily answered and the whole idea remains very controversial. Less controversial is the d = 26 origin of the 496 gauge bosons of SO(32) or $E_8 \times E_8$. Although a traditional Kaluza-Klein compactification on T^{16} could yield only the 16 gauge bosons of $[U(1)]^{16}$, the fact that we are dealing with a string rather than a field theory means that the string can wrap itself around the torus to form topologically non-trivial configurations, and the extra 480 gauge bosons also emerge as "solitons" (Frenkel & Kac 1980; Goddard & Olive 1985a).

7 KALUZA-KLEIN APPROACH TO STRINGS

In Section 3, we saw that the most appealing feature of Kaluza-Klein theories is that what we perceive to be internal symmetries are really space-time symmetries in a higher dimension. In particular, it is not necessary to postulate the separate existence of Yang-Mills fields; they are an automatic consequence of gravity.

It is ironic, therefore, that the recent spectacular successes of superstrings seem to ignore this beautiful concept. First, as discussed in Section 4, the Yang-Mills fields of $E_8 \times E_8$ or SO(32) are present as primary fields already in the d = 10 equations of motion. The bosonic sector of the massless states consists of the graviton $g_{\mu\nu}(x)$, the anti-symmetric tensor potential $B_{\mu\nu}(x)$ and the dilaton $\phi(x)$ coupled to the Yang-Mills potential $A_\mu{}^i$ (i=1,...,496). Secondly, as discussed in Section 5, the favoured compactification to d = 4 yields no Kaluza-Klein gauge fields since Calabi-Yau manifolds have no continuous symmetries.

What has happened to the traditional Kaluza-Klein ideas that

* Yang-Mills $\subset$ gravity?
* Gauge group G $\subset$ general co-ordinate group (i.e., given by the isometry group of M_k)?
* All symmetries are space-time symmetries?

As we have seen in Section 4, this traditional Kaluza-Klein idea fell out of favour largely because of its inability to produce chiral fermions. In order to have chiral fermions in four dimensions, we needed both gravity <u>and</u> Yang-Mills already in the higher dimension. Since the whole idea of Kaluza-Klein was to get internal symmetries from space-time symmetries, having both Yang-Mills and extra dimensions is, to say the least, gilding the lily.

However, if one is stubborn about Kaluza-Klein one can simply cut the Gordian knot and dispense with fermions altogether! Indeed, in Sections 8 and 9, this is exactly what we are going to do. We shall attempt to derive properties of d = 10 superstrings from spontaneous compactification of the old <u>bosonic</u> string on the group manifold $E_8 \times E_8$ (Duff et al. 1985; Duff et al. 1986b). Since $E_8 \times E_8$ has dimension 496 this means starting with the bosonic string in

$$d = 10 + 496 = 506 \tag{38}$$

dimensions! It may seem odd to talk about the bosonic string in d = 506 because we saw in Section 6 that its critical dimension was 26. However, this is strictly true only when the string world-sheet moves through a flat space-time. If we allow space-time to be curved as on a group manifold, then as discussed in Section 8, the critical dimension can change.

Of course, we have only swapped one problem for another. Having reintroduced Kaluza-Klein in the context of the purely bosonic string we are left with the question "where do the fermions come from?" To resolve this problem we shall appeal to the Freund idea of Section 6. At the same time as we presented the Kaluza-Klein compactification of the bosonic string from d = 506 to d = 10 on the group manifold $E_8 \times E_8$ (Duff et al. 1985), Casher et al. (1985) independently presented the Frenkel-Kac compactification of the bosonic string from d = 26 to d = 10 on the torus T^{16}. Although ostensibly very different, the aim of these two approaches was the same: to derive properties of the d = 10 superstrings starting from only the bosonic string. In fact, one can show that, using non-Abelian fermionization (Witten 1984) or by exploiting the uniqueness of representations of Kac-Moody algebras (Goddard and Olive 1985b) that these two approaches are in fact equivalent! We shall now describe the Kaluza-Klein approach in more detail.

8 EINSTEIN'S EQUATIONS FROM VANISHING CONFORMAL ANOMALY

Our starting point is the off-shell effective action Γ of Fradkin and Tseytlin (1985a,b) for the infinite set of local fields corresponding to the modes of the free bosonic string spectrum. Γ is the generating functional of all possible off-shell scattering amplitudes on arbitrary backgrounds. In the case of closed oriented strings Γ is defined (in Euclidean signature) by

$$\Gamma[\hat{\phi}(x),\hat{g}_{MN}(x),\hat{B}_{MN}(x),\ldots] = \Sigma_{\chi=2,0,-2\ldots} e^{\sigma\chi}\int[d\gamma_{ab}]_{\chi}[dx^{M}]e^{-S} \tag{39}$$

where σ is a dimensionless parameter and where

$$S = \int_{M^2} \frac{d^2\zeta}{4\pi\alpha'} \{\sqrt{\gamma}\, \gamma^{ab}\partial_a x^M \partial_b x^N \hat{g}_{MN}(x) + \alpha'\sqrt{\gamma}\, R(\gamma)\hat{\phi}(x)$$
$$+ \varepsilon^{ab}\partial_a x^M \partial_b x^N \hat{B}_{MN}(x) + \ldots\} \tag{40}$$

Here $x^M(\zeta)$ defines the embedding of the two-dimensional world-sheet M_2 in a space-time M_d $(M,N = 1,\ldots,d)$; $\zeta^a = (\tau,\sigma)$ are the co-ordinates on M_2; $R(\gamma)$ is the curvature scalar of the world-sheet metric γ_{ab}. The graviton $\hat{g}_{MN}(x)$, the antisymmetric tensor $\hat{B}_{MN}(x)$ and the dilaton $\hat{\phi}(x)$ correspond to the massless modes of the bosonic string spectrum. As in Section 3, the hats "$\wedge$" are to remind us that these are d-dimensional quantities, the $n < d$ dimensional quantities will have no hats. The dots refer to terms describing the higher spin massive modes and the scalar "tachyon". [By ignoring these higher modes in (40) we are implicitly assuming that, in the correct vacuum state of the theory, these fields have vanishing vacuum expectation values. Thus, we are ignoring the possibility of a "space-invaders scenario" in the sense of Duff et al. (1986a).]

For string consistency, the two-dimensional theory must be conformally invariant and hence the two-dimensional world-sheet stress tensor must be traceless. The general structure of the trace is given by

$$\theta^a_a = \hat{\beta}^\phi \sqrt{\gamma}\, R(\gamma) + \hat{\beta}^g_{MN}\sqrt{\gamma}\, \gamma^{ab}\partial_a x^N + \hat{\beta}^B_{MN}\varepsilon^{ab}\partial_a x^M \partial_b x^N \tag{41}$$

where $\hat{\beta}^\phi$, $\hat{\beta}^g_{MN}$ and $\hat{\beta}^B_{MN}$ are local functionals of $\hat{\phi}(x)$, $\hat{g}_{MN}(x)$ and $\hat{B}_{MN}(x)$. As noted by Polyakov (1981), the first term is just the two-dimensional gravitational conformal anomaly (Capper and Duff 1974; Deser et al. 1986; Duff 1977) while the second and third terms are present even for a flat world-sheet and arise from the self-interactions. The β functions in the tree approximation (in the string expansion) have been calculated by Callan et al. (1985). They find:

$$16\pi^2\hat{\beta}^\phi = \frac{d-26}{3\alpha'} + \left[4(\partial\hat{\phi})^2 - 4\,\hat{\Box}\,\hat{\phi} - \hat{R} + \frac{1}{12}\hat{H}^2\right] + O(\alpha') \tag{42}$$

$$\hat{\beta}^g_{MN} = \hat{R}_{MN} - \frac{1}{4}\hat{H}_M{}^{PQ}\hat{H}_{NPQ} + 2\,\hat{\nabla}_M\partial_N\hat{\phi} + O(\alpha') \tag{43}$$

$$\hat{\beta}^\beta_{MN} = \hat{\nabla}_P\hat{H}^P{}_{MN} - 2\,\hat{H}^P{}_{MN}\partial_P\hat{\phi} + O(\alpha') \tag{44}$$

where $\hat{H}_{MNP}$ is the field strength of $\hat{B}_{MN}$

$$\hat{H}_{MNP} \equiv 3\,\partial_{[M}\hat{B}_{NP]} \tag{45}$$

The absence of a trace anomaly, i.e., the vanishing of $\hat{\beta}^\phi$, $\hat{\beta}^g$ and $\hat{\beta}^B$ is therefore just equivalent to the Einstein matter field equations obtained from the effective Lagrangian

$$\mathcal{L}_d = \sqrt{-\hat{g}}\, e^{-2\hat{\phi}}\left[\frac{d-26}{3\alpha'} - \hat{R} - 4(\partial\hat{\phi})^2 + \frac{1}{12}\hat{H}^2 + \alpha'\hat{R}_{MNPQ}\hat{R}^{MNPQ}\right] + O(\alpha'^2) \tag{46}$$

where we have written the curvature squared term explicitly. Apart from the cosmological term, this action was first calculated by Tseytlin (1985).

As a historical aside, I remember giving a talk on conformal anomalies at the GR7 Conference in Tel Aviv in 1974. The talk immediately after mine was by John Schwarz explaining how Einstein's theory emerged from strings. I wish I had known then that the latter follows from the former!

Of course, these Einstein-matter field equations are not quite the familiar ones, first of all because they include a dilaton and an antisymmetric tensor and secondly because they involve an infinite series in α' with higher and higher derivatives. Nevertheless, it is now of interest to re-examine in the new string framework all the old questions of GR: exact solutions, black holes, singularity theorems, cosmology, etc. Since most of the people working on strings are particle physicists, this presents a golden opportunity to the general relativist, once he can liberate himself from the shackles of classical four-dimensional physics.

9 SPONTANEOUS COMPACTIFICATION OF THE BOSONIC STRING

We now turn to the question of spontaneous compactification. One obvious vacuum solution to the field equations corresponds to $\langle\hat{\phi}\rangle$ = constant, $\langle\hat{H}_{MNP}\rangle = 0$ and $\langle\hat{g}_{MN}\rangle$ the flat metric, but this is valid only for $d = 26$. In this case, the possible ground states are given by

$$M_d = M^n \times T^r \tag{47}$$

where M^n is n-dimensional Minkowski space and T^r is the r torus with

$$n = 26 - r \tag{48}$$

However, for $d > 26$ the cosmological term in (46) obliges us to look for solutions in which some of the dimensions are compactified on a curved manifold, and we can now follow the traditional Kaluza-Klein interpretation. Accordingly, we split the indices

$$x^M = (x^\mu, y^m) \tag{49}$$

where x^μ ($\mu = 1,\ldots,n$) refers to space-time and y^m ($m = 1,\ldots,k$) to the extra dimensions. One solution which suggests itself corresponds to the case

$$M_d = M^n \times G \tag{50}$$

where G is a non-Abelian group manifold of dimension k given by

$$n = d - k \tag{51}$$

In this case

$$\langle\hat{g}_{mn}\rangle = \frac{1}{c_A} f_{mpq} f_n{}^{pq} = g_{mn}(y) \tag{52}$$

where f_{mnp} are the structure constants of the group, converted to world indices and c_A is the second-order Casimir in the adjoint representation. This will indeed be a solution provided

$$\langle\hat{\phi}(x,y)\rangle = \text{constant} \tag{53}$$

$$\langle\hat{H}_{mnp}(x,y)\rangle = \hat{H}_{mnp}(y) = m\, f_{mnp} \tag{54}$$

where m is a constant with dimensions of mass. Thus, the three-form in the extra dimensions corresponds to the parallelizing torsion on the group manifold, i.e., the curvature calculated from the connection $\hat{\Gamma}_{mnp} \pm \frac{1}{2}\hat{H}_{mnp}$ vanishes. This ensures that $\langle\hat{\beta}^g\rangle$ and $\langle\hat{\beta}^B\rangle$ vanish identically from (43) and (44), whereas substitution into the $\hat{\beta}^\phi = 0$ equation yields

$$d-26 = k\alpha' m^2 c_A/2 + O(\alpha'^2) \tag{55}$$

Fortunately, this equation has already been solved to all orders in α' because if we substitute the ground state values of $\hat{g}_{MN}$, $\hat{B}_{MN}$ and $\hat{\phi}$ into the Fradkin-Tseytlin action (40), the result is nothing but a non-linear σ-model on $M^n\times G$ with a Wess-Zumino term provided by $\langle\hat{H}_{mnp}\rangle$. [The dilaton term in (40) reduces to the Euler number $\chi = (1/2\pi)\int d^2\xi\sqrt{\gamma}\, R(\gamma)$ and renormalizes the parameter σ.] Such σ-models have been well studied in the literature (Goddard and Olive 1985a,b,1986; Curtright and Zachos 1984; Witten 1984; Nemeschansky and Yankielowicz 1984; Bergshoeff et al. 1986; Altschuller and Nilles 1985; Jain et al. 1985; Gepner and Witten 1986; Redlich and Schnitzer 1986). In fact we know that $\langle\hat{\beta}^g\rangle$ and $\langle\hat{\beta}^B\rangle$ vanish identically to all orders in α' and that $\langle\hat{\beta}^\phi\rangle$ also vanishes provided we satisfy the critical dimension formula

$$d - 26 = \frac{kc_A}{2p+c_A} \tag{56}$$

and that the radius of the group manifold is quantized in units of $\alpha'^{-\frac{1}{2}}$

$$p^2m^2\alpha' = 1 \tag{57}$$

where p is an integer. The appearance of the integer p follows from the fact that the coefficient of the Wess-Zumino term is quantized. In what follows we shall confine our attention to the case p = 1 for which the non-linear σ-model is equivalent on the one hand to a theory of free fermions and on the other to a theory of free bosons defined on $M^n\times T^r$

where r is the rank of G. In this case, the only restriction on G is that it be simply laced (the root vectors having the same length) as shown in Table III. However, we shall later insist that G also be self-dual (root lattice equal to the weight lattice) which restricts us to $G = E_8 \times E_8$ or $G = SO(32)$, i.e., $k = 496$, $r = 16$, $c_A = 60$ and hence $d = 506$ and $n = 10$. It is no coincidence that these are the two anomaly-free groups of $n = 10$ superstrings. Thus one of the goals of this investigation is to establish a dictionary between compactification of the $d = 506$ bosonic string on the $k = 496$ group manifold and compactification of the $d = 26$ bosonic string on the $r = 16$ torus.

Table III: The simply-laced groups

G	dim G	rank G	c_A
SU(r+1)	r(r+2)	r	2r+2
SO(2r)	r(2r-1)	r	4r-4
E_6	78	6	24
E_7	133	7	36
E_8	248	8	60

The bi-invariant metric on the group manifold is, as its name suggests, invariant under both left and right actions and hence has isometry group $G_L \times G_R$. Since the VEVs of $\hat{\phi}$ and $\hat{H}_{MNP}$ in Section 8 are also $G_L \times G_R$-invariant, the massless sector of the compactified theory will contain the Yang-Mills gauge bosons of $G_L \times G_R$ denoted $(A_\mu{}^i, \tilde{A}_\mu{}^i)$ with field strengths $(F_{\mu\nu}{}^i, \tilde{F}_{\mu\nu}{}^i)$. If we now substitute the Kaluza-Klein ansatz into the d-dimensional $\hat{\beta}$ functions (42)-(44), we obtain corresponding n-dimensional β functions which are themselves just equivalent to the Einstein-matter field equations obtained from the Lagrangian

$$\mathcal{L}_n = \sqrt{-g}\, e^{-2\phi}[-R - 4(\partial\phi)^2 + \frac{1}{12}H^2 + \frac{\alpha'}{2}(2R_{\mu\nu\rho\sigma}R^{\mu\nu\rho\sigma} - F^i_{\mu\nu}F^{\mu\nu i} - \tilde{F}^i_{\mu\nu}\tilde{F}^{\mu\nu i})] + O(\alpha'^2) \tag{58}$$

where now, instead of $d\hat{H} = 0$, we have

$$dH = \alpha' \mathrm{Tr}(\tilde{F} \wedge \tilde{F} - F \wedge F) \tag{59}$$

i.e., we have acquired n-dimensional Chern-Simons terms even though there were none in d-dimensions.

The reason why we bother substituting into the equations of motion rather than directly into the Lagrangian is to check Kaluza-Klein consistency, which is an important topic in its own right which I shall not have time to discuss in detail. The basic problem is that the massless Kaluza-Klein ansatz which singles out the low-energy sector of the compactified theory is not guaranteed to solve the higher-dimensional field equations. Thus setting the massive states to zero could impose unacceptable constraints on the surviving massless states which are not implied by the field equations. It is a remarkable property of the bosonic string compactified on the group manifold (which is also shared by the S^7 compactification of d = 11 supergravity) that the massless ansatz is completely consistent, provided we include all the massless modes, in particular the massless scalars which for simplicity have not been included in Eq. (58). See Duff et al. (1986b) for details. This consistency problem also arises if one tries to derive the effective low-energy supergravity theory obtained from Calabi-Yau compactification of the 10-dimensional heterotic string by simply discarding the massive states (Duff et al. 1986a). See Font & Quevedo (1986) for a recent detailed discussion.

10 THE HETEROTIC AND TYPE II STRINGS

To obtain the corresponding Lagrangian for the heterotic string, we

* Choose $G = E_8 \times E_8$ for which $k = 496$, $r = 16$, $c_A = 60$, $d = 506$ and hence $n = 10$
* Decompose $G_R \supset SO(8)$, i.e, $\tilde{A}_\mu{}^i$ $(i=1,..,496) \to \tilde{A}_\mu{}^{ab}$ $(a,b,=1,..,8)$;
* Following Charap & Duff (1977), identify the right-handed Yang-Mills gauge potential with the gravitational spin connection

$$\tilde{A}_\mu{}^{ab} = \omega_\mu{}^{ab} \tag{60}$$

From (58) and (59), we obtain

$$\mathcal{L}_{\text{heterotic}} = \sqrt{-g}\, e^{-2\phi}\Big[-R - 4(\partial\phi)^2 + \frac{1}{12}H^2 + \frac{\alpha'}{2}\big(R_{\mu\nu\rho\sigma}R^{\mu\nu\rho\sigma} - F^i_{\mu\nu}F^{\mu\nu i}\big)\Big] + O(\alpha'^2) \tag{61}$$

where

$$dH = \alpha' \mathrm{Tr}(R \wedge R - F \wedge F) \tag{62}$$

But this is just the heterotic string Lagrangian of Section 5, together with both the Yang-Mills and Lorentz Chern-Simons terms!

It should be admitted, however, that in common with Casher et al. (1985) and Englert et al. (1986), we have as yet no dynamical understanding of the decomposition $G_R \supset SO(8)$ and the identification (60). Nor do we see any justification for the truncation of the string spectrum which seems to be entailed in reproducing that of the heterotic string. The idea is that states whose G_R index i $(i = 1,..,496)$ runs over the 8_s spinor index α $(\alpha = 1,..,8)$ of $SO(8)_{\text{internal}}$ transform as fermion representations of the diagonal SO(8)

$$SO(8)_{\text{heterotic}} \subset SO(8)_{\text{bosonic}} \times SO(8)_{\text{internal}} \tag{63}$$

which is identified as the transverse space-time group of the heterotic string. Hence, in some sense, the G_R Yang-Mills boson $\tilde{A}^\alpha_\mu$ is really the gravitino

$$\tilde{A}_\mu{}^\alpha \to \psi_\mu{}^\alpha, \tag{64}$$

the scalars $S_i{}^\alpha$ are really the gauginos

$$S_i{}^\alpha \to \chi_i{}^\alpha \tag{65}$$

and the Yang-Mills parameter $\tilde{\alpha}^\alpha$ is really the supersymmetry parameter

$$\tilde{\alpha}^\alpha \to \varepsilon^\alpha \tag{66}$$

but the origin of the Fermi statistics remains obscure. We would like to be able to say that there are distinct vacua of the bosonic string relative to one of which all states transform as bosons, but relative to the other some states transform as fermions. Unfortunately, we are not yet in a position to make this more precise. Nevertheless, we glimpse the beginnings of the explanation for supersymmetry by using (60), (64) and (66) to convert the G_R Yang-Mills transformation rule

$$\delta\tilde{A}_\mu{}^i = \partial_\mu\tilde{\alpha}^i + f^i{}_{jk}\,\tilde{A}_\mu{}^j\tilde{\alpha}^k \tag{67}$$

into the gravitino transformation rule

$$\delta\psi_\mu^\alpha = \partial_\mu\varepsilon^\alpha - \frac{1}{4}\,\omega_\mu{}^{ab}[\gamma_{ab}]^\alpha_\beta\varepsilon^\beta + \ldots \tag{68}$$

on using the property of the E_8 structure constants

$$f^\alpha{}_{[ab]\beta} = -\frac{1}{4}\,[\gamma_{ab}]^\alpha_\beta \tag{69}$$

Similarly the Type IIA and Type IIB theories are obtained by

* decomposing both $G_R \supset SO(8)$ and $G_L \supset SO(8)$
* identifying both the left- and right-handed Yang-Mills gauge potentials with gravitational spin connection

$$A_\mu{}^{ab} = \tilde{A}_\mu{}^{ab} = \omega_\mu{}^{ab} \tag{70}$$

to obtain

$$\mathcal{L}_{\text{Type II}} = \sqrt{-g}\, e^{-2\phi}\left[-R - 4(\partial\phi)^2 + \frac{1}{12}H^2\right] + O(\alpha'^2) \tag{71}$$

$$dH = 0 \tag{72}$$

The extra 64 bosonic degrees of freedom are then provided by $S^{\dot{\alpha}\beta}$ or $S^{\alpha\beta}$ corresponding to the embeddings $(8_c,8_s)$ in the case of Type IIA or $(8_s,8_s)$ in the case of Type IIB. The second supersymmetry of Type II has the same origin in G_L as did the first in G_R, i.e., with $A_\mu{}^\alpha$ and α^α playing the part of gravitino and supersymmetry parameter.

Thus we arrive at the bizarre picture of a three-in-one world that can be described equivalently in 10, 26 or 506 dimensions as in Table IV.

TABLE IV: A three-in-one world described equivalently by 10, 26, or 506 dimensions.

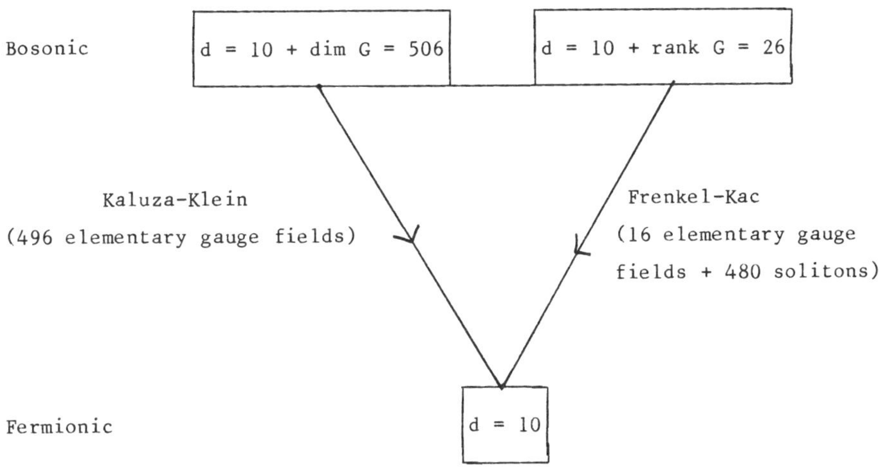

11 CONCLUSIONS

I began my talk in Stockholm by posing a little quiz for the audience, and my talk ended (as these talks often do) by the audience posing one for me. The questions centred around the (perhaps understandable) reaction of the average classical general relativist : you particle physicists promised us that the way forward lay in supergravity, then you said it was Kaluza-Klein, now you tell us it is strings. Is it safe to ignore string theory, and hope that it will go away?

This is just the question posed at the end of Weinberg's (1986) closing talk at the Oregon Meeting, so I will let him answer it.

> "In trying to answer these questions, I must say right away that there is not the slightest shred of experimental evidence for string theory. The same was also true of the other theories that we developed in our desperate attempt to go beyond the standard model, in particular for supersymmetry and Kaluza-Klein theories, which have now been incorporated into superstring theory. Never has so much brilliant mathematics been done by physicists with so little encouragement from experiment. Furthermore, just as for supersymmetry and Kaluza-Klein theories, the string theories have still not settled down so that they could make very definite predictions which could be tested experimentally. However, in this respect I think string theories are really different from the Kaluza-Klein and supersymmetry theories. Supersymmetry is a symmetry like Lorentz invariance; it allows a tremendous variety of possible dynamical theories. Likewise, Kaluza-Klein theory is just a general idea that there might be some higher number of dimensions which are compactified; again, this idea allows a great many specific theories. String theories, on the other hand, are very rigid. There are almost no string theories at all, and of the few possibilities only one at present (the "heterotic" superstring of Gross, Harvey, Martinec and Rohm) seems at all promising phenomenologically."

> "Relying on a general sense that nothing ever in this life works out the way we want it to ("we" here means theorists),

my guess is that there are many surprises that will be provided to us both by imaginative theorists and by enterprising experimentalists before we finally get the solution to our problem. Nevertheless, I would argue that it is not safe to ignore string theories and wait for the next change in fashion. These theories are much too promising and too beautiful for us not to take them very seriously and explore their consequences for as long as it takes."

ACKNOWLEDGEMENTS

I have benefited from conversatioons with Ali Chamseddine, Chris Isham, Bengt Nilsson, Chris Pope, Douglas Ross and Nick Warner.

REFERENCES

Altschuler, D. & Nilles, H.P. (1985). Phys. Lett., 154B, 226.

Alvarez-Gaumé, L., Ginsparg, P., Moore, G. & Vafa, C. (1986). HUTP 86/A013.

Awada, M., Duff, M.J. & Pope, C.N. (1983). Phys. Rev. Lett., 50, 294.

Bergshoeff, E., Randjbar-Daemi, S., Salam, A., Sarmadi, H. & Sezgin, E. (1986). Nucl. Phys. B269, 77.

Bergshoeff, E., de Roo, M., de Wit, B., van Nieuwenhuizen, P. (1982). Nucl. Phys., B195, 97.

Candelas, P., Horowitz, G.T., Strominger, A. and Witten, E. (1985). Nucl Phys. B258, 46.

Capper, D.M. & Duff, M.J. (1974). Nuovo Cimento, 23A, 173.

Callan, C., Friedan, D., Martinec, E. & Perry, M. (1985). Nucl. Phys., B262, 593.

Casher, A., Englert, F., Nicolai, H. & Taormina, A. (1985). Phys. Lett., 162B, 121.

Chamseddine, A.H. (1981). Phys. Rev., D1724, 3065.

Chapline, G.F. & Manton, N.S. (1983), Phys. Lett.,B120, 105.

Charap, J.M. & Duff, M.J. (1977), Phys. Lett., 69B, 445.

Coleman, S. (1975). Phys. Rev., D11, 1088.

Cremmer, E., Julia, B. & Scherk, J. (1978). Phys. Lett., 76B, 409.

Christofano, C., Maiella, G., Musto, R., Nicodemi, F. & Pettorino, R. (1986). Univ. of Naples preprint DFNA/24.

Curtright, T. & Zachos, C. (1984). Phys. Rev. Lett., 53, 1799.

Deser, S., Duff, M.J. & Isham, C.J. (1976). Nucl. Phys., B114, 29.

Dixon, L., Harvey, J.A., Vafa, C. & Witten, E. (1985). Nucl. Phys.,B261, 678.

Dixon, L. & Harvey, J. (1986). Nucl. Phys. B274, 93.

Duff, M.J. (1977). Nucl. Phys., B125, 334.

Duff, M.J., Nilsson, B.E.W. & Pope, C.N. (1985). Phys. Lett., 163B, 343.

Duff, M.J., Nilsson, B.E.W. & Pope, C.N. (1986a). Physics Reports, 130, 1.

Duff, M.J., Nilsson, B.E.W., Pope, C.N. & Warner, N. (1986b). Phys. Lett., B171, 170.

Duff, M.J. (1986). Phys. Lett. B773, 289.

Einstein, A. & Bergmann, P. (1938). Ann. Math., 39, 685.

Englert, F., Nicolai, H. & Schellekens, A. (1986). CERN preprint TH.4360/86.

Font, A. & Quevedo, F. (1986). CERN preprint TH.4569/86 (1986).

Fradkin E.S. & Tseytlin, A. (1985a). Phys. Lett., B158, 316.

Fradkin, E.S. & Tseytlin, A. (1985b). Nucl. Phys., B261, 1.

Freeman, M.D., Pope, C.N., Sohnius, M.F. & Stelle, K.S. (1986). Phys. Lett., 178B, 199.

Freeman, M.D. & Pope, C.N. (1986). Phys. Lett., 174B, 48.

Frenkel, I. & Kac, V.G. (1980). Inv. Math., 62, 23.

Freund, P.G.O. & Rubin, M.A. (1980). Phys. Lett., 97B, 233.

Freund, P.G.O. (1984). Phys. Lett., 151B, 387.

Gepner, D. & Witten, E. (1986). Princeton preprint.

Goddard, P. & Olive, D. (1985a). In Workshop on Vertex Operators in Mathematics and Physics, Berkeley.

Goddard, P. & Olive, D. (1985b), Nucl. Phys., B257, 226.

Goddard P. & Olive, D. (1986). DAMTP preprint.

Green, M.B. & Schwarz, J. (1981). Nucl. Phys., B181, 502.

Green, M.B. & Schwarz, J. (1984). Phys. Lett., B149, 117.

Grisaru, M.T., Van der Ven, A.E.M. & Zanon, D. (1986). Phys. Lett., 173B, 423.

Gross, D., Harvey, J., Martinec, E. and Rohm, R. (1985). Phys. Rev. Lett., 54, 502.

Jain, S., Shankar, R. & Wadia, S.R. (1985). Phys. Rev., D32, 2713.

Kaluza, Th. (1921). Sitzungber. Preuss. Akad. Wiss. Berlin Math. Phys., K1, 966.

Klein, O. (1926a). Z. Phys., 37, 895.

Klein, O. (1926b), Nature, 118, 516.

Knizhnik, V.G. & Zamolodchikov, A.B. (1984). Nucl. Phys., B247, 83.

Nahm, W. (1979). Nucl. Phys., B135, 149.

Nambu, Y. (1970). In "Proc. Int. Conf. on Symmetry and Quark Models", (Gordon & Breach).

Narain, K.S. (1986). Phys. Lett. B169, 41.

Neveu, A. & Schwarz, J. (1971). Nucl. Phys., B31, 86.

Nemenchansky, D. & Yankielowicz, S. (1984). Phys. Rev. Lett., 54, 620.

Nepomechie, R.I. (1985). Seattle preprint.

Polyakov, A.M. (1981). Phys. Lett., 103B, 207.

Ramond, P. (1971). Phys. Rev., D3, 2415.

Redlich, A.N. & Schnitzer, H.J. (1986). Phys. Lett., 167B, 315.

Scherk, J. & Schwarz, J. (1974). Nucl. Phys., B81, 118.

Tseytlin, A. (1985). In Quantum Field Theory and Quantum Statistics, Ed. C.J. Isham (Adam Hilger, Bristol, 1986).

Veneziano, G. (1968). Nuovo Cimento 57A, 190.

Witten, E. (1984), Commun. Math. Phys., 92, 455.

Witten, E. (1985). In Shelter Island II, Eds. R. Jackiw, N. Khuri, S. Weinberg & E. Witten (MIT Press 1985).

Weinberg, S. (1986). Proceedings of the Oregon Meeting, Eugene, Oregon (Aug. 1985), ed. R.C. Hwa (World Scientific, Singapore, 1986).

FOLKLORE IN RELATIVITY AND WHAT IS REALLY KNOWN

Jürgen Ehlers
Max-Planck-Institut für Physik und Astrophysik
Institut für Astrophysik
8046 Garching b. München, FRG

Before I start with the subject of this talk, I wish to say something about its **title** and its **purpose**. When the title was proposed to me by the scientific organizing committee – I did not invent it – I, perhaps carelessly, accepted it, thinking it may be worthwhile to you and also possible for me to separate what is **believed** from what is **established** in classical general relativity. I still feel that it is worthwhile to make that effort, but at the same time I have to admit that I am unable to present to you now the result of such an effort in a way that is in any sense complete or systematic. Thus the sole purpose of the following remarks is to indicate in a few instances the difference between belief and knowledge as I see it, being aware that there are differences of opinion on what constitutes knowledge. My role in this endeavour is somewhat like that of the child in the story of the Emperor's New Clothes. As in that story, it may bring some relief to some of you if things are spelled out which are seen by all, but usually not admitted. Comments and corrections will, of course, be much appreciated.

I shall restrict attention to what is now called classical GR. In view of what we have heard yesterday from Chris Isham I doubt that anyone would dare try to separate belief from knowledge in quantum gravity.

To put you into the appropriate mood for this lecture I should like to quote John Leighton Synge. In a 1964 Boyle medal lecture given at the Royal Irish Academy he said [1]: "As science advances, it seems to have a sort of scorched earth policy. The advancing army is full of enthusiasm for its advancing into the unknown, and the unknown is always exciting. If it glances back at the territory it has overrun, it sees little but dullness, the dullness of what seems to be completely known, with little prospect of adding to that knowledge by a deeper understanding."

So, let us now turn to the dullness of what seems to be completely known. Perhaps it is neither dull nor, in fact, really known.

Let me begin with some rather general remarks. Very often it is asserted that the gravitational force between an electron and a proton in a hydrogen atom is smaller than the Coulomb force by a factor of about 10^{40}. Does that statement represent knowledge or folklore? We have certainly no direct experimental evidence for it, nor is it part of a relativistic theory covering consistently electromagnetic and gravitational interactions between subatomic particles governed by quantum mechanics. On the other hand the statement **is** part of non-relativistic quantum mechanics, and within that framework it makes successful, if indirect, contact with observed facts of stellar structure via the appropriate thermodynamic limit (see W. Thirring [2] and the references given there). This simple example shows that a decision "folklore or knowledge?" cannot be made unambiguously; what matters is in each case a clarification of the context and sense in which the statement is one or the other.

What about the statement "The metric ceases to be well-defined below the scale of the Planck length because of quantum fluctuations"? Is that folklore or knowledge? The answer depends on the criteria you wish to use.

How about the **SR/GR correspondence principle** in the form: "Non-gravitational laws of physics can be taken over from SR to GR by the rule $\eta \to g$, $\partial \to \nabla$"? Useful as it is as a heuristic guide, it is also well-known that a) for laws of higher than first differential order the rule is ambiguous, because of "$\nabla_{[a}\nabla_{b]} \neq 0$", and b) even for SR-laws which are of first differential order, the minimal coupling rule in some cases (wave equations with spin larger than two) does not lead to a sensible GR-generalization of a consistent SR-theory [3].

As another example let me mention one version of **Mach's Principle** in the form: "The distribution and motion of matter uniquely determines the fields, in particular the metric." In GR that assertion is neither true nor false, but **meaningless**, since so far any (classical or quantal) description of the structure and the state of matter, in particular its "distribution and motion", already involves a metric as an essential prerequisite. (This has been pointed out already in 1924 in unsurpassed clarity in a "popular" essay by

Hermann Weyl [4].) $T^{\alpha\beta}$ (or $T_{\alpha\beta}$ or $T^{\alpha}{}_{\beta}$) alone, for example, does not describe matter.

Let me now go on to more specific statements in GR concerning the **field equation** and its relation to **equations of motion**. To be concrete, let us idealize matter as a perfect fluid, and let us choose as its thermodynamic potential the pressure p as a function of the specific enthalpy h and the specific entropy s. We really know the following [5]:

Theorem: A (sufficiently differentiable) solution $(g_{\alpha\beta}, U^{\alpha}, h, s)$ of

$$G^{\alpha\beta} = 8\pi\{(\mu+p)U^{\alpha}U^{\beta} + p\ g^{\alpha\beta}\}, \tag{1}$$

$$(\rho U^{\alpha})_{;\alpha} = 0 \tag{2}$$

and

$$g_{\alpha\beta}U^{\alpha}U^{\beta} = -1, \tag{3}$$

where $p = p(h, s)$ and

$$\rho = \frac{\partial p}{\partial h}, \quad \mu + p = h\rho, \tag{4}$$

is locally uniquely (up to diffeomorphisms) determined by appropriate initial data obeying constraints and satisfying the inequalities $\rho > 0$, $h > 0$, $\partial p/\partial s < 0$. This solution depends causally on the data provided

$$0 < \left[\frac{\partial p}{\partial \mu}\right]_s \leqslant 1, \tag{5}$$

i.e. if the speed of sound $(\partial p/\partial\mu)^{1/2}$ of the fluid does not exceed the fundamental speed $c = 1$.

(An analogous, spatially global theorem for a compact initial 3-space or for asymptotically flat initial data on R^3 presumably holds, but does not appear to have been stated in the literature.)

According to this theorem, for a fluid (and presumably for other kinds of continuous media) equations of motion in addition to (1)-(4) are not needed. The field equation combined with a suitable source model determines the source's motion as well as the evolution of the gravitational field. The reason for this is, of course, that Einstein's equation implies the local energy-momentum law

$$T^{\alpha\beta}{}_{;\beta} = 0 \ . \tag{6}$$

However, for two or more material bodies or black holes separated by empty space no corresponding theorem is known. Since Einstein's field equation is incompatible with point sources and relativistic causality also excludes rigid bodies (which would violate the solid-body analog of (5)), relativistic dynamics have to deal, at least at a fundamental level, with extended, deformable bodies. (Although this circumstance implies considerable "technical" complications for the theory, it may basically be considered rather as one of its assets; for i) macroscopic bodies **are** extended, and ii) in relativistic microscopic theories starting with electron theory the point particle idealization has led to at least as many difficulties as successes, viz. all kinds of infinities.)

The simplest objects of classical GR-dynamics which correspond to point particles in Newtonian dynamics appear to be, in fact, black holes, as is shown by the Israel-Carter-Robinson theorems [6] and is further supported by the membrane description [7]. The essential unsolved problem in establishing mathematically the existence of relativistic N-body space-times is to find a way to determine the motion of the **boundaries** of the bodies, which are characteristics of the evolution equations along which the pressure (or normal stress) vanishes and the density may be discontinuous. The existence of initial data obeying the constraints which may be compatible with N-body problems has been established ([8], [9], [10]), but for the reason just mentioned it has not been possible to show that there actually exist, for these or similar initial data, solutions to the evolution equations. (The existence theorem for fluids stated above requires $\mu+p$ to be strictly positive everywhere on the initial hypersurface and thus excludes fluid balls separated by empty regions.) In this respect the situation is better for black holes, a fact which supports the view that black holes are the simplest objects of GR-dynamics. Even in Newton's theory no existence theorems for non-stationary systems of gravitationally interacting deformable bodies appear to be known. – Giving up the idealization of sharp boundaries is no cure since one does not know how to model bodies without boundaries.

In spite of the facts just reported, folklore has it that Einstein's field equation implies equations of motion

a) for test bodies in external fields, and

b) for heavy, i.e. gravitationally interacting bodies.

Such claims are based on formal, so far not rigorously justifiable, approximation methods. Within such approaches, **approximate equations of motion** for point-like bodies - nearly spherical objects which are small compared to their distances - have been obtained as ordinary post-Galilei-invariant, respectively retarded, Poincare-invariant differential equations for world lines l, l', ... representing the (centers of mass of the) bodies. Their formulation is based on and involves a flat, non-dynamical, Galilean respectively Lorentzian background metric, and they arise as conditions which those world-lines have to satisfy in order that there exists a dynamical, "physical" Lorentz metric $g_{\alpha\beta}$ and a matter tensor $T^{\alpha\beta}$ such that

a) outside some world tubes containing l, l', ..., the vacuum field equation holds approximately for $g_{\alpha\beta}$;

b) $g_{\alpha\beta}$ obeys some asymptotic flatness (and possibly initial-) condition which implies that $g_{\alpha\beta}$ equals asymptotically at infinity the background metric, and an absence-of-incoming-radiation condition;

c) ($g_{\alpha\beta}$, $T^{\alpha\beta}$) obey approximately within world tubes containing l, l', ..., interior field equations for a specified matter model, or alternatively $g_{\alpha\beta}$ asymptotically equals, near any one of these lines, a Schwarzschild-, Kerr-, or similar exterior one-body metric.

In brief: Equations of motion are (approximate) conditions which are necessary and sufficient in order that the fields of various bodies which have, near the bodies, a form characteristic of the bodies in question, fit together so as to satisfy, outside of and between the bodies, the vacuum field equation and obey, at large distances, certain asymptotic conditions. To transform this very suggestive "field theoretic interpretation of the laws of mechanics", already indicated by G. Mie in 1913 and explicitly spelled out for GR first by H. Weyl in 1920, into precise mathematical statements has so far not been possible, in spite of many efforts.

I have indicated the somewhat subtle logical status of approximate equations of motions in GR in order to emphasize:

1) Although equations of motion (as opposed to "laws of motion"; for this distinction see the contribution of P. Havas in [11]) arise as conditions of solvability of the field equation for the dynamical, curved metric, they do not contain this metric, but instead contain and presuppose a given, non-dynamical background metric.
2) Observationally testable predictions about the relative motions of the bodies, e.g. about angles between light rays as functions of proper times require, besides the knowledge of the world lines l, l', ... as solutions of the equations of motion, also the physical metric $g_{\alpha\beta}$ "between" these lines.
3) Although each one of the world lines l, l', ... is a geodesic of a certain metric (even when gravitational radiation reaction forces are taken into account), which arises from the total metric $g_{\alpha\beta}$ by subtracting a "self-potential" $h_{\alpha\beta}$ – this "geodetic law" is useless for computing that world line except at the first (linear) post-Minkowskian order, since at higher orders the metric of which l is a geodesic depends itself functionally on l. Only in a test-body limit does the geodetic law become useful even at higher orders.

Approximate equations of motion in this sense have been derived now by various, mutually agreeing approximation methods, even up to the 2½ post-Newtonian order where gravitational radiation reaction terms appear for the first time (see, e.g., [12], [13] and the references therein). It is far from clear, however, whether any of the formal approximation methods converges or provides metrics which indeed approximate actual solutions, and whether exact equations of motion for point-like bodies or black holes exist according to GR. (Formal approximation methods at best provide fields $(g_{\alpha\beta}, T^{\alpha\beta})$ such that Synge's "error tensor" $(8\pi)^{-1}G^{\alpha\beta}-T^{\alpha\beta}$ is "small" in some sense. Nobody has shown, however, that that implies the existence of an exact solution $(\bar{g}_{\alpha\beta}, \bar{T}^{\alpha\beta})$ such that $g_{\alpha\beta}-\bar{g}_{\alpha\beta}$ and $T^{\alpha\beta}-\bar{T}^{\alpha\beta}$ are "small".)

Further progress concerning the general-relativistic N-body problem requires presumably a combination of the strengths of various approximation methods with each other and with rigorous parts of GR such as the ordinary and characteristic initial value problems (see J. Winicour [14]), asymptotics, and Dixon's rigorous treatment of extended bodies as further elaborated in [15], [16]; In this context compare the remarks made in [24] and the references cited there. I consider it as unfortunate, in view of the central importance of the N-body problem for a theory of gravitation and for observational tests of its field equation, that these challenging problems attract relatively little attention compared to many speculative or purely formal research topics in General Relativity. From the point of view of physics it also seems to be inappropriate that most mathematical research on Einstein's equation focusses on the vacuum equation. What is observed is matter and (so far non-gravitational) radiation; after all the vacuum field was introduced to account for interactions between bodies.

Perhaps surprisingly, even the simpler problem of deriving the **geodetic law for test bodies** has not been rigorously solved. It is true that if, in a given smooth space-time $(M, g_{\alpha\beta})$, a distributional stress-energy-momentum tensor $T^{\alpha\beta}$ is supported by a time-like world line $l \equiv z(s)$, obeys the dominant energy condition and satisfies (6), then l is geodetic and $T^{\alpha\beta}$ is monopolar,

$$T^{\alpha\beta}(x) = m \int \dot{z}^{\alpha}\dot{z}^{\beta} \; \delta(x-z(s)) \; ds,$$

and m is a positive constant [15]. **But** since such a $T^{\alpha\beta}$ is incompatible with Einstein's equation, it is not a valid representation of a body in GR. Statements about test bodies ought to be considered as limits of statements about heavy bodies (including black holes), so that they may be applied, as approximations, whenever actual bodies satisfy nearly the conditions characterizing the limit. The formally convenient textbook definition of a test body as one which has no self-field is inadequate since it excludes most cases to which one would like to apply the test body concept. Not only does that definition fail in the case of the Earth viewed as a test body in the Schwarzschild field of the Sun - even in the case of an inch-sized ball out of heavy metal falling freely near the Earth's surface the curvature within and near the ball is essentially determined by the ball, and not by the Earth. Nevertheless one conjectures the geodetic law to apply to both cases. Newtonian theory and approximation methods (compare remark 3) above) indicate that that law holds approximately provided i) the size of the body in question is small compared to the scale on which the external field changes, ii) the world line used to represent the body is chosen appropriately, and iii) the connection with respect to which that world line is geodetic is an external one. It still remains a challenge to derive a corresponding limit theorem from the field equation, perhaps by means of Dixon's [17] description of bodies in GR or by "rigorising" the Infeld-Schild [18] argument. The crucial problem is to find a suitable definition of a self-field, perhaps by solving the constraint equations in leading order near the body in question. (In this context see the penetrating remarks by Dixon in [11], section 6 which are in agreement with Damour's approximate treatment quoted above.) It is perhaps worth recalling here that the dominant energy condition on $T^{\alpha}{}_{\beta}$ implies causal motion [19], and ensures the existence of a center-of-mass line with reasonable properties [20], and that it precludes the existence of models of bodies compatible with Einstein's field equation and having a finite number of multipole moments only [15], such as pole-dipole particles.

Another statement which seems to be generally believed, usually on rather flimsy arguments, is: **Newton's theory of gravity is a "limit" of Einstein's**. To understand this limit relation is important since i) Newtonian theory successfully explains many gravitational phenomena and ii) the approximation methods on which the comparison of GR with observations is based assume such a limit relation and even use Newtonian concepts such as masses and linear momenta of gravitationally interacting bodies, which have no precise meaning in GR.

About this limit relation the following is known (see, e.g. [21], [22], [23].):

1. There exists a reformulation of the theories of Newton (N) and of Einstein (E) in terms of common, generally covariant spacetime concepts and laws such that N and E differ only in the value of one constant which appears in one of the laws and which equals 0 in N while it is c^{-1} (the inverse speed of light) in E. (This holds for isolated systems. Otherwise, N differs from E by one additional local law.)
2. Using the common formulation of N and E alluded to in 1) one can give a precise meaning to the statement that a one-parameter family of models ("solutions") of E converges to a model of N, where $\epsilon = c^{-1}$ serves as parameter.
3. One can exhibit many examples of families having Newtonian limits such as the suitably parametrized Schwarzschild interior plus exterior solution with test bodies moving around the central body, the Schwarzschild and

Kerr black holes, Weyl solutions, Friedmann-Lemaitre cosmological models, Gödel's model, the Oppenheimer-Snyder collapsing star model, and even plane waves. This shows that the limit definition is not empty and seems to capture what it is supposed to. (The NUT-family of solutions admits, interestingly and perhaps not surprisingly, a limit which is not quite Newtonian; in the limit $c \to \infty$ a non-Newtonian gravitomagnetic term survives.)

4. M. Lottermoser has shown (to be published) that there exist ϵ-dependent, in ϵ-smooth, asymptotically flat initial data obeying the constraints for Einstein's equation with perfect fluid sources and reducing for $\epsilon = 0$ to Newtonian initial data which may represent bodies separated by empty space.
5. J. Winicour has shown how Newtonian initial data for a self-gravitating compactly supported perfect fluid system can be used to determine uniquely characteristic initial data on an outgoing (physical, not background) null-cone for a corresponding asymptotically flat relativistic system which appears to be free of incoming gravitational radiation. The relativistic data reproduce the Newtonian data in the limit $c \to \infty$. He has used such data to prove the **first rigorous version of a far-field quadrupole radiation law**; see [14] and the references therein. (As I pointed out in my talk at the Jena Conference, the only possibility to formulate that law as a <u>theorem</u> is to phrase it as a statement about the $c \to \infty$ -limit of a relation for a family of GR-spacetimes having a Newtonian limit, since that law combines the Newtonian quantity $\ddot{Q}$ (Q: quadrupole moment of the source, time derivatives taken according to Newtonian dynamics) with the GR-quantity N (the news function); Winicour's result confirms that expectation.)
6. Approximate solutions of Einstein's field equation representing isolated gravitating systems constructed via post-Minkowskian methods without input from Newtonian theory **do** have Newtonian limits.

These facts give evidence in favour of, but do not prove that "lim E = N" ; for the following question has <u>not</u> been answered:

7. Do solutions of the field equation evolving from ordinary or characteristic initial data which have Newtonian limits, also have such limits for some finite, ordinary or retarded time? (Conceivably one Einsteinian solution is in osculating approximation, at different times (space-like

hypersurfaces), not with one, but with the members of a family of different Newtonian solutions, in analogy to the way in which in the classical theory of "perturbations of the elements" (or variation of constants) one perturbed solution of an N-body problem is in osculating approximation with a family of unperturbed solutions each member of which has different values of the "elements" which would correspond to parameters characterizing different Newtonian solutions. Such an osculating limit is, in fact, already suggested by the anomalous periastron advance if the Newtonian limit is required to be valid uniformly in time. In this case the varying element would be the time of passage through the periastron.)

Note that a positive answer to the last question is assumed in post-Newtonian approximation methods. If this assumption were incorrect, the comparison of GR with observations of the solar system, the binary pulsar or cataclysmic binaries would lose its theoretical basis. Unfortunately, there seems to be little hope for answering that question rigorously in the near future.

In attempting to separate folklore from knowledge I have focussed on just a few problems which I consider on the one hand as important for general relativity as a part of physics and which, on the other hand, lead to reasonably well-defined mathematical questions. It would seem to be meritorious to consider other areas of GR in a similar spirit. More often than not it may turn out that a problem which, according to published statements, has been solved, still poses significant questions.

References

1. Synge, J.L., Proc. Roy. Dublin Society A4, 253 (1964)
2. Thirring, W., Vorlesungen über Mathematische Physik, Bd. IV, Springer, Berlin (1980)
3. Buchdahl, H.A., Nuov. Cim. 10, 96 (1958);
   Bell, P. and Szekeres, P., Int. J. Theor. Phys. 6, 111 (1972)
4. Weyl, H., Die Naturwissenschaften, 12. Jahrg. (1924)

5. Lichnerowicz, A., Relativistic Hydrodynamics And Magnetohydrodynamics, Benjamin, N.Y. (1967)
6. Carter, B., p. 294 in General Relativity, S.W. Hawking et al. (eds.), CUP (1979)
7. Thorne, K.S. et al. (eds.), Black Holes: the Membrane Paradigm, Yale Univ. Press, New Haven & London, (1986)
8. Lichnerowicz, A., Journ. de Math. 23, 37 (1944)
9. Choquet-Bruhat, Y., p. 17 in Aspetti Matematici Della Teoria Della Relativita, Atti Dei Convegni Lincei 57, (1983)
10. Lottermoser, M. (to be published)
11. Ehlers, J. (ed.), Isolated Gravitating Systems in General Relativity, North-Holland, Amsterdam (1979)
12. Damour, T., p. 59 in Gravitational Radiation, N. Deruelle et al. (eds.), North-Holland, Amsterdam 1983; see also General Relativity and Gravitation, B. Bertotti et al. (eds.), Reidel, (1984)
13. Futamase, T., p. 81 in Gravitational Collapse And Relativity, H. Sato et al. (eds.), World Scientific, Singapore (1986)
14. Winicour, J., Institute Report MPA 253, Max-Planck-Institut für Astrophysik, Garching bei München, (1986)
15. Schattner, R.H., The Structure of Extended Bodies in General Relativity, Ph.-Thesis, University of Munich (1980) (unpublished)
16. Schattner, R. and Lawitzky, S., Ann. Inst. Henri Poincare 40, 291 (1984)
17. Dixon, W.G., in ref. 11
18. Infeld, L. and Schild, A., Rev. Mod. Phys. 21, 408 (1949)
19. Hawking, S.W. and Ellis, G.F.R., The Large Scale Structure of Spacetime, CUP (1973)
20. Schattner, R., GRG 10, 377; 395 (1979)
21. Dautcourt, G., Acta Phys. Polon. 65, 637 (1964)
22. Künzle, H.P., GRG 7, 445 (1976)
23. Ehlers, J., p. 65 in Grundlagenprobleme der modernen Physik, J. Nitsch et al. (eds.), BI Mannheim, (1981); also p. 387 in Logic, Methodology and Philosophy of Science VII, B. Marcus et al. (eds.), Elsevier (1986)
24. Ehlers, J. and Walker, M., p., 125 in General Relativity and Gravitation, B. Bertotti et al. (eds.), Reidel, Amsterdam, (1984)

Superstring Theory

Michael B. Green,
Dept. of Physics, Queen Mary College, University of London,
Mile End Road, London E1 4NS

In this talk I shall review the structure of superstring theory and the reasons why particle physicists are enthusiastic about this somewhat novel kind of theory for explaining the unity of the particles and their interactions. One of the remarkable features of the recent developments is the close connection with various areas of modern mathematics.

String theories have a curious history since they were originally developed in the early 1970's in attempts to explain the structure of the strong interaction. In those days the strings were supposed to describe strongly interacting particles so that, for example, a meson was thought of as a quark and an antiquark tied together by a string. The need for such theories appeared to evaporate with the coming of QCD, the theory of quarks interacting with gluons, which provided an apparently satisfactory and more conventional field theoretic basis for the strong interactions. The early string theories were inconsistent. Following the incorporation of supersymmetry, the symmetry which relates fermions and bosons, these inconsistencies disappear. String theories of this type are called superstring theories.

It has been realized for a long time that the inconsistency of general relativity and quantum mechanics probably requires a radical modification of theories at short distances. This is indicated by a simple application of the Heisenberg uncertainty relation at the incredibly small distance scales around the Planck length ($\sim 10^{-35}$ metres). At this scale the energy fluctuations are so great that according to general relativity microscopic virtual black holes form, in which case the structure of the theory ceases to have much meaning. In string theories this problem is avoided by a radical modification of the description of space-time which is no longer described in terms of a collection of points.

The conventional quantum field theories on which elementary particle theories are based, such as quantum electrodynamics and general relativity, describe the interactions of point particles such as photons, quarks, electrons, etc. String theories differ in a fundamental manner since they describe the interaction of string-like relativistic quanta, which have one-dimensional extension. The motion of a single string can be described in terms of the infinite number of normal modes which have unlimited frequencies, ν_i. The separation between these frequencies depends on a parameter, the string tension T. Quantum mechanically, these modes become states of definite energy corresponding to particles of definite mass.

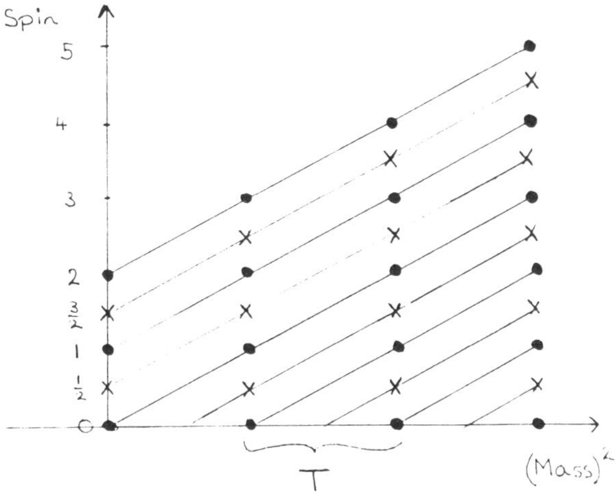

Figure 1. The spectrum of the heterotic string theory, showing the angular momenta and masses of the states. Dots represent boson states and crosses fermion states (each of which is typically very degenerate). The labelling of the angular momentum refers to the theory in four dimensions.

The figure illustrates a typical spectrum that emerges from a superstring theory. The states lie on straight-line trajectories (which are known as *Regge* trajectories) in a plot of the angular momentum versus $(\text{mass})^2$. There are an unlimited number of particles in the theory including particles of arbitrarily high spin.

The fact that string theories consistently describe particles of arbitrary spin is one of the reasons for the importance of these theories. There is no consistent description of high-spin particles in conventional quantum field theories with a finite number of species of particles. In particular, I have already alluded to the problems of quantum gravity, a theory with a spin 2 graviton.

The need for a consistent description of high-spin particles is experimentally very evident in the data on *strong interaction* physics. Many strongly interacting particles – mesons and baryons – were discovered during the 1960's and found to lie on straight-line Regge trajectories in plots such as the one in the figure. Experimentally, the slope of these trajectories was found to be around $1Gev^2$ and there are no massless hadronic particle states so the intercept of the trajectories is shifted from that of the figure. It was the similarity of this hadronic data which gave rise to the original string theory based on the Veneziano model which was formulated in 1968. The inability of string theories to reproduce the observed trajectory intercepts as well as the fact that the early string theories were inconsistent because they contained tachyons (particles with negative $(\text{mass})^2$) led to the demise of this application of string theory.

In its modern incarnation string theory is supposed to encompass a unification of all the interactions. The figure shows that among the states of the string are massless states of spin 1 and spin 2. These are interpreted as Yang–Mills gauge particles and the graviton of general relativity. The string tension sets the dimensional scale of the theory – so it must be of the same order of magnitude as the dimensional parameter in a theory containing gravity, namely, the Planck scale. This means that the slope of the trajectories (*i.e.*, the inverse of the string tension) is tiny so that the first massive states have a mass of around the Planck scale ($10^{19}Gev$!). This is somewhat higher than the mass scale of the strong interactions. Another important feature is that superstring theories do not have the tachyon states which killed the original string theories.

The elementary particles that are observed in experiments (the quarks, leptons and gauge particles) are massless compared to the Planck scale and are expected to arise from among the superstring ground states. The massive string states are not directly observable but their existence is crucial for the consistency of the theory since they affect the very short distance (or high energy) behaviour of the theory, which is just the regime in which Einstein's theory has quantum mechanical problems.

The quantum-mechanical behaviour of superstrings restricts the possible theories of this kind dramatically. A particularly striking restriction is on the dimensionality of space-time, which is forced to be ten. The idea of extra spatial dimensions goes back to the work of Kaluza and Klein but string theory is the first example of a theory which demands higher dimensions for its consistency. Obviously, if the theory is to make contact with observed physics, these extra dimensions must be very curled up. The hope is that the equations of the theory determine this *compactification* of dimensions.

One aspect of superstring theories which makes them most appealing as a fundamental description of physics is the fact that they very naturally incorporate the non-conservation of parity, or *chirality*, of observed physics. There are well-known problems, known as *chiral anomalies*, in incorporating chirality into quantum theories. Generically the conservation laws (such as the conservation of electric charge), which are crucial for the consistency of the theory, break down. The standard model of elementary particles (the union of QCD and the electroweak theory) is chiral and the fact that it must be free of anomalies provides experimentally observed constraints on the charges of the quarks and leptons. In higher dimensions, such as the ten dimensions in which superstring theories are defined, there can also be anomalies in general relativity. Ensuring the absence of all anomalies is then much more difficult. The consequent restrictions on the possible theories lead to very few possibilities. The only possible ones contain unifications of all the forces and particles associated with symmetry groups which are subgroups of either $SO(32)$ or $E_8 \times E_8$.

Since string theory contains a radical modification of general relativity it must contain a principle that generalizes the geometrical principle embodied in Einstein's theory. Unfortunately, such a principle is not yet understood – almost all the calculations are based on a formulation of the theory which is analogous to using a perturbative expansion of the Einstein theory around a particular fixed classical background geometry. Even at this level certain superstring theories avoid the familiar problems of quantum gravity such as ultraviolet infinities and anomalies. Requiring this consistency selects only a very few possible theories. In these theories the Yang–Mills symmetry associated with the forces other than gravity is determined in an almost unique manner. There are certainly further, unexplored, constraints that may well restrict the possible theories even further which raises the possibility that there is only one possible consistent quantum string theory. In order to make contact with physics such a theory has to explain a wealth of experimental data.

The formulation of string theories

The dynamics of a classical relativistic string can be described by an action principle that generalizes that of a relativistic point particle. In the case of the particle the simplest Lorentz covariant action is simply the length of the world-line traversed as the particle moves through space-time. As a string moves through D-dimensional space-time it sweeps out a two-dimensional *world-sheet*. Denoting the parameters of the world-sheet by σ, τ the space-time coordinate of any point on the string at a given time, $X^\mu(\sigma,\tau)$ ($\mu = 0, \dots, D-1$), is a mapping of the sheet into D-dimensional space-time. The obvious generalization of the point particle action to the case of a string is to choose the action to be proportional to the area of the world-sheet

$$S = -T \int d\tau d\sigma \sqrt{\dot{X}^2 X'^2 - \left(\dot{X} \cdot X'\right)^2}, \tag{1}$$

where $\dot{X} = \partial X/\partial\tau$ and $X' = \partial X/\partial\sigma$ and the world indices, μ, are contracted by using the metric $g^{\mu\nu}(X)$ of the space in which the world-sheet is embedded. It is important for the interpretation of the theory that the area is a quantity that is reparametrization invariant, *i.e.*, it is invariant under arbitrary redefinitions of the parameters σ and τ. This is general coordinate invariance in *two* dimensions. The action can be rewritten in a manner that makes this connection with *two-dimensional* general relativity very explicit by introducing a two-dimensional metric tensor $h^{\alpha\beta}(\sigma,\tau)$ $(\alpha,\beta = \tau,\sigma)$ and interpreting the coordinates $X^\mu(\sigma,\tau)$ as D *fields* on the world-sheet which transform as scalars under reparametrizations. This form of the action is given by

$$S = -\frac{T}{2} \int d\sigma d\tau \sqrt{h} h^{\alpha\beta} \partial_\alpha X^\mu \partial_\beta X^\nu g_{\mu\nu}(X). \tag{2}$$

The two-dimensional metric tensor, $h^{\alpha\beta}$, is a sort of Lagrange multiplier in two dimensions. In the classical theory $h^{\mu\nu}$ can be eliminated by substituting the solution of its equation of motion back into the action (2). This converts it into the action (1) so that the two formulations are equivalent. In the quantum theory it is very convenient to use the action of (2) in which case many well-known geometric properties of two-dimensional surfaces can be applied. A particularly important feature of (2) is that in addition to the obvious two-dimensional coordinate invariance the action is invariant under arbitrary Weyl rescalings of $h^{\alpha\beta}$

$$h^{\alpha\beta} \to h'^{\alpha\beta} = e^{\phi(\sigma,\tau)} h^{\alpha\beta}. \tag{3}$$

It is possible to choose a gauge (at least locally) in which the metric has the form $h^{\alpha\beta} = e^{\phi}\eta^{\alpha\beta}$, where $\eta^{\alpha\beta}$ is the flat two-dimensional metric. In this gauge the theory is invariant under *two-dimensional conformal transformations* corresponding to the transformations $\phi \to \phi'$.

The conformal invariance of the theory is of paramount importance in understanding the consistency of string theory for the following reason. In ordinary gauge field theory (such as Maxwell's electrodynamics) the vector potential, A^μ, describes four modes of oscillation including an obviously unphysical one in the time-like direction. The fact that this mode actually decouples from physical processes is a result of the gauge symmetry of Maxwell's equations – we know that photons only have two independent modes of polarization. String theory has a potentially infinitely worse problem with time-like modes since the coordinate $X^\mu(\sigma)$ has an infinite number of time-like degrees of freedom corresponding to the normal modes of X^0. In fact string theory has an infinite number of gauge invariances, analogous to the single gauge symmetry of electrodynamics, which guarantee that all these unphysical modes decouple from physical processes. This enormous symmetry is a special feature of the two-dimensional conformal group which is associated with an infinite dimensional algebra known as the Virasoro algebra. The generators of this algebra, L_m, are the modes of the two-dimensional energy-momentum tensor, $T_{\alpha\beta}$, which satisfy

$$[L_m, L_n] = (m-n)L_{m+n} + \frac{D-26}{12} m^2(m-1). \tag{4}$$

An important feature of this algebra is the anomalous term, proportional to $D-26$. This is a quantum-mechanical effect – a quantum breakdown of the conformal symmetry. In superstring theory there are extra, fermionic, coordinates which embody the supersymmetry of the theory and also describe the internal quantum numbers which are charge densities smeared on the world-sheet. In that case the anomaly is proportional to $D-10$. The term proportional to 26 in (4) (or 10 in the case of superstring theory) is a feature of the quantum mechanical theory that arises from a proper treatment of the fixing of the conformal gauge in (2) which requires the introduction of Fadeev–Popov ghost coordinates.

From (4) it is obvious that the conformal symmetry only holds when the dimension of space-time is restricted to specific values. For the Veneziano model this *critical* dimension is 26 whereas for superstrings it is ten. It is also at this point that the spectrum is restricted so that the open-string sector contains a massless vector state (a gauge boson) and the closed-string sector includes the massless spin 2 state (the graviton). In these dimensions only the transverse polarizations of the string are physical, just as only the transversely polarized photons appear in Maxwell's electrodynamics. There is a suggested procedure for dealing with string theories in dimensions lower than these critical values but it is only in the critical dimension that the physical modes are purely transverse, which corresponds most closely to the case of Yang–Mills theory.

Two open strings interact by touching at their endpoints and fusing into one string. Conversely, an open string can split into two open strings. A theory containing open strings necessarily also contains closed ones since the two ends of a single open string can join to form a closed string. Since the closed string includes within its states the massless spin 2 graviton this means that gravity must automatically be included in any theory that has open strings. Certain theories only have closed strings. Two of these can interact by touching at an internal point and joining to form a single closed string.

String scattering amplitudes can be calculated by a generalization of Feynman's path integral approach to quantum mechanics. The amplitude for a process is given by an integral over all the possible world-sheets that join the incoming and outgoing states. Each path is weighted by the exponential of the action for the path (which is the area of the world-sheet). The surface swept out by interacting strings can be thought of as a Riemann surface. This yields a series of terms corresponding to the topological classification of the surface. These diagrams are the string version of the Feynman diagrams of ordinary perturbation theory. The classification of these diagrams is simplest for the closed-string theories since the surfaces have no boundaries. In that case the lowest-order (or 'Born') diagrams are given by summing over surfaces which are topologically equivalent to a sphere.

By making use of the conformal invariance of the theory, an arbitrary tree diagram can be represented on a world-sheet which is a Riemann sphere. The corrections to these lowest-order diagrams are surfaces with handles.

These are the loop corrections which represent the many-string effects present in a quantum field theory of strings. For example, at one loop the world-sheet for the scattering of closed strings is topologically a torus. Unlike in the case of the sphere, not all tori are conformally equivalent. A torus is parametrized by a single complex number, or 'modulus', τ. The expression for the scattering amplitude includes an integral over all conformally inequivalent values of τ. A feature of crucial importance is that different domains of the τ

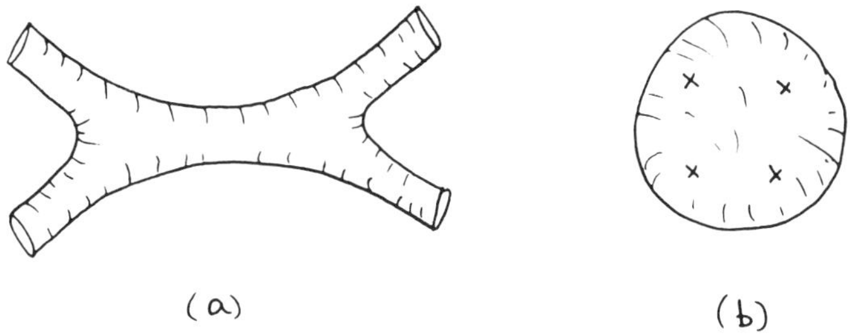

Figure 2. The world-sheet for the scattering of two incoming closed strings into two outgoing closed strings. (a) is mapped by a conformal transformation into the Riemann sphere (b). The incoming and outgoing strings become singular points on the surface of the sphere.

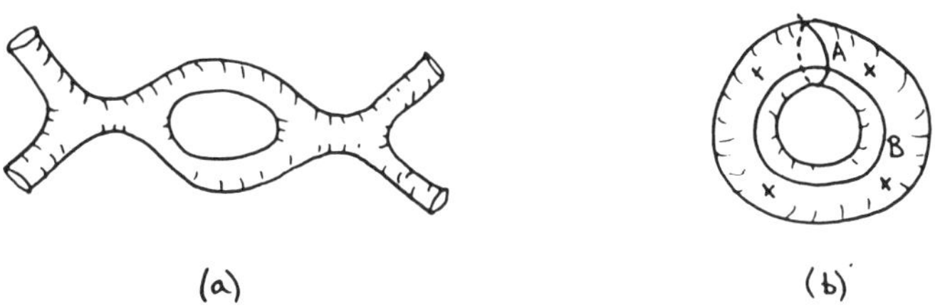

Figure 3. The one-loop correction to the scattering amplitude of closed strings is a world-sheet (a) which is conformally equivalent to a torus (b). Not all tori are conformally equivalent, however. Inequivalent tori are labelled by a complex parameter, τ (the modulus). The calculation of the amplitude involves an integration over τ, taking into account the discrete large diffeomorphisms corresponding to modular transformations.

integration are related by *modular transformations*,

$$\tau \to \tau' = \frac{a\tau + b}{c\tau + d}, \tag{5}$$

(where a, b, c and d are integers satisfying $ad - bc = 1$). These transformation are globally non-trivial diffeomorphisms of the torus which are not continuously connected to the identity. It is essential for the consistency of the theory that these transformations are invariances of the theory. This is another feature that restricts the dimension of space-time to the critical one ($D = 26$ for the bosonic theory and $D = 10$ for superstring theories).

Modular invariance is also a crucial constraint on the formulation of string theory in a curved space-time background.

Quantum numbers

The original way of incorporating quantum numbers into string theory was to attach charges to the endpoints of open strings. Originally, these charges were required to lie in the fundamental representation of the groups $SO(n)$, $Sp(n)$ or $U(n)$. It was then found that the only choice that is free of anomalies when the quantum loop amplitudes are considered is the group $SO(32)$. In this case there are 496 massless gauge particles and their 496 massless fermion supersymmetric partners corresponding to the adjoint representation of the group.

A second group which is also known to give a cancellation of anomalies is the group $E_8 \times E_8$. This is not one of the groups that can be obtained from open strings. There is, however, a much more elegant way of incorporating quantum numbers into closed string theories by means of a current density on the world-sheet. This is used in the construction of the *heterotic* string theory (invented by Gross, Harvey, Martinec and Rohm). In that case the action (2) must have an additional term involving 32 world-sheet fermions that carry the internal quantum numbers. These fermions are chiral in the two-dimensional sense. Such a theory can only be made consistent with modular invariance if the gauge groups are $E_8 \times E_8$ or $spin(32)/Z_2$ (which has the same Lie algebra as $SO(32)$). It is remarkable that the only theories that can be constructed in this way are precisely those with symmetry groups which were known to be free of anomalies and that these theories also give *finite* one-loop amplitudes. More recently a third theory has been added to the list of anomaly-free chiral string theories. This one has a symmetry group $O(16) \times O(16)$ which is a subgroup of both $SO(32)$ and $E_8 \times E_8$. The latter theory has divergent loop amplitudes, however.

It is interesting that the richest of the string theories, the heterotic string, has many properties in two dimensions that we also want in the ten-dimensional theory, namely, coordinate invariance, supersymmetry, chirality and freedom from anomalies.

The emergence of four-dimensional physics

We have seen that conformal invariance imposes strong restrictions which arise even when the background space-time is flat (*i.e.*, $g^{\mu\nu} = \eta^{\mu\nu}$, the D-dimensional Minkowski metric in (2)). Obviously, if string theory is to explain the world in which we live, $D-4$ dimensions must be very curved so that in these dimensions the space-time metric is far from that of flat space. A theory defined by an action of the form of (2) is known as a two-dimensional nonlinear σ model. In this case the requirement of conformal invariance imposes very strong restrictions on the possible metrics. These constraints on the possible structure of the D-dimensional space-time geometries are quite remarkable since they arise from a *two*-dimensional principle, namely, that of conformal invariance. There is no analogue of these constraints in conventional relativity based on the Einstein equations.

The condition that the σ model be conformally invariant can be translated into the condition that the renormalization group β functions of the theory vanish. These conditions had been studied in their own right before the connection with string theory had been appreciated. The result is that there is a single function $\beta^g_{\mu\nu}(g)$ which is a function of the Riemann curvature made out of the metric tensor in the usual way. In perturbation theory in powers of $1/T$ the condition for conformal invariance is

$$\beta^g_{\mu\nu} = R_{\mu\nu} + O(1/T) = 0. \tag{6}$$

In this equation $R_{\mu\nu}$ is the Ricci tensor formed from the metric, $g_{\mu\nu}$. Remarkably, this is Einstein's equation together with corrections which are negligible at low energies. This demonstrates the fact that string theory contains Einstein's equations at low energies. Of course, the series of terms in (6) is infinite, containing the structure of the complete string theory (and generalizing the action of general relativity).

There can be other terms in the σ model associated with background fields in addition to the metric. One example of such a term is one of the form

$$\int d\sigma d\tau \epsilon^{\alpha\beta} \partial_\alpha X^\mu \partial_\beta X^\nu B_{\mu\nu}(X), \tag{7}$$

where $\epsilon^{\alpha\beta}$ is the two-dimensional Levi-Civita symbol and $B_{\mu\nu}(X)$ is an antisymmetric tensor. The condition for the theory to be conformally invariant now requires the vanishing of the β function, $\beta^B = 0$, which is the equation of motion for the antisymmetric massless tensor field $B_{\mu\nu}$ which is another of the massless states of the string. More generally, there are similar terms with a field corresponding to each of the infinite number of component states of the string. In practice, most of the work on this subject considers terms in which the nonzero background fields correspond to all the massless states of the string. There is a β function for each of these fields which is required to be zero if the theory is conformally invariant. These conditions on the β functions are just the field equations for these massless fields.

There has been a great deal of activity in trying to find solutions of the full equations for the vanishing of the β functions to all orders in $1/T$. One possible class of curved spaces which satisfy these equations are the so-called *Calabi–Yau* spaces. These are Kähler spaces of vanishing first Chern class. The interesting case is when six dimensions are curved, leaving four flat dimensions. In this case the simplest Calabi–Yau spaces are Ricci-flat (the Ricci tensor vanishes) which implies that they have $SU(3)$ holonomy (*i.e.*, the Riemann curvature, $R_{\mu\nu}^{mn}$, is an $SU(3)$ matrix in the tangent-space indices, m, n). It is remarkable that these are precisely the spaces which had been conjectured to be required in order for the effective four-dimensional theory to have any chance of explaining observed four-dimensional physics.

Physics from Calabi–Yau Spaces

There is still a long way to go before the standard model can be derived from the superstring theory. It is nevertheless interesting that there is a fairly clear scenario (devised by Candelas, Horowitz, Strominger and Witten) for how the huge symmetry groups associated with the ten-dimensional theory may break when the extra dimensions curl up and many realistic features emerge.

One procedure is to study the approximate equations that emerge from the low energy expansion of the superstring theory in terms of the fields that correpond to the massless states of the theory. There are very strong topological constraints which relate the Yang–Mills field strength, $F_{\mu\nu}^{ab}$, to the space-time curvature, $R_{\mu\nu}^{mn}$. These constraints suggest that the non-zero curvature in the six extra dimensions should be identified with the Yang–Mills field in these dimensions, *i.e.*, $R = F$. However, as usual a non-zero value for a field, $< F > \neq 0$, is the sign of the breaking of a symmetry. The symmetry group breaks to a group which commutes with the subgroup in which the field strength lies. Since the curvature has $SU(3)$ holonomy in the present case, $< F >$ lies in a $SU(3)$ subgroup of either of the large groups $SO(32)$ or $E_8 \times E_8$.

Starting from the $E_8 \times E_8$ group in ten dimensions (the $SO(32)$ case seems not to be related to realistic physics) the effective four-dimensional theory has the smaller symmetry group $E_6 \times E_8$ and is still supersymmetric. (The field strength, $< F >$, is in the $SU(3)$ subgroup of E_8 that commutes with the E_6 subgroup.) This is an effective supersymmetric grand unified theory for the four-dimensional theory at high energy. The particle content includes fermions lying in **27**'s which are crucial for obtaining chiral physics. The number of generations of quarks and leptons is the number of **27**'s minus the number of $\bar{\mathbf{27}}$'s. This is determined by a simple topological invariant of the internal six-dimensional Calabi–Yau space, namely its Euler characteristic. Since there are very many Calabi–Yau spaces an arbitrary choice must be made. Phenomenological considerations, including astrophysical abundances, suggest that there are, at most, four generations. This severely restricts the possible spaces of interest.

At this stage the symmetry group looks like one of the popular grand unified groups, E_6, multiplied by an extra E_8 factor. This extra factor is associated with a sector of matter which is neutral under the E_6 charges and therefore is only observable via its gravitational interactions. If this exists in significant abundance it constitutes a novel form of dark matter. Unfortunately, it is premature to predict its properties.

The E_6 grand unified group breaks to the standard model, $SU(3) \times SU(2)_L \times U(1)$ at low energy. This can happen by a topological mechanism that is very different from the usual mechanism involving Higgs particles in the adjoint representation of E_6, which is just as well since the required Higgs particles are not present. The new mechanism makes use of the fact that those Calabi–Yau spaces which are plausibly relevant are not simply connected (*i.e.*, they have holes). As a result E_6 flux loops can get trapped in the holes which leads to a breaking of the symmetry (but not supersymmetry). There are discrete choices for this breaking but the arbitrariness is much less than in the conventional Higgs mechanism. It is very easy to get the standard model from this breaking but almost inevitably there are extra symmetry factors such as $U(1)^2$ or $SU(2)$. This suggests that there are extra low energy gauge symmetries together with associated gauge particles.

There are other plausible experimental consequences of this scheme which are only understood at a qualitative level. These include: The existence of at least one axion. The existence of Planck-mass, unconfined, fractionally-charged particles. The existence of Planck-mass, multiple-charged magnetic monopoles.

I have described this ingenious scheme, based on the low energy effective field theory obtained from the string, to illustrate that there is a plausible way in which observed physics can emerge from the apparently very abstract ten-dimensional starting point. The scheme has many difficulties and may ultimately be completely wrong. Many of the severe problems are associated with the fact that perturbative calculations are not justifiable. Crucially missing from the scheme is an understanding of how supersymmetry is broken and the associated breaking of $SU(2) \times U(1)$ together with the fact that the observed particles have non-zero masses which are tiny numbers when measured in Planck units. There are indications that these effects are due to a fermion condensate similar to that in a superconductor but there is no quantitative understanding of this.

Other theoretical ideas

A very attractive alternative approach to the study of the curled-up, effectively four-dimensional theory is to study the complete string theory in backgrounds which are highly curved. It is not possible to take this to be a Calabi–Yau space since this is too complicated. However, it is possible to take the background to be a singular limit of a Calabi–Yau space in which all the curvature is concentrated at points while the rest of the space is flat. Such a space is an *orbifold* rather than a manifold since it has conical singularities. The orbifolds of relevance are six-dimensional tori with a discrete isometry group divided out. The singular nature of an orbifold is not a problem for the string theory while it would be in a conventional theory of gravity. Orbifolds can also be constructed which are not limits of ordinary manifolds. The fact that exact statements can be made about string theory in an orbifold background suggests that they may play a specially important rôle in the theory.

There has been a lot of activity in studying the multi-loop diagrams. This involves calculating the properties of string scattering amplitudes described in terms of Riemann surfaces of arbitrary genus. The resulting expressions depend on the geometric properties of the space of moduli (or super-moduli in the case of superstrings). The simplest quantities to be calculated are the amplitudes with no external particles, *i.e.*, the vacuum loop diagrams. For a g-loop amplitude these are expressed as integrals over the $(3g-3)$-dimensional space of complex moduli of a Riemann surface of genus g. There has been a lot of progress on this problem for the bosonic (26-dimensional) theory. In the case of superstring theory the integration also involves the $(2g-2)$-dimensional space of the supermoduli. An important issue that should be resolved by such considerations is whether the cancellation of infinities and chiral anomalies, seen at one loop, persists to all orders of perturbation theory.

A major subject of research is the formulation of a field theory of strings. The present description of strings in terms of a given background space is not satisfactory for a fundamental understanding of the theory. Just as in general relativity, we expect that the theory should have a geometrical formulation in which the particles (the strings) emerge as some kind of fluctuations in something like a metric field. The favourite approach is to generalize the idea of a field to a string field, $\Psi[X(\sigma)]$, which takes its values in the space of all string configurations. Various versions of such a string field theory have been successfully formulated for open bosonic strings. The generalization to superstring theories and to closed strings is, as yet, unknown. The hope is that since the closed-string theories contain general relativity as a low energy approximation there must be a generalization of the principle of relativity to a larger principle in the much larger space of string configurations. This is not obvious from the formulation of string theory in terms of two-dimensional world-sheet principles described above. Understanding such a principle might lead to a much deeper understanding of the predictions of the theory.

Bibliography

I have not included a detailed list of references to the subject of superstring theory but a very extensive list can be found in the two-volume book, 'Superstring Theory', by M.B. Green, J.H. Schwarz and E. Witten (C.U.P., 1986).

GRAVITY-WAVE ASTRONOMY

L.P. Grishchuk
Sternberg Astronomical Institute
119899 Moscow V-234, USSR

The chairman of the GR11 program committee, Malcolm MacCallum, has said that the published version of a plenary lecture should contain "what was actually said" at the conference, and not "what might have been said". Following this advice, I present a short account of my lecture making just minor modifications and additions.

1 Introduction. Gravity-wave astronomy in action

Gravity-wave astronomy is usually treated as a very fascinating branch of science promising great discoveries and achievements, but ... in the very distant future. However, during the last few years the situation has considerably changed; these days gravity-wave astronomy actually exists and is developing successfully. We still have to wait some time before an undeniable registration of cosmic gravitational radiation occurs in laboratory conditions. However, the convincing observations of the indirect manifestations of gravitational waves (secular change of orbit in the binary pulsar PSR 1913+16 [1]); the recognition of the decisive role of gravitational radiation in concrete, existing astrophysical events (such as evolution of close binaries resulting in the supernovae type I explosions [2]); the derivation of essential limitations on the early Universe parameters from gravity-wave considerations (confrontation of various theoretical predictions on spectra and energy density of relic gravitons with observational data [3], [4]) – all these demonstrate that gravity-wave astronomy already acts as a valuable means of exploring nature.

2 Theory and some new results

The theoretical basis for gravity-wave astronomy is general relativity (GR). If there existed some fundamental defects in GR they would also make doubtful the conclusions referring to gravitational waves. Much discussion arises on how to describe the energy and momentum of gravitational fields, especially in view of the application of an energy-momentum pseudotensor. I will present some new results on this topic.

GR is chiefly distinguished from other physical theories such as electrodynamics in that it uses one and the same quantities as variables of the gravitational field and as components of the space-time metric. In the ordinary "geometrical" formulation of GR this double role is played by the metric tensor $g_{\mu\nu}(x)$ of a curved space-time. In order to make GR similar to usual field theories, it is necessary to introduce the notion of a dynamical gravitational field $h^{\mu\nu}(x)$ given on some background space-time (for instance, flat) endowed with a metric tensor $\gamma^{\mu\nu}(x)$. A convenient relation between these quantities is given by $\sqrt{(-g)}g^{\mu\nu} = \sqrt{(-\gamma)}(\gamma^{\mu\nu} + h^{\mu\nu})$. In this way one can develop an entirely equivalent "field-theoretical" formulation of GR. Such a theory is strict and possesses all the necessary attributes. The theory contains a Lagrangian and an action; the variational principle generates field equations totally equivalent to the Einstein equations; the standard rules define the energy-momentum tensor (*not* a pseudotensor!) of the dynamical field, together with the common conservation laws reflecting the symmetries of the background space-time; the theory admits coordinate and gauge transformations. Whilst identical to the "geometrical" formulation of GR in all experimentally verifiable conclusions, such a "field-theoretical" formulation of GR has the form of a field theory and is convenient for theoretical investigations. This approach gives additional support to some standard results which have been previously obtained with the help of the energy-momentum pseudotensor. (More details on "field-theoretical" formulations of GR are given in [5], [6]).

Another subject generating a lively discussion in recent years deals with the equations of motion of gravitating bodies including the gravitational radiation reaction force. The most detailed results have been obtained for a pair of spherically symmetric non-rotating bodies. It is assumed that the relative velocity of the bodies is small, $v/c \ll 1$, that is the slow motion approximation is used. This problem has now been solved to the degree of rigour and completeness which is acceptable in usual celestial mechanics. The equations of motion for a body as a whole are derived taking account of all relativistic corrections up to the terms of order $(v/c)^5$ inclusive. All periodic and secular corrections to the Keplerian orbit have been found. The explicit expression for the radiation reaction force has been derived and dE/dt, dp^i/dt, dL^i/dt, computed, where E is the total energy of the system, and p^i and L^i are its momentum and angular momentum. The explicit and complete expression for the Lagrangian of the system, valid in $(v/c)^4$ approximation inclusive, where the system is still conservative, has been derived, and it has been shown that the theory so constructed is applicable to extended as well as to compact bodies (such as neutron stars or black holes) irrespective of their internal structure. The reason for this is that the equations of motion include only a single parameter of each body, namely its relativistic (Tolman) mass. It is worth emphasising that the conclusions regarding the secular change of the orbital period are exactly the same as those derived by the more common technique of using the Einstein quadrupole formula. The exhaustive analysis of this problem is a consequence of much independent work done by different methods but reaching partially overlapping or coinciding results (see for example, [7] -[9]).

3 Burst and periodic sources of gravitational waves

Gravity-wave fields are usually classified as *burst, periodic* and *stochastic*. We will consider them successively. Only those astrophysical sources of gravitational radiation which seem to be the most interesting and definitely exist, will be analysed.

Currently, the greatest attention is being paid to the impulsive gravitational radiation associated with cosmic catastrophes. The majority of experimental programmes are aimed at detecting astrophysical burst sources. The predicted strengths of the burst sources together with current experimental limits and expected sensitivity of some detectors are presented in Fig.1. The gravity-wave field at the Earth is characterised by a dimensionless amplitude h as a function of frequency ν (Hz.).

A typical source of impulsive gravitational radiation is a supernova explosion. We take the mass of a star $M \approx 10M_{\odot}$ and assume that $10^{-5}Mc^2$ is converted into gravitational radiation . The wave amplitude produced by a single event decreases with distance r as r^{-1}. The expected event rate (e.r.) grows linearly with the spatial volume considered. In Fig.1 are given the values of h and e.r. for events occurring in the center of the Galaxy ($r = 10$ kpc), in the Virgo cluster ($r = 20$ Mpc) and in remote galaxies lying at cosmological distances ($r = 10^3 - 6.10^3$ Mpc). Unfortunately, there are large uncertainties both in h and in e.r.

Much better known is the shape and amplitude of the gravitational signal emitted when a pair of compact objects (white dwarfs, neutron stars or even black holes) spiral together and subsequently coalesce. An almost periodic wave is emitted at the early stages of the process while the final stage looks like a burst.

Coalescence of neutron stars is a powerful but very rare event. The number of events becomes acceptable from the observational viewpoint only if one is able to include in the sphere of observation the very distant galaxies. An estimate of the e.r. is given in Fig.1. In the frequency range $10^2\,\mathrm{Hz} \lesssim \nu \lesssim 10^3$ Hz, the gravitational radiation has the form of individual events since not more than one emitting binary gets into the bandwidth $\Delta\nu \approx \nu$. The signal produced by an individual event in the center of the Galaxy is indicated by the wavy line *n* coalescence* in Fig.1. However, many systems radiate at frequencies $\nu < 10^2$ Hz. Their radiation overlaps and forms a sort of stochastic background. A characteristic dimensionless r.m.s. amplitude $h(\nu)$ yielded by all such sources confined to the Hubble sphere is depicted by a dashed line *n* coalescences* in Fig.1 [10].

A signal emitted by one compact binary is quasi-periodic. This property can be used to increase the sensitivity of the detector. One can repeat the measurements over some tens of periods before the frequency changes substantially. In terms of h, the sensitivity increases by a factor $\sqrt{n}$, where $n = \nu' / \nu^2$ and ν is a slowly varying frequency [12]. At the final stage of

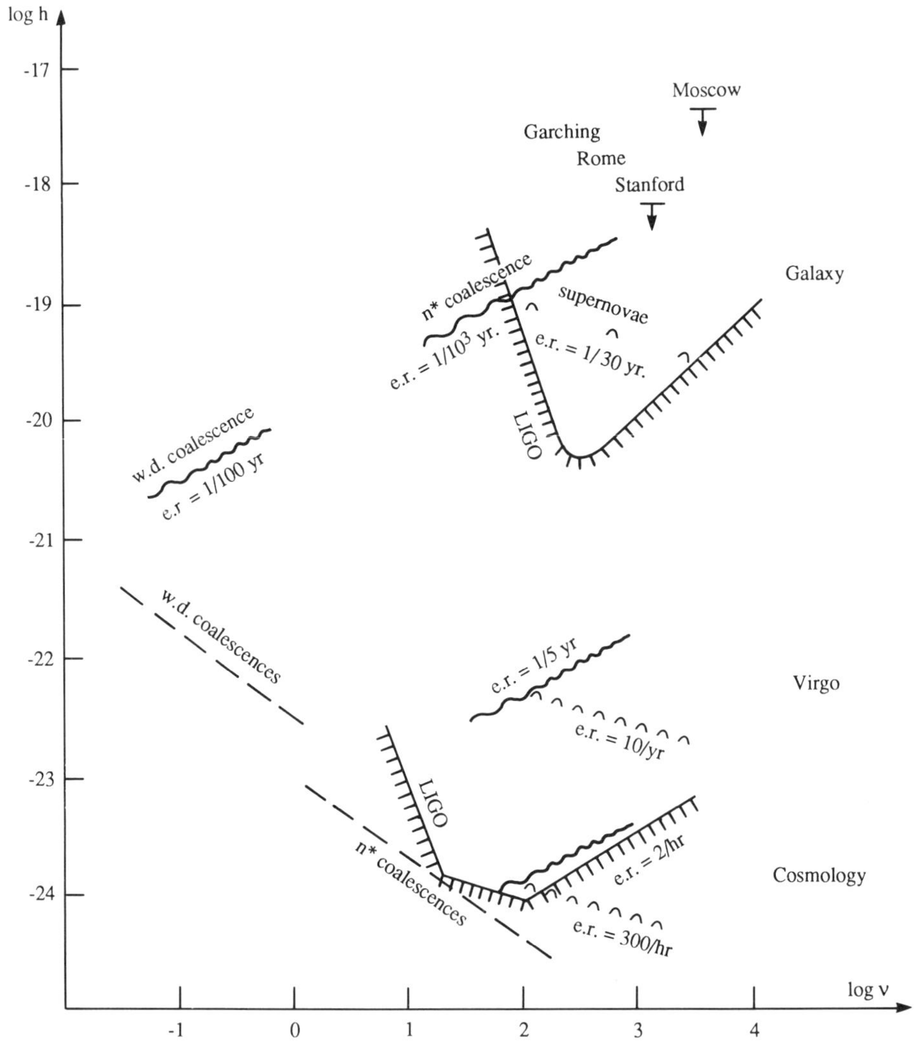

Figure 1: Burst sources

coalescence $n \approx 20$–30.

A theory of spiralling which could permit one to extract additional important information from observations is well worked-out. By measuring $h(t)$ and $\nu(t)$ simultaneously one can find the absolute distance to the source since it turns out to be possible to get rid of the unknown masses of the stars forming the binary [13].

Coalescence of white dwarfs occurs more often than coalescence of neutron stars. However, because of the larger radius of a dwarf, the coalescence ceases at a lower frequency ($\nu \approx 1$Hz) and smaller amplitude of the emitted gravitational wave. A single event in the centre of our Galaxy is shown in Fig.1 by the wavy line *w.d. coalescence* . All such events in the Hubble sphere form a stochastic background *w.d. coalescences* [10].

The experimental limits achieved are shown by arrows in Fig.1. The expected sensitivity of the proposed Laser Interferometric Gravitational Observatory (LIGO) [33] is also given. It is seen that possible later experiments (the lower curve) should permit detection of many astrophysical sources of gravitational radiation. It is important to note that similar projects are under consideration in several countries. (Experimental possibilities are discussed in more detail in section 5.)

Now let us turn to periodic sources (see Fig.2). A typical source is a nearby double star, for instance, ι Boo. All the double stars in the Galaxy produce a stochastic background. Its characteristics depend on the distribution of the stars over masses, orbit parameters, evolutionary scenarios. The outcome of a new calculation of $h(\nu)$ is given in Fig.2 [10] (see also [11]). The maximum of the spectrum is formed by contributions from about $N \approx 3.10^7$ systems. The spectrum ceases to be continuous around the frequency $\nu \approx 10^{-2}$ Hz. At higher frequencies there is no more than one binary in the bandwidth $\Delta\nu \approx \nu$.

In Fig.2 are shown also the expected values of h from pulsars deformed by various possible amounts. The dashed line indicates the performance of an accreting neutron star whose angular velocity has risen above some critical value. As a result, the highly deformed star becomes an efficient source of gravitational radiation [14], [15]. The actual position of the dashed line in Fig.2 takes into account the factor $\sqrt{N}$, where $N, \approx 100$, is the expected number of such stars in the Galaxy.

Periodicity of a signal makes it possible to extract the signal from the noise over a long time. The expected sensitivity of facilities like LIGO is given in Fig.2 under the assumption that the integration time is $\tau \approx 10^6$ sec. One very impressive project is that of a spaceborne laser interferometer with arm length of order $10^6 - 10^8$ km [16]. This detecting system is most sensitive to those frequencies where double stars radiate. The detector can be used to detect a single binary or a stochastic background. In the latter case its sensitivity given in Fig.2 assumes $\tau \approx 10^6$ sec.

It is worth mentioning that, in principle, Doppler tracking for a pair of spacecraft can also be arranged in an interferometric scheme. In this case, one can considerably weaken the requirements for the stability of the frequency standard which lies at the very basis of this technique and determines its sensitivity.

4 Cosmological stochastic background: the early Universe

As we saw above, a stochastic background can be generated by the totality of periodic or impulsive sources existing in the present epoch. However, the most fundamental and interesting is the primordial, or cosmological, background of gravitational waves whose origin is related to the processes in the very early Universe. The observation of these relic gravitons promises unique information on the physical conditions at the Planck time.

Let us recall the main features of the graviton creation mechanism resulting in the formation of the primordial stochastic background ([17],[3]). This recollection is especially proper now since literally the same principles are actively used in modern inflationary cosmology as an explanation for the origin of those primordial scalar field fluctuations capable of producing the density perturbations and eventually the observed large scale structure of the Universe (see, for example, [18]).

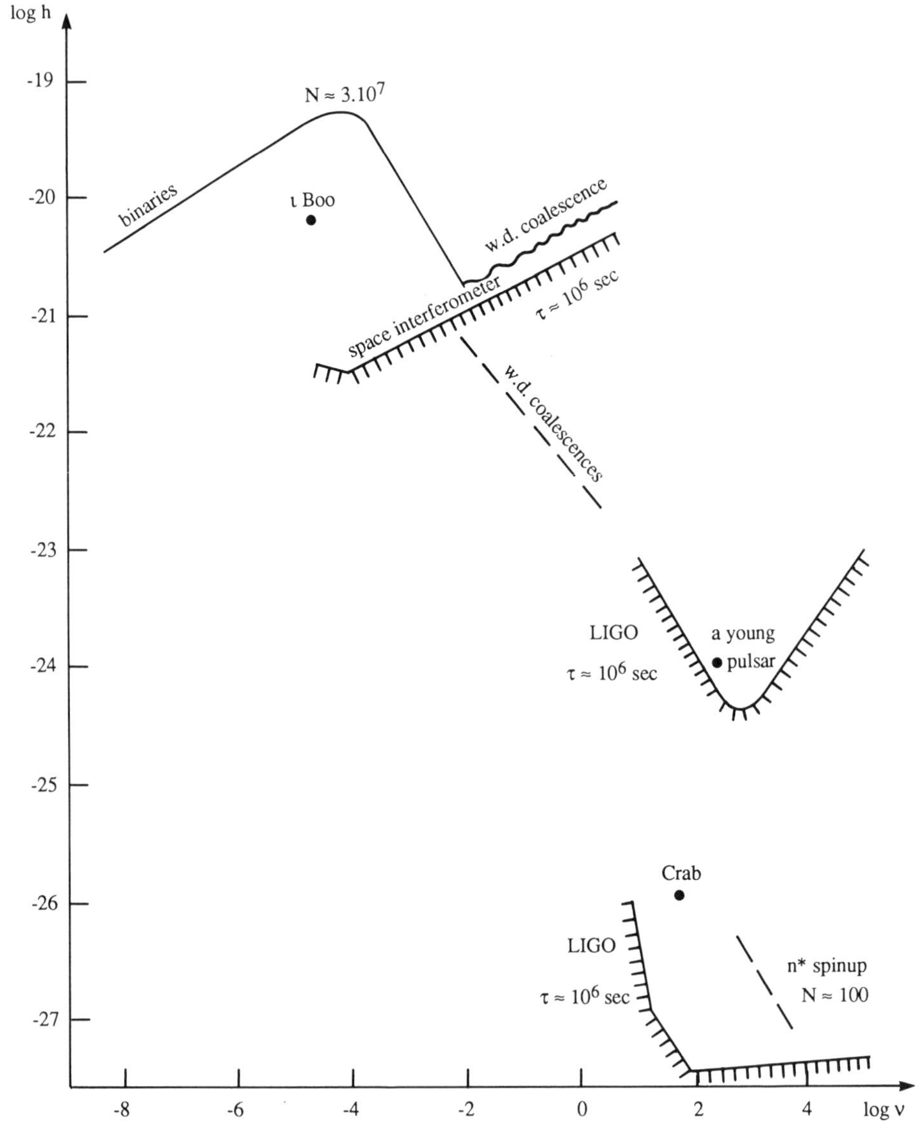

Figure 2: Periodic sources

The gravity-wave field $h_i^k = a^{-1}\mu(\eta)\, G_i^k(x)$ interacting with the gravitational field of the homogeneous isotropic Universe obeys the equation:

$$\mu'' + \mu(n^2 - a''/a) = 0, \qquad (1)$$

where $a(\eta)$ is the cosmological scale factor and n is the dimensionless wave number, the wavelength being $\lambda_g = 2\pi a/n$.

The physical interpretation of (1) is obvious — it is the equation for an oscillator parametrically excited by a variable frequency. This equation can also be regarded as the Schrödinger equation for a particle with energy n^2 in a potential $U(\eta) = a''/a$, with the η-time derivative playing the role of a spatial derivative.

A typical potential $U(\eta)$ is drawn in Fig.3. In the domains where a wave with a given n satisfies the condition $n^2 \gg |U(\eta)|$, the amplitude of the wave decreases adiabatically, that is according to the law $h \sim e^{in\eta}/a$. However in the domain where $n^2 \ll |U(\eta)|$, the amplitude of the wave remains almost constant despite the growth of the scale factor $a(\eta)$. After penetrating the barrier and emerging in the domain where again $n^2 \gg |U(\eta)|$, the wave will have a larger amplitude than it would have had if it had behaved according to the adiabatic law. This is the essence of the superadiabatic amplification of classical waves and quantum zero-point fluctuations. (The adiabatic behaviour of the wave under the barrier is schematically shown in Fig.3 by a broken line.)

One can see from Fig.3 that waves with different wave numbers n are amplified by different amounts. In this way an initial spectrum of the waves (for instance, the spectrum of the zero-point fluctuations) transforms into a final one. Various assumptions about $a(\eta)$ in the early Universe lead to different predictions about the mean square amplitude $\langle h^2 \rangle$, the energy density ε_g, and the spectrum of the cosmological stochastic background in the present epoch.

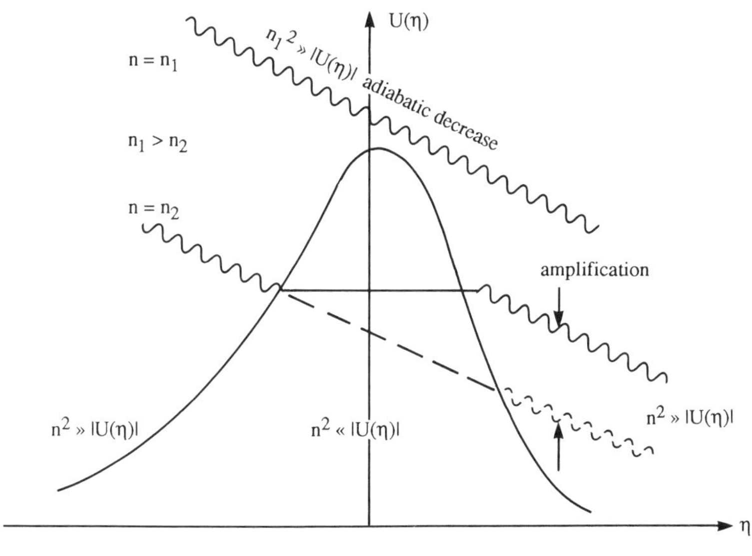

Figure 3: Amplification of waves

In order to make contact with observations easier, it is convenient to use the following definitions:

$$h(x,t) \sim \int h_k e^{ikx} d^3k, \quad \langle h_k h_{k'} \rangle = h_\nu^2 \delta(k-k')$$

$$\langle h^2 \rangle \sim \int h_\nu^2 \nu^2 d\nu \sim \int h^2(\nu) d\nu/\nu, \quad h(\nu) \equiv \sqrt{(h_\nu^2 \nu^3)}$$

$$\varepsilon_g = \int \varepsilon_\nu d\nu = \int \varepsilon(\nu) d\nu/\nu \sim \int h^2(\nu)\nu^2 d\nu/\nu \, .$$

A spectrum can be characterised by $h(\nu)$ or $\varepsilon(\nu)$. It is also common to use the parameter $\Omega_g(\nu) = \varepsilon(\nu)/\varepsilon_{cr}$, where ε_{cr} is the critical cosmological density.

By using the Einstein equations, the η-time dependence of the scale factor $a(\eta)$, and hence the form of $U(\eta)$, can be viewed in terms of the effective equation of state $p(\varepsilon)$. The deviations from $p = \varepsilon/3$ toward stiffer equations of state, that is $p > \varepsilon/3$, lead to the "violet" spectra of relic gravitons where the total energy density ε_g is determined by the high-frequency end of the spectrum. On the other hand, negative effective pressure, $p < 0$, leads to "red" spectra. Some possible spectra are represented in Fig.4 by thin unbroken lines (see also [19]). They describe the spectral flux density F_ν (erg sec^{-1} cm^{-2} Hz^{-1} ster^{-1}) where $\varepsilon_g = 4\pi/c(\int F_\nu d\nu)$. The spectrum of the 3°K blackbody radiation is presented for comparison. Its total energy density is ε_γ.

Too stiff an equation of state, say $p \approx \varepsilon$, could not hold during the hadronic epoch (10^{-43} sec $< t < 10^{-6}$ sec), since it predicts an unacceptably high energy density of relic gravitons, $\Omega_g \gg 1$, in the high-frequency band, $\nu \sim 10^{12}$ Hz. On the contrary, the parameters of the inflationary models (whose effective equation of state is $p = -\varepsilon$), are constrained by the predicted amplitude of the lowest frequency gravitational waves, $\nu \sim 10^{-18}$ Hz.

The spectrum of $\langle h^2 \rangle$ generated immediately after the end of the inflationary period follows easily from the above-mentioned elementary considerations. Indeed, the spectrum of the quantum zero-point fluctuations is defined by $h_n \sim \sqrt{(2n)^{-1}}$, the scale factor being $a(\eta) = -\eta^{-1}$ (η grows from $-\infty$). A wave with a given n begins to amplify from the time η_i determined by the condition $n^2\eta_i^2 \approx 1$. The longer waves, having smaller n, stay under the barrier for a longer η-time . The amplification factor is equal to the ratio $a(\eta_f)/a(\eta_i)$, where η_f marks the end of the inflationary period and the end of the amplification. In the case considered this reduces to the multiplication of h_n by n^{-1}. As a result, by the end of inflation one gets the spectrum $h_n \sim \sqrt{(2n)^{-1}}.n^{-1} \sim n^{-3/2}$, which means that $h^2(n) \sim$ constant and $\langle h^2 \rangle \sim \int n^{-1} dn$. This spectrum is called the *flat Zeldovich-Harrison spectrum* [20], [21]. (This is an analog of the $1/f$ noise spectrum known in radiophysics.)

After the inflation, at the radiation-dominated stage, the waves neither amplify nor attenuate: they behave adiabatically since $p = \varepsilon/3$ and $U(\eta) \equiv 0$. Only later, at the matter-dominated stage, will there again be $U(\eta) \neq 0$. At that stage the lowest frequency waves whose present-day frequencies are below $\nu \approx 10^{-15}$ Hz underwent additional amplification. The rest of the waves experienced the usual transition from $\lambda_g \gg R_H$ to $\lambda_g \ll R_H$ where R_H is the Hubble radius. (For more details about gravitons in inflationary models see [22]-[26], [4].) A possible inflationary spectrum limited from above by observations of $\Delta T/T$ is given in Fig.4. For waves

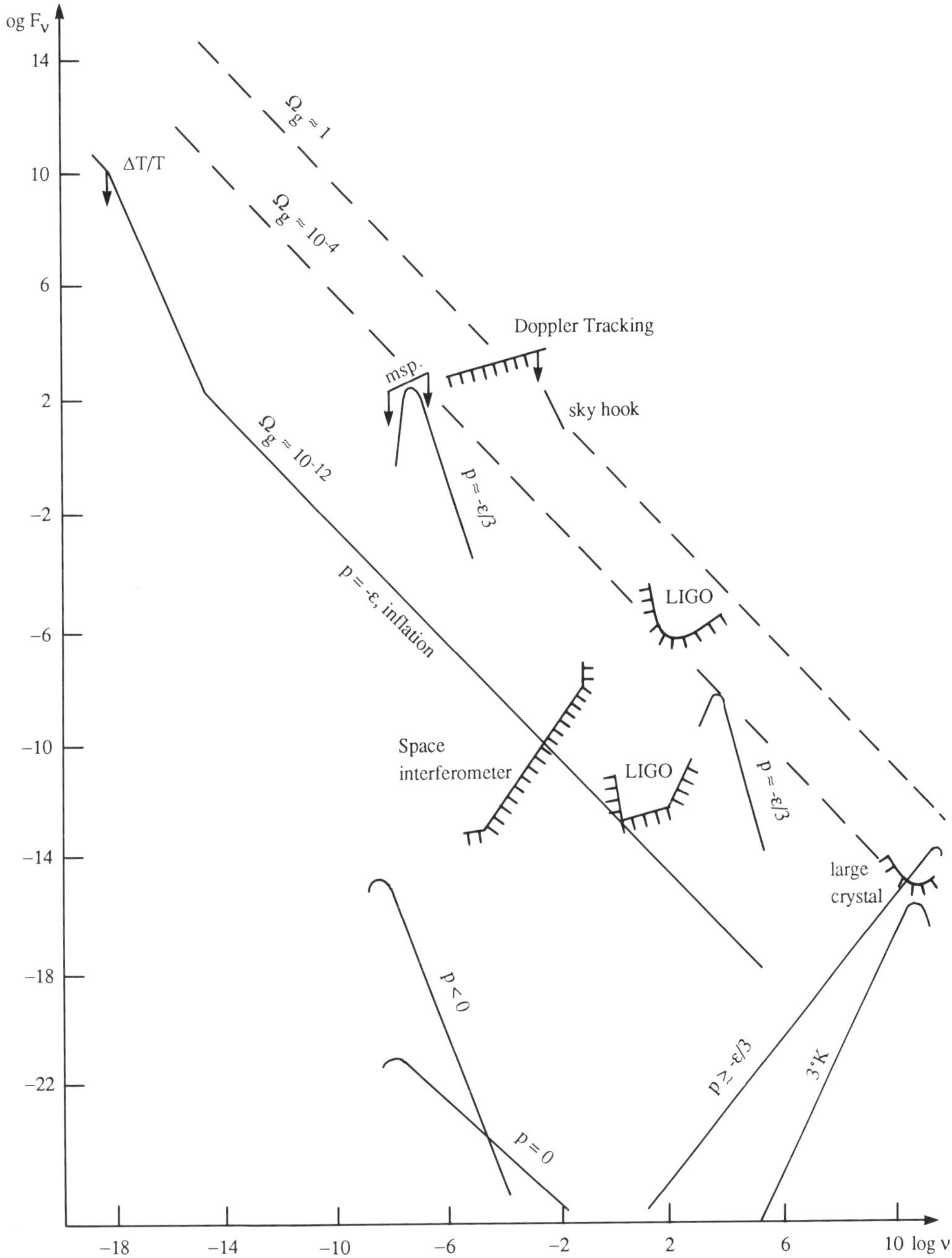

Figure 4: Stochastic background

with $\nu \gtrsim 10^{-15}$ Hz a corresponding Ω_g is $\Omega_g \approx 10^{-12}$.

For illustration, two spectra produced in models with $p = -\varepsilon/3$ are also given. They differ in the assumed values of the time t_m when $p = -\varepsilon/3$ is replaced by the usual $p = \varepsilon/3$. In the first case it is assumed that $t_m \approx 10^{-6}$ sec, in the second, $t_m \approx 10^{-27}$ sec. In effect, we are dealing with the "injection" of gravitons possessing energy density equal to the energy density of the background matter. From t_m up to the present epoch, the gravitons cool adiabatically.

The broken lines indicate the position of maxima in the set of broad spectra, $\Delta\nu \approx \nu$, having the integral $\varepsilon_g \approx \varepsilon_{cr}$ ($\Omega_g \approx 1$) or $\varepsilon_g \approx \varepsilon_\gamma$ ($\Omega_g \approx 10^{-4}$). In the latter case one has $h(\nu) \approx 10^{-20}/\nu$.

Now let us turn to the experimental limits on the stochastic background. In Fig.4 they are marked by arrows. In the lowest frequency band the limits follow from the absence of the angular variations $\Delta T/T$ in the microwave 3°K radiation. The information about shorter waves, $\lambda_g \approx$ a few light years, follows from pulsar timing observations (msp). This technique already gives the essential limits in terms of Ω_g. According to the new data [1], $\Omega_g \lesssim 10^{-6}$ for $\nu \approx (10^{-7} - 10^{-8})$Hz. The prolongation of these observations will make the limits on Ω_g much more stringent provided the experimenter possesses a very good phase-keeping standard. In the course of longer observations the sensitivity increases not only due to the statistical reduction of noise, but, mainly, because of shifting toward longer gravitational waves. Finally, some limits of the order of $\Omega_g \lesssim 1$, follow from the Doppler- tracking technique in the frequency band $\nu \approx (10^{-2} - 10^{-4})$Hz [27].

The prospects for stricter limitations to Ω_g look promising. As is seen from Fig.4 an especially impressive project is the one for the laser interferometer in space [16]. However, in addition to technical difficulties, it will encounter the problem of avoiding the stochastic gravitational radiation noise produced by binary stars in the Galaxy. This will probably necessitate a search for the appropriate frequency region and may also require us to think of the suitable orientation of the antenna's directivity pattern.

The planned sensitivity of other techniques is also presented in Fig.4. They include: the Doppler tracking of spacecraft [28]–[31], the "sky-hook" detector [32], ground-based interferometers like LIGO [33]. The possibility of using specially constructed crystal antennae for detecting the high-frequency stochastic background deserves particular attention [34]. The expected sensitivity is shown by the curve *large crystal* . This suggestion will be discussed in more detail in section 6. It is important to stress that the high-frequency part of the non-thermal spectra as well as the maximum of the envisaged 1°K thermal gravitational radiation may concentrate in the considered region $\nu \approx (10^{10} - 10^{11})$Hz.

Stochastic gravitational radiation can also be generated by other physical processes in the early Universe, [35]–[37]. However, they seem to be more problematical.

5 Detection

Gravity-wave antennae can be divided into two large groups: oscillators and free masses. The traditional solid-state antennae (Weber's bars) belong to the first group. Their current best sensitivity to burst signals lies at the level $h \approx 5.10^{-18} - 10^{-19}$. The gradual increase of the sensitivity is slow but constant [38].

The second group includes man-made as well as natural detectors. Their common feature is the essential role played by the electromagnetic field propagating through the weak gravity-wave field. To this group one can assign the measurement of the angular variations $\Delta T/T$ in the electromagnetic microwave background, the pulsar timing observations, the Doppler tracking of spacecraft, the ground-based and cosmic interferometers, the "full sense" electro-magnetic detectors-resonators, etc. (For a review see, for example, [19], [39].)

In the geometric optics approximation we deal with very simple formulae. The electromagnetic signal emitted by one free body at the time t_e and received by another free body at the time t_0 experiences the frequency variation:

$$\frac{\Delta\nu}{\nu} = z(t) = \frac{1}{2}\int_{t_e}^{t_0} \frac{\partial h_{ik}}{\partial t} n^i n^k \, dt$$

where n^i is a unit 3-vector pointing to the source. The variation of the phase, or of the arrival times of individual pulses, is described by :

$$R(t) = \int_0^t z(t')\, dt' .$$

In a stochastic gravitational radiation field one has $\langle R\rangle = 0$, but $\langle R^2\rangle \neq 0$.

A typical example of this technique is pulsar timing. In principle, this technique is very sensitive (in terms of Ω_g) to long gravitational waves, $\lambda_g \gg cT$, where T is the observation time span. However, long waves cause only very slow, secular variations in the arrival times. Similar variations are also caused by a simpler reason, namely the secular change of the pulsar's frequency. Not having a way to separate these effects, one is forced to attribute to the pulsar itself all the low-frequency components and to get rid of them by "polynomial fitting". As a consequence, this technique turns out to be most sensitive to waves of wavelength $\lambda_g \approx cT$, [40]–[41].

The brightest hopes are related to the building of ground-based laser interferometers. Experiments in the USA, Great Britain (Glasgow), Germany (Munich), and France (Orsay), are at different stages of realization. We consider briefly one of these projects – LIGO [33].

The observatory will be located at two sites, one on the West coast of the USA and another on the East coast. One, two, or even several, interferometers will be established at each site. The maximal arm length will be 4 km. The whole system will be maintained under vacuum and seismic isolation. The interferometers will be furnished with very low loss mirrors. Later development is expected to utilize some new concepts such as "light recycling". This method increases the light flux within the interferometer and enhances its sensitivity. The first detectors are expected to operate in 1990 or later. The maximal sensitivity will be reached perhaps by 2000. The corresponding two levels of sensitivity are represented by the curves marked LIGO in Figs.1, 2 and 4.

6 New Ideas and prospects

Although several gravity-wave detectors have been operating for a long time and others are in a state of technical realization, gravity-wave detection is not yet purely an engineering science. There is still plenty of room for new ideas and suggestions even at the level of first principles.

As an example I can mention two new effects in the motion of free test bodies – *kinematic resonance* and the *memory effect* [42]. Both effects refer to the final state of a pair of bodies initially at rest at a time t_0 and then subject to the action of a weak gravitational wave. And both effects follow from the elementary equation of mechanics: $mx'' = F(t)$, to which the equation of motion for a particle subjected to the action of a gravity-wave force reduces in the simplest circumstances.

The kinematic resonance effect states that the averaged distance between two particles made free at some t_0 will linearly increase or decrease depending on the initial phase of the wave. Of course, the usual oscillations are superimposed on this systematic motion. By time τ, the systematic change in the distance Δl is expressed by the formula:

$$\frac{\Delta l}{l} = \frac{1}{2} h\,(\omega\tau)\cos\psi$$

where l is the initial distance and h, ω and ψ are the amplitude, frequency and phase of the wave. It is important to note that τ may be very large: $\omega\tau \gg 1$. (This effect is a kind of drift effect considered earlier, [43].)

The memory effect states that the final distance between two bodies which have experienced the action of a gravity-wave burst is, in general, different from the initial one. The net variation of the distance is determined by the simple formula (see also [44], [45]):

$$\Delta l = \frac{1}{m}\int\!\!\int (F(\tau)\, d\tau)\, dt \, .$$

It is shown in [42] that both effects can be used to increase the sensitivity in the ground-based and cosmic interferometers as well as in the observations undertaken in the program POINTS, [46].

Another direction of research deals with the feasibility study of possible experiments in detecting the high-frequency part of the relic stochastic radiation, $\nu \approx (10^{10} - 10^{11})$Hz. This suggestion implies the use of a specially constructed crystal antenna ("large crystal") and is based on the calculation of the cross-section for such an antenna.

Let a gravitational wave impinge on a long rod of length l (a one-dimensional crystal). The gravitational wavelength λ_g is assumed to be much longer than l. It is known that if the wave excites the lowest acoustic mode, $n = 1$, the absorption cross-section integrated over frequencies is equal to $\sigma_1 \approx GMv_s^2/c^3$, where M is the mass of the rod and v_s the speed of sound, [47], [48]. Now, let the rod be excited by a high-frequency wave (with $l \ll \lambda_g$, as before) that excites the n-th mode of oscillations, $n \gg 1$. The assumption usually made is that the ends of the rod are free. Then one obtains the cross-section for the n-th mode $\sigma_n \approx \sigma_1/n^2$, [48]. In other words, under these boundary conditions, the higher n, the smaller the cross-section.

However, for a crystal and, in fact, for any system where volume effects are more important than the boundary ones, the appropriate boundary conditions are periodic conditions, not free end conditions. By carrying out calculations with the periodic boundary conditions we obtain $\sigma_n \approx \sigma_1$, that is the same cross-section for each n. In the case of a broad-band gravitational signal, with $\Delta\nu \approx \nu$, one has to take into account the number of excited modes, Δn, and calculate the total cross-section, σ_{tot}. For a one-dimensional system one has $\Delta n \approx n$ and:

$$\sigma_{tot} \approx \frac{G}{c^3} M v_s^2 \, . \, n = \frac{G}{c^3} m v_s^2 \, . \, n^2 \, .$$

In the last equality we have introduced the mass m of one small element of the rod, of length of the order of the n-th mode acoustic wavelength λ_s, so that $M = mn$. This element can be considered as an elementary quadrupole. Thus we see that σ_{tot} is proportional to the squared number of these quadrupoles. For a three-dimensional system with a characteristic $l \ll \lambda_g$ one can also expect that we will find $\sigma_{tot} \approx Gmv_s^2 n^2/c^3$, where m is the mass of an element with linear size of order λ_s, and $M = mn^3$.

These formulae are valid as long as l is smaller than λ_g. In order to get a large composite antenna with total volume larger than λ_s^3 we must use many such crystals. The cross-section of the composite antenna is proportional to the total number of the crystals. Let us apply these estimates to the problem of detecting the relic stochastic background. We assume that in the considered frequency band, $\varepsilon_g \approx \varepsilon_\gamma$. As a criterion of detectability we accept the condition that the whole system has absorbed energy of the order of one quantum, $h\nu$. Then it turns out that the total mass of the antenna should be of the order $M_{tot} \approx 10^4$ g and the observation time $\tau \approx 10^5$ sec. Of course, these estimates respect only the principal demands and ignore completely the various technical problems. However, taken at face value, they show that the detection of relic background does not seem hopeless.

7 Conclusion

I hope that this lecture demonstrates the importance, the fascination and the fruitful development of gravity-wave astronomy. However there is one additional motivation that makes research in this field so attractive. Gravitational radiation is so weak that I believe, at least I hope, it can be used only for peaceful aims.

References

1. Taylor, J.H. (1986). Invited talk at GR 11 Conference (see this volume).
2. Iben, I., Tutukov, A.V. (1984). Ap. J. Supp., **54**, *2*, 335.
3. Grishchuk, L.P. (1977). Ann. N.Y. Acad. Sci., **302**, 439.
4. Rubakov, K.A., Sazhin, M.V. & Veryaskin, A.V. (1982). Phys. Lett., **115B**, 189.
5. Grishchuk, L.P., Petrov, A.N. & Popova,A.D. (1984). Comm. Math. Phys., **94**, 379.
6. Zeldovich,Ya.B. & Grishchuk, L.P. (1986). Uspekhi Fiz. Nauk., **149**, 4, 694.
7. Damour,T. (1983). Article in *Gravitational radiation,* p.59, N.Deruelle and T.Piran (eds.), North-Holland, Amsterdam.
8. Grishchuk, L.P., Kopejkin, S.M. (1986). In *Relativity in Celestial mechanics and Astrometry*, edited by J.Kovalevsky and V.A.Brumberg, North-Holland, Amsterdam.
9. Shäfer, G. (1986), GRG, **18**, 255.
10. Lipunov, V.M., Postnov, K.V. & Prokhorov, M. (1986) Astrophysics & Space Science (submitted for publication).
11. Hills, D., Bender, P.L., Faller, J.E. & Webbink, R.E. (1986). Abstracts of contributed papers, GR 11, p. 509.
12. Thorne, K.S. (private communication).
13. Schutz, B.F. (1986). Cardiff preprint.
14. Friedman, J. & Schutz, B.F. (1978). Ap.J., **222**, 281.
15. Wagoner, R.V. (1984). Ap.J., **278**, 345.
16. Faller, J.E., Bender, P.L., Chan, Y.M., Hall, J.P., Hills, D. & Hough, J. (1983). 10th International Conference on GRG, Contributed Papers. Edited by B.Bertotti, F. De Felice, A. Pascolini, Rome, p. 960.
17. Grishchuk, L.P. (1974). Zh.E.T.F. **67**,825 (Sov. Phys, JETP, **40**, 409); (1975). Lett. Nuovo Cim., **12**, 60; (1977). Uspekhi Fiz. Nauk., **121**, 629 (Sov. Phys. Uspekhi, **20**, 319).
18. Turner, M.S. (1986). Invited talk at GR 11 Conference. See this volume.
19. Grishchuk, L.P. & Polnarev, A.G. (1980). In: *General Relativity and Gravitation*, volume II, page 393, edited by A. Held, Plenum Press, New York.
20. Zeldovich, Ya. B. (1972). M.N.R.A.S., **160**,1.
21. Harrison, E.R. (1970). Phys. Rev., **D1**, 2726.
22. Starobinsky, A.A. (1979).Pis'ma Zh.E.T.F., **30**, 719 (Sov.Phys.JETP Lett, **30**, 682).
23. Fabbri, R. & Pollock, M.D. (1983). Phys. Lett., **B125**, 445.
24. Abbott, L.F. & Wise, M.B. (1984). Nucl Phys., **B244**, 541.
    Abbott, L.F. & Harari, D.DS. (1986). Nucl. Phys., **B264**, 487.
25. Witten, E. (1984). Phys. Rev., **D30**, 272.
26. Krauss, L.M. (1986). GRG, **18**, 723.
27. Hellings, R.W., Callahan, P.S., Anderson, J.D. & Moffet, A.T. (1981). Phys. Rev., **D23**, 844.
28. Anderson, A.J. (1983). 10th International Conference on GRG, Contributed papers. Edited by B.Bertotti, F. De Felice, A. Pascolini, Rome, p. 945.
29. Anderson, J.D., Lau, E.L. & Estabrook, F.B. (1986). Abstracts of contributed papers, GR 11, p.130.
30. Bertotti, B. & Iess, L. (1985). GRG, **17**, 1043.

30. Bertotti, B. & Iess, L. (1985). GRG, **17**, 1043.
31. Mashhoon, B. & Grishchuk, L.P. (1980). Ap. J., **236**, 990.
32. Braginsky, V.B. & Thorne, K.S. (1985). Nature, **316**, 610.
33. Drever, R.W.P. & Weiss, R. Proposal to NSF for LIGO (in preparation).
34. Grishchuk, L.P. & Il'tchenko, A.S. (1984). Abstracts of VI Soviet Gravitational Conference, Moscow, M.G.P.I. Volume 1, p. 303.
35. Carr, B.J. (1980). Astron. Astrophys., **89**, 6.
36. Rees, M.J. (1983). Article in *Gravitational radiation,* p.296, N.Deruelle and T.Piran (eds.), North-Holland, Amsterdam.
37. Hogan, C. (1986). M.N.R.A.S., **218**, 629.
38. Fairbank, W.M. (1986). Workshop summary talk at GR 11 Conference. See this volume.
39. Grishchuk, L.P.(1983). Proceedings of 9th International Conference on General Relativity and Gravitation, p. 255, edited by E.Schmutzer, Cambridge University Press, Cambridge.
40. Bertotti, B., Carr, B.J. & Rees, M.J. (1983). M.N.R.A.S. **203**, 945.
41. Blandford, R., Narayan, R. & Romani, R.W. (1984). J. Astrophys. Astr., **5**, 369.
42. Braginsky, V.B. & Grishchuk, L.P. (1985). Zh.E.T.P., **89**, 744.
43. Grishchuk, L.P. (1974). Zh.E.T.P., **66**, 833 (Sov. Phys. JETP, **39**, 402).
44. Zeldovich, Ya.B. & Polnarev, A.G. (1974). Astron. Zhurn., **51**, 30.
45. Kovac, S. & Thorne, K.S. (1978). Ap. J., **224**, 62.
46. Reasenberg, R.D. & Shapiro, I.I. (1982). Acta Astronautica. **9**, 103.
47. Weber, J. (1961). *General Relativity and Gravitational Waves,* Wiley, New York.
48. Misner, C.W., Thorne, K.S. & Wheeler, J.A. (1973). *Gravitation,* Freeman and Co., San Francisco.

QUANTUM GRAVITY

C.J.Isham
The Blackett Laboratory, Imperial College
University of London, UK.

1 INTRODUCTION

The current generation of quantum gravitationalists lives in stirring times. The enthusiasm and intensity of effort engendered by the recent developments in superstring theory has surpassed anything that I have previously encountered and, when preparing my quantum gravity review talk, I wondered at first whether there was anything to be said other than "Come to the string theory lecture on Thursday afternoon". But on reflection I realized that, paradoxically, there has been more published in "straight" quantum gravity since the last GRG meeting than in any earlier span of three years; and it is on this work that I shall report.

In analysing the rather complicated status of quantum gravity studies one needs to identify and distinguish the structural foundations of the various approaches to the subject. There is no unique way of doing this but a good start may be made by considering how any particular approach views the a priori and/or a posteriori roles of the concepts of (i) spacetime, and (ii) quantum theory. Indeed, much of the disagreement in the past about how one should approach the quantization of gravity has been grounded in differing expectations of what a quantum theory of gravity should look like. And this is turn reflects the multifarious answers to the question of whether the ideas of classical general relativity and/or quantum physics should be built into the very foundations of the subject or assigned a more "phenomenological" role in which they are recoverable from the theory only in some appropriate limit.

The idea that quantum theory is itself derivable from a higher structure is attractive; but there has been little direct progress in this direction. However, the situation regarding spacetime concepts is somewhat different and in this context it is useful to distinguish between "primary" and "secondary" quantum theories of gravity:

(P) A "primary" quantum theory of gravity is one whose starting point is the classical action of general relativity (or some simple extension, for example adding (Riemann curvature)^2 terms) together with the associated geometrical and spacetime concepts. This theory is then "quantized"; rather as the classical Maxwell equations can be quantized, but with the metric tensor replacing the electromagnetic potential. Such a programme might involve Hilbert space operators, or functional integrals; it might use a covariant or a canonical framework; or one might even try to extend the quantization algorithm to the fabric (topology, differential structure, etc.) of spacetime itself. The common factor in all this is the view that quantization is an "active" process: it is a soteriological action ("quantize" is a verb!) that one performs on a classical theory to elevate it into the world of quantum physics.

(S) A "secondary" quantum theory of gravity is a quantum theory whose starting point is *not* classical general relativity, but which nevertheless has the property that this particular classical system can be recovered from the theory in some suitable limit (eg. at low energies). Such a quantum theory might be obtained by an active quantization of some other classical system; and it is very likely that it will assume certain concepts pertaining to space and time. But, a priori, this system and these concepts are *not* those of classical general relativity. Note that, given a scheme of type (P), one can study its classical limit and presumably recover the classical theory from which one started. But the situation for schemes of type (S) is different. True, the classical theory of general relativity is recoverable in some limit, but one cannot reverse the procedure by actively quantizing this system to recover the original quantum theory.

This distinction between primary and secondary quantum theories of gravity is not meant to be sacrosanct (many philosophical problems arise in attempting to make precise the idea that one theory/conceptual scheme is the limit of another), but it is a useful working tool for distinguishing between some approaches to the subject. A good example of a secondary scheme is the "induced gravity" programme proposed by Sacharov and Zeldovich, and much developed in the early 1980's by Adler, Zee etc. In this theory, the conceptual spacetime structure of general relativity *is* imposed a priori, but only the matter fields (not the gravitational metric) are quantized . The action principle includes the coupling of the metric to these matter fields but *not* the Riemann curvature scalar of gravity itself; indeed, the central thesis of induced gravity is that this will arise as a counterterm in the quantization of the matter fields.

The induced gravity programme has some attractive features, not least of which is the prospect of predicting a value for Newton's constant. However, it has recently received what seems to be a mortal blow at the hands of David (1984) who has shown that the non-Borel summability of the ultra-violet divergences in the bare coupling constants results in an intrinsic, non-perturbative ambiguity in these predictions. The argument was extended by David & Strominger (1984) to scale-invariant gravity (which also has no Riemann scalar term in the action), and research activity in these areas now seems to have ceased.

A more striking example of a secondary theory is afforded by the superstring programme in which the classical theory of relativity is recovered in the zero-slope/infinite-tension limit. The underlying classical theory (in so far as there is one) has no connection with general relativity and there is currently much interest in the question of precisely how the familiar spacetime concepts do emerge. This is closely connected with the ground state/compactifications of the theory and is one of the most important problems confronting the superstring programme. In so far as the string is deemed to move in a background space (albeit of 10 or 26 dimensions) it might be felt that a number of spacetime concepts have already been built into the theory. But there is a growing feeling that this simple physical picture is not the last word on the matter and that the final form of the string theory may not make any specific reference at all to what we would regard as "physical" space or time. A good example of this philosophy can be found in the recent work of Friedan & Shenkar (1986) who seek to express the entire string theory in terms of holomorphic vector bundles on the universal moduli space of all Riemann surfaces, and with the expectation that physical spacetime must be reconstructed from within this framework. String theory is a fascinating and important branch of theoretical physics, and it is discussed at length in the article by Mike Green. However, my task is to report on recent developments in the primary quantum gravity programmes, and to this we must now turn.

Viewed from the perspective of standard quantum field theory, the obvious way of quantizing the gravitational field is: (i) expand the four-dimensional metric $g_{\mu\nu}(x)$ around a Minkowski space background $\eta_{\mu\nu}$ by writing $g_{\mu\nu}(x) = \eta_{\mu\nu} + h_{\mu\nu}(x)$ and (ii), quantize the spin-2 field $h_{\mu\nu}(x)$ using conventional, Poincaré covariant techniques. However, the ensuing weak-field perturbation theory is power counting non-renormalizable and only a set of miraculous cancellations could remove the unwanted ultraviolet divergences. That no such miracle occurs has finally been confirmed by Goroff & Sagnotti (1985,1986) who have calculated the explicit 2-loop divergences; and they do not vanish. This result was widely expected, and for some time the emphasis in primary quantum gravity research has been either on modifying the classical theory of general relativity to improve the ultra-violet behaviour (eg. supergravity, higher-derivative theories), or in developing non weak-field perturbative techniques (eg. Regge calculus).

In this review talk/paper I shall attempt to summarize some of the research programmes that have been particularly active since the last GRG meeting and which are likely to be developed further in the future. I will not be discussing Kaluza-Klein theory or supergravity as these topics are covered in the article by Mike Duff. Regrettably, there is also not enough time/space to report on quantum field theory in a curved, but fixed background. One especially active area is quantum field theory in the early universe and/or DeSitter space and this will perhaps be mentioned in the article by Mike Turner. An impressive example is the work by Parker & Toms (1985) and Jack & Parker (1985) in which the large curvature regime is studied using a non-perturbative summation of the DeWitt-Hadamard-Schwinger series for the propagator. There have also been interesting developments in the rigorous formulation of quantum field theory in a curved spacetime (e.g. Haag et al 1984; Dimock 1985; Dimock & Kay 1985; Kay 1985) and details of some of this will be found in the report by Peter Hajicek.

In addition to my review talk at the GRG meeting there was a quantum gravity workshop organised by David Boulware. He has very kindly given me his "share" of pages in these Proceedings and I have included the relevant material in this article.

2 HIGHER-DERIVATIVE GRAVITY

2.1 Introduction.

One long-standing response by the particle physics community to the perturbative non-renormalizability of gravity, is to replace the simple Einstein lagrangian with one that involves squares of the Riemann curvature. The most general such form is

$$L = \sqrt{-g}\ \{ L + aR + bR^2 + cC^{\alpha\beta\gamma\delta} C_{\alpha\beta\gamma\delta} \} \tag{2.1}$$

where L,a,b and c are constants, and R and $C_{\alpha\beta\gamma\delta}$ are the Riemann scalar and Weyl curvature respectively . The motivation for such a change is the softening of the short distance behaviour of the graviton propagator induced by the presence of fourth-order derivatives of the metric tensor. In fact the quantized theory becomes renormalizable (Stelle 1977). Unfortunately, this is at the expense of unitarity - a major problem to which we shall return shortly. However, on the bonus side, it has been known for some time that the theory is "asymptotically free". More precisely, the energy dependence of the effective coupling constants is such that the theory is dominated at high energies by (Riemann curvature)^2 effects while, at low energies, the usual Einstein theory is recovered. The precise behaviour of the R^2-term is in doubt following a recent calculation of the relevant β-function by Avramidy & Barvinsky (1985) which disagrees with the earlier result of Fradkin & Tseytlin (1981,1982). This does not affect any of the major implications of the asymptotic freedom, and the question will be doubtless be settled definitively if part II of the very careful work of Barth & Christensen (1983) ever appears.

At a more classical level, there have been several useful developments over the last few years in our understanding of (Riemannian curvature)^2 theories of gravity. A careful analysis of the canonical Hamiltonian structure was made by Boulware (1984), and extended to higher-derivative supergravity by Boulware and Deser (1984). Higher-derivative versions of the celebrated positive energy theorems have also featured prominently . Thus Boulware et al (1983) considered scale-invariant gravity (obtained by putting L = a = 0 in eq.(2.1)) and showed that any asymptotically flat solution to the field equations necessarily has zero total energy. A variety of positive energy and stability theorems were found by Strominger (1984) and Whitt (1984) for the special case when $L = \sqrt{-g}\ \{ aR + bR^2 \}$. These results were rederived and extended in a comprehensive analysis by Boulware et al (1986) who studied the complete lagrangian of eq.(2.1) by constructing a supersymmetric version of the classical theory and then invoking the manifest positivity of the energy in such systems.

This progress in understanding the classical features of the lagrangian (2.1) is pleasing, but of course the central question is the existence or otherwise of a satisfactory quantum theory. And here we encounter the crucial difficulty mentioned above: the renormalizability coming from the $[p^2(p^2 - a)]^{-1}$ form of the graviton propagator is achieved at the expense of acquiring a massive spin-2 "ghost" partner for the physical massless graviton, and hence a violation of unitarity. Within the context of perturbation theory this simultaneous gain and loss forms an inseparable syzygy, and the only hope is to find an intrinsically non-perturbative way of performing the quantization. Past attempts have used expansions in the inverse of the number of matter fields, or in the dimension of spacetime, and correspond to summing certain infinite sets of Feynman diagrams. This has the desirable effect of freeing the leading order S-matrix elements from unwanted singularities because the offending ghost pole splits into a complex pair that moves away from the real axis. However, attempts to extend this to higher orders usually invoke the Lee-Wick mechanism which involves an indefinite metric quantization with undesirable zero norm states being removed with the aid of much "adjustment by hand". This procedure introduces an intrinsic acausality into the theory and it is often criticised for that reason. However, in the case of quantum gravity, this pathology is only seen at Planck size distances $(Gh/c^3)^{1/2} \approx 10^{-33}$ cms., and it can be argued that anything is permitted at this scale! More serious perhaps is the general inelegance of the Lee-Wick mechanism and the difficulties that have been encountered in attempting to incorporate it within a genuine quantum field theoretic framework. For example, Boulware & Gross (1984) tried hard to reexpress it in the language of euclidean space functional integrals, but to no avail.

The failure to incorporate a euclidean spacetime is particularly disturbing as this forms the backbone of so much modern work on quantum field theory. Of course, the issue of Riemannian v. pseudo-Riemannian manifolds raises many deep questions in quantum gravity and we will return to some of these later. But for the moment we are concerned with those approaches in which a flat background is imposed a priori, and in this context it should be noted that Antoniadis & Tomboulis (1986) have shown how the formal Ward identities for the euclidean space, BRS-quantized higher-derivative theory suggest a gauge-dependence for the mass of the ghost particle. They view this as an indication that the ghost is an artifact of the weak-field perturbation methods rather than an intrinsic feature of the system itself. It thus becomes all the more imperative to seek new ways of defining the quantum theory, and to some of these we now turn.

2.2 Lattice gauge theory

One of the most fruitful approaches to the functional integral quantization of Yang-Mills theory involves discretizing the system by replacing euclidean spacetime with a cubical lattice and approximating the functional integral by a finite dimensional integral over group-valued functions defined on the 1-simplices ("links") of the lattice. Such methods have been employed extensively in rigorous studies of the existence of quantum field theories and, in the case of QCD, of the existence of phase transitions that could separate the quark confinement and high energy, asymptotically free, regimes. The hardest problem is to prove the existence of a continuum limit as the lattice spacing is taken to zero; for a general review see Halliday (1984).

Several attempts have been made to apply these methods to quantum gravity with, not surprisingly, a tendency to emphasise those formulations of general relativity (or related theories) that most mimic the non-abelian gauge structure of Yang-Mills theory. To prove unitarity for the Lorentzian continued, continuum limit (if it exists), the lattice theory must be shown to satisfy reflection positivity (which concerns the way the euclidean fields behave under reflections in a lattice plane), and most recent work has been directed at this goal.

A good example is afforded by Kondo (1984) who developed the earlier work of Mannion & Taylor (1981) on the vierbein, first-order form of lattice gravity and showed that the theory is indeed reflection positive. This scheme is marked by an asymmetric treatment of the spin-connection and vierbein. Only the former is a genuine "gauge variable" and, as such, assignable to the links of the lattice; the vierbein is treated as a matter field and is specified at the vertices . A similar approach was used by Tomboulis (1984) in a study of the general (Riemann curvature)^2 lagrangians with which we are concerned in this chapter. He found reflection positivity for all values of the coupling constants and argued forcibly that the continuum limit should be unitary. Other authors have removed the vierbein/spin-connection asymmetry by using the modified versions of general relativity which arise from gauging the deSitter or conformal groups. A recent example is the work of Kaku (1986) which also includes a "random lattice", ie. the positions of the lattice points are extra integration variables in addition to the gauge fields on the links. He succeeds in proving reflection positivity, and advances the hope that the use of a random lattice will assist in the thorny problem of recovering the spacetime diffeomorphism group invariance in the continuum limit (see below).

One of the most comprehensive recent discussions is that of Menotti & Pelissetto (1985,1986) who have studied a variety of "gauged" versions of general relativity and, in all cases, succeeded in proving reflection positivity. However, they also demonstrated the existence of a gravitational analogue of the notorious "fermion doubling" problem that occurs in conventional lattice gauge theories. More precisely, they performed a weak field expansion of the discretized quantum gravity theory and showed that the hypercubical symmetry of the lattice results in the appearance of 16 independent gravitons! Great care would need to be taken if attempts were made to remove these spurious modes by modifying the lattice action: this works in Yang-Mills theory, but in gravity it could play havoc with the recovery of a diffeomorphism group invariance in the continuum limit. Indeed, it is clear that in all these lattice approaches the crucial questions to be answered are,

(i) Does the continuum limit exist at all?

(ii) If so, is general relativity reproduced at low energies? In particular, how does the diffeomorphism group invariance of the classical theory reappear as the lattice structure is removed?

These questions are extremely difficult and it is not surprising that very little is currently known about the matter, beyond perhaps a general tendency for those not working on this particular approach to quantum gravity to assume that the answer to question (i) is "No"! The use of a fixed hypercubical lattice does seem at variance with the geometrical and topological ideas underlying the classical theory of general relativity and it has often been suggested that the alternative use of Regge calculus (in which spacetime is approximated by a dynamical, triangulated simplicial complex) stands more chance of reproducing these features.

2.3 Regge calculus and $(\text{Riemann})^2$

There are a number of reasons why one should be interested in Regge calculus and I will discuss the subject in detail in section 3.3. For the moment however we are only concerned with the application of these techniques to the higher-derivative versions of general relativity. Careful work by Hamber (1984) and Hamber & Williams (1984,1985,1986b) has produced a set of analogues for $(\text{Riemann curvature})^2$ of the well-known Regge form of the scalar curvature lagrangian, and the stage is thereby set for analytical and/or numerical investigations of the resulting discretized classical and quantum systems. A first attempt is contained in Hamber & Williams (1985,1986a,b) who report the results of computer calculations for the "Reggeized" functional integral in two- and four-dimensional spacetimes. In the latter case they use the simplified lagrangian

$$L = \sqrt{g}\,\{ L + aR + bR^2 \} \tag{2.2}$$

in which the (perhaps rather crucial!) Weyl curvature term in eq.(2.1) has been dropped. One cannot deduce too much from these initial calculations although it is clear that the rapidly increasing power of "supercomputers" is destined to affect quantum gravity studies as much as any other branch of physics. In this context, we should also note the recent work of Berg (1985, 1986) using Monte Carlo techniques on the Reggeized functional integral in a four-dimensional spacetime with just a Riemann curvature scalar lagrangian. His results suggest the presence of a phase transition, and it is possible that the continuum limit may actually exist, albeit only for a limited range of the coupling constant. This is a remarkable conclusion and further work along these lines is awaited with much interest.

Returning to the higher-derivative theories, it should be noted that the papers cited above do not discuss the question of unitarity and/or the Regge analogue of the reflection positivity found in the fixed, hypercubical lattice schemes. Indeed, this question has been rather neglected in Regge calculus based discussions of quantum gravity, and the only reference of which I am aware is Fröhlich (1981). From all perspectives, the problem of the continuum limit is extremely difficult, and one needs to be quite an optimist to believe that gravity can really be quantized by the simple expedient of adding $(\text{Riemann curvature})^2$ terms to the lagrangian. I suspect that the majority of those who still hold that gravity can be quantized at all in the primary sense expect that something far more dramatic will need to be added to the formalism. Indeed, one of the main attractions of the Regge calculus scheme is its potential extendibility to incorporate the idea that the topology of spacetime must be made part of the dynamical structure of the quantum theory. However, the question of spacetime topology and quantum gravity has implications well beyond just the Regge programme, and justifies a chapter in its own right.

3 QUANTUM SPACETIME STRUCTURE

3.1 Quantum Foam

Almost by definition, any "active" quantization of general relativity will involve some type of quantum fluctuation in the metric tensor. Far more speculative however is the not infrequently expressed feeling that any successful quantum theory of gravity must also incorporate quantum fluctuations in the fabric of spacetime itself. This sounds exciting, but the meaning of such a concept is unclear, and a number of interpretations have been proposed over the years. One of the more productive ideas has evolved from Hawking's remarkable discovery of the thermal, quantum mechanically driven, production of particles by a black hole. Using the theory of thermal Green's functions, this process can be rederived in a functional integral framework where the background spacetime is Riemannian, rather than pseudo-Riemannian. This suggested to Hawking that the quantum gravity analogue of the instanton structure of Yang-Mills theory would involve a functional integral over all Riemannian metrics on all possible four-manifolds. It is this appearance of many topologically inequivalent spacetimes that gives rise to the concept of "spacetime foam" - a term first coined by Wheeler in his investigations of the "quantum wormhole" structure of three-space in canonically quantized gravity.

Much of the work on the Hawking process has involved quantum field theory on a *fixed* spacetime background, and I will not be reporting on that here. However, within the context of quantum foam, one relevant result is due to York (1985) who showed that, contrary to earlier expectations, it *is* possible to construct a canonical ensemble of black-holes, provided that one allows for holes of two different masses, the lighter one being unstable. Another interesting development in the study of the Hawking effect is the recent revival of interest in the so-called "elliptic interpretation" of the spacetime structure of a black hole or uniformly accelerating observer. This employs a global identification of the left and right event horizons, which has a significant effect on a quantum field propagating in such a space. The subject is reviewed in Gibbons (1986) and Sánchez & Whiting (1986) and extended to deSitter space in Sánchez & Folacci (1986). Freese et al (1985) claim that this is the explanation of the mysterious factor of 2 that t'Hooft (1985) found in the Hawking temperature - but this still seems to be a contentious issue.

A striking feature of the Hawking process is the loss by an external observer of the information that crosses the event horizon. This results in the transformation of an incoming pure quantum state into an outgoing state that is mixed, and Hawking (1982,1984b) suggested that a similar process might occur in quantum gravity proper if virtual blackholes appeared as part of the spacetime foam. He constructed a set of axioms for quantum gravity which are based on the familiar Wightman axioms but with the S-matrix replaced by a "$-matrix" (to accomodate the pure-to-mixed transition) and a corresponding slackening of the condition of asymptotic completeness. This aspect was developed by Alvarez-Gaume & Gomez (1983) and the general idea was studied by Banks et al (1984), and by Gross (1984) in the context of a Kaluza-Klein model. Ellis et al (1984) considered the experimental implications and decided that, perhaps not unsurprisingly, they lie outside the reach of present-day technology. Nevertheless, the Hawking axioms are a good illustration of the fundamental way in which the ideas of quantum field theory might be modified under the influence of spacetime concepts (such as event horizons) drawn from classical general relativity.

Another intriguing application of the idea of quantum foam is due to Crane & Smolin (1986). They study perturbative quantum gravity in the presence of such a background with the condition that the quantum fluctuations of the metric tensor should vanish on the event horizons of the virtual black holes. Rather remarkably, this leads to a theory that is ultra-violet cut-off, and hence finite! One weakness of this work is the use of a fixed density distribution $\rho(\text{mass})$ of black holes in the background, whereas one would like this also to be part of the dynamical structure. But there is a compensating feature: the particular choice $\rho(\text{mass}) = (\text{mass})^{-4}$ generates a *fractal* spacetime - an intriguing concept that has arisen recently in other contexts (see for example Page 1984; Svözil & Zeilinger 1985; Englert et al 1986; Müller & Schäfer 1986).

There is no doubt that the generation of mixed states via the Hawking process is something that continues to fascinate workers in both classical and quantum gravity. For example, Penrose (1986) has been developing the idea that the black hole process may relate in some deep way to the pure-to-mixed state transition that arises in the interpretation of the measurement process in conventional quantum theory (see also Károlyházy et al 1986). Such a situation would imply a radical revision of our current understanding of both quantum theory and general relativity; it affords a good illustration of the profundity of the a priori concepts that could lie at the heart of quantum gravity.

3.2 Lorentzian spacetime effects

Hawking's ideas on spacetime foam have been developed almost entirely within the framework of Riemannian geometry, and it is still a debatable point as to how this should be related to the use of pseudo-Riemannian/Lorentzian metrics. Wheeler's original introduction of spacetime foam was within the context of (intrinsically Lorentzian) canonical quantization, and the emphasis there was on the topological structure of the spatial 3-manifold Σ rather than the four-dimensional spacetime. Many of the quantum topological effects in canonical quantization derive from the global structure of the classical configuration space which, in the case of general relativity, is Riem Σ/Diff Σ, where Riem Σ and Diff Σ are respectively the set of all Riemannian metrics on Σ and the diffeomorphism group of Σ. The most relevant topological information is contained in the first homotopy group,

$$\pi_1(\text{Riem } \Sigma/\text{Diff } \Sigma) \approx \pi_0(\text{Diff } \Sigma) \tag{3.1}$$

which gives rise to a gravitational analogue of the Yang-Mills θ-vacua. Following earlier work of Friedman & Witt (1983), Witt (1986) has performed extensive calculations of π_0(Diff Σ) for the special case when Σ is the quotient of the 3-sphere by some finite group action.

These effects arise from a fixed topology for Σ and are relatively easy to understand. A much harder question within this Lorentzian picture of spacetime is how quantum induced *changes* in the 3-topology could arise. Any two Riemannian 3-manifolds Σ_1 and Σ_2 are topologically cobordant and hence can be "filled in" with a 4-manifold M whose boundary is the disjoint union of Σ_1 and Σ_2. However this does not necessarily mean that they are *Lorentz* cobordant, ie. there may not exist a smooth Lorentzian metric on M that can induce the Riemannnian metrics on Σ_1 and Σ_2 . The necessary and sufficient condition (Sorkin 1985a,b) for the existence of such a metric is purely topological: if the dimension of M is even (resp. odd) then the Euler number of M must vanish (resp. the Euler numbers of Σ_1 and Σ_2 must be equal).

In the contrary case, the "metric" will necessarily have points of degeneracy which, as shown by Anderson & DeWitt (1985), can have striking effects on the quantum theory. As Bryce DeWitt explained at the GRG11 quantum gravity workshop, he and Anderson have studied scalar quantum field theory in a fixed two-dimensional spacetime whose topological structure (a "pair of trousers"; see Fig.1) is chosen precisely because it does *not* admit a globally defined, smooth Lorentzian metric.

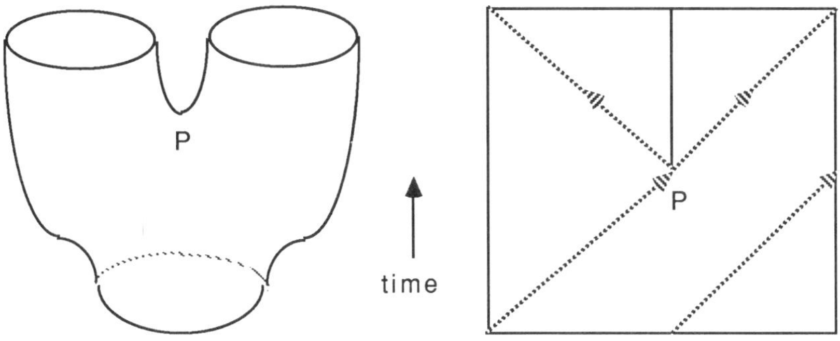

Fig. 1

In order to quantize the field it suffices to decide on the propagation of classical waves on this singular spacetime - the (linear) quantum field theory then follows from the usual techniques of Bogolubov transformations. Anderson and DeWitt tackled this question by mapping the 2-manifold onto the "cut-square" in the right of Fig.1 and then specified that all waves should propagate in the obvious way, except that a wave incident on the singular point P should split and pass symmetrically up each leg of the trousers. The ensuing quantum field theory predicts an infinite production of particles from P, and the authors suggested that if this were to occur in a full quantum theory of gravity, the resulting gravitational field would cause the space to dilate violently, and the incipient topology change would be aborted!

There are a number of uncertainties surrounding this work: for example it is not clear if the effect is really due to the changing topology with its degenerate metric, or to the singular nature of the "split-square" representation of the spacetime. And in either case, the dependence of the results on the details of the assumed classical propagation is unknown. But the whole programme is most intriguing and it is to be hoped that this line of study will be developed further.

Note that the problem of a degenerate Lorentzian metric is directly relevant to the theory of the first-quantized string, which involves a non-linear σ-model defined on a two-dimensional space Ξ and taking its values in a 10- (or 26-) dimensional spacetime. Much of the work on string theory employs a Riemannian metric on Ξ (which thus becomes one of the classical Riemannian surfaces) and is often accompanied with the implicit assumption that it is a trivial matter to adapt the results to the case where Ξ is Lorentzian. But the work of Anderson and DeWitt might suggest otherwise.

At this point it is worth emphasising that, from a topological perspective, there is a lot to be said in general for investigating lower dimensional versions of general relativity, and this has been the subject of a mini research programme in its own right. In particular, Deser & Jackiw (1984), Deser et al (1984), and Jackiw (1985) have studied the classical theory in two and three dimensions, and results on the canonical quantization of lower dimensional gravity have been reported by Henneaux (1984,1985), Martinec (1984) and Banks et al (1985).

3.3 Regge Calculus

One of the more striking features of quantum gravity research over the last few years has been the steady increase in the use of Regge calculus. This has been applied mainly to the Riemannian, functional integral approach and involves approximating the functional integral as,

$$< O > = \Sigma\, \nu(K) \int d\mu(l)\, O(l)\, \exp\text{-}S^{Regge}(l) \qquad (3.2)$$

where K is a simplicial complex with N_0 vertices and O(l) is an observable written as a function of the edge-lengths l of the simplices. The Regge approximation to the Einstein action $S^{Regge}(l)$ is usually given as a function of the area of the 2-simplices and of the defect angles of the 4-simplices containing these as faces. However, as the notation suggests, it can be rewritten as a (complicated!) function of the edge-lengths alone. It is assumed /hoped that the continuum limit will be achieved as N_0 goes to infinity.

Many intriguing questions and problems arise within this primary approach to quantum gravity. For example:

(1) What is the measure $d\mu$? Should it preserve the generalized triangle inequalities satisfied by the simplices in a piece-wise flat complex? Or should we admit singular limits that might correspond in some way to the inclusion of complexes that are not approximations to smooth manifolds? Should $d\mu$ be scale invariant? Or, following the recent suggestion of Jevicki & Ninomiya (1986), should it contain a "non-local" part associated with the conformal anomaly? There is also the question of the measure/weight factor $\nu(K)$ which might be determined by a Regge analogue of reflection positivity, and hence ultimately by the requirement of unitarity (Fröhlich 1981). Note that many of these questions arise also in the conventional (ie. continuum) functional integral approach to quantum gravity, and a useful recent reference in this respect is Toms (1986).

(2) What about the conformal modes of the metric? A familiar defect of the continuum Riemannianized action is its failure to be positive definite because of these modes, and this is complicated in the Regge case by the difficulty reported in Rocek & Williams (1984) of defining conformal transformations in a way that respects the triangle inequalities.

(3) Is there a continuum limit? The manner in which classical general relativity (with a single spacetime manifold M) can be reached from the Regge action has been discussed in detail by Cheeger et al (1984), Freidberg & Lee (1984), Feinberg et al (1984), and Barrett (1986). This work makes very precise the sense in which the curvature tensor on a N-dimensional Regge simplex is a δ-function concentrated on the faces of dimension N-2. Rocek & Williams (1984) and Feinberg et al (1984) have studied the simultaneous weak-field and continuum limits of the quantum theory and have shown that the correct graviton propagator is obtained. A similar result for a Lorentzian signature is given in Williams (1986). These weak-field results confirm the suspicion that Regge complexes are better than hypercubical lattices (there is only one graviton!), but of course this is a far cry from proving the existence of a continuum limit for the full quantum theory.

(4) Can we recover the spacetime diffeomorphism group? This question is of vital significance in discussing the continuum limit and can be approached from several perspectives. For example, a triangulation of a Riemannian manifold M can be viewed as a continuous map τ from an abstract simplicial complex K to M; and then each diffeomorphism ϕ of M generates a new triangulation $\phi\circ\tau : K \to M$. Unfortunately, the discretized Regge action is *not* invariant under the changes in edge-lengths thus induced, although Rocek & Williams (1984) and Hartle (1985a) have suggested that some sort of "approximate" invariance group may exist. Whether or not such a structure would suffice for quantization is not known. But, in any event, there is clearly a great danger of over-counting the metric modes (eg. Piran & Strominger 1986) and this has lead Römer & Zähringer (1986) to develop a type of "gauge-fixing" for the triangulations. They advance a scheme whereby, given the complex K, there corresponds to each Riemannian metric on M just one triangulation $\tau : K \to M$ obtained, roughly speaking, by minimising the deviations of the edge-lengths from their mean. This raises the spectre of the Regge analogue of Faddeev-Popov ghosts, BRS-invariance etc; but the idea is intriguing and deserves further study.

(5) Is there a version of Regge calculus adapted to the canonical quantization of gravity? This is a long-standing problem in which Piran & Williams (1986) have recently made some progress. Unfortunately, they were obliged to neglect the shift vector in their formalism and, as a consequence, the transverse constraints are missing and there is an inconsistency in the time evolution of the longitudinal constraint. Friedman & Jack (1986) provided partial remedies for these defects, but the Poisson brackets of their set of longitudinal and transverse constraints fail to close on an algebra. This is a major impediment to implementing a successful quantization and it is clear that, as in the "covariant" approaches, the crucial task is to find a Regge version of the diffeomorphism group. It is unlikely that much further progress will be made until this is achieved.

(6) Over what class Γ of complexes K should the functional integral in eq.(3.2) be summed? This is one of the deepest questions that can be posed within the Regge framework, and its answer requires a profound insight into the a priori role in quantum gravity of the classical concepts of space and time. The simplest choice might be to restrict the complexes K to those that are simplicial approximations to genuine smooth manifolds (but note that not all non-compact 4-manifolds are triangulable), and such a view is implicitly adopted in much of the work on spacetime quantum foam. But it is very questionable. For example, it is well-known that the measure in a path integral is concentrated on distinctly non-smooth paths; perhaps the same is true for topologies? Hartle (1985b) has considered these problems in some depth and has raised a number of important issues. He emphasises that, in choosing the class Γ of complexes, one should require that:

(i) the Regge action be extendible to all members of Γ;

(ii) there exists an algorithm for deciding whether or not any particular space belongs to Γ;

(iii) smooth manifolds dominate in the classical limit (in order to recover the equivalence principle).

In the special case of two-dimensional complexes (with the action chosen as the Euler number of the complex), Hartle shows how these restrictions do indeed suggest a particular choice for Γ: the set of "pseudo-manifolds". However, it is not clear how this discussion should be extended to spaces of higher dimension where the terms in the action are no longer purely topological and the examination of the classical limit involves metric as well as topological considerations. Evidently, there is still a long way to go before we acquire the necessary deeper understanding of what type of spacetime topology can/should occur inside the functional integral.

Many of the problems considered above have analogues in the well-developed field of random surface theory which has a variety of applications in theoretical physics (Fröhlich 1985). This is a subject in which students of Regge calculus should certainly take an interest.

4 QUANTUM COSMOLOGY

4.1 The Hartle-Hawking programme

One major motivation for studying quantum gravity is to gain insight into the dynamical structure of the universe at times less than the characteristic Planck time $(Gh/c^5)^{1/2} \approx 10^{-42}$ secs. Problems involving time evolution are most naturally studied within the framework of canonical quantization which, in the case of general relativity, means solving the

Wheeler-DeWitt equation,

$$H_L(g_{ab}, \pi^{cd})\,\Psi[g] = 0 \tag{4.1}$$

where H_L is the longitudinal constraint (which generates deformations of the spatial 3-surface Σ normal to itself) and $g_{ab}(x)$ and $\pi^{cd}(x) = -i\delta/\delta g_{cd}(x)$ are the canonical operators representing the 3-metric and its conjugate variable. In many respects, eq.(4.1) is the gravitational analogue of the time-dependent Schrödinger equation. But note the absence of any explicit time label on the state vector Ψ since "time" is deemed to be implicitly present already in the 3-geometry argument of Ψ (I am assuming that Σ is compact with no asymptotic regions). This feature has generated much discussion over the years, and I will return to it later. Note also that the state vector $\Psi[g]$ is a functional on either Riem Σ/Diff Σ or Riem Σ, according to whether or not we have previously imposed the transverse constraints,

$$H_i(g_{ab}, \pi^{cd})\,\Psi[g] = 0 \quad , \; i = 1,2,3 \tag{4.2}$$

which are the generators of diffeomorphisms of Σ.

The first extended burst of activity in quantum cosmology took place in the early 1970s and was conducted within the framework of this operator based canonical scheme. However, much recent work has used functional integration, motivated by the observation that in ordinary quantum mechanics a solution to the time-dependent Schrödinger equation can be obtained by defining

$$\psi(x,t) := \int dq()\ \exp -i\int^t L[q(s)]\,ds \tag{4.3}$$

where the path integral is taken over all paths $q(s)$, $t_0 \leq s \leq t$, such that $q(t) = x$. The gravitational analogue of this expression has been studied recently by Berger (1985), Berger & Vogeli (1985) and Barvinsky & Ponomariov (1986). Note however that there are many solutions to the Schrödinger equation: the particular one generated by eq.(4.3) is determined by the behaviour that must be specified for the paths $q(s)$ at the initial time $s=t_0$. In the case of quantum cosmology this means we are faced with the prospect of specifying boundary conditions at very early times, perhaps even at the initial singularity. This situation is sometimes felt to be unsatisfactory from a philosophical and/or a psychological viewpoint. We appear to live in only *one* universe, and it is desirable to have a theory that can predict this unique configuration, rather than being obliged to feed it in by hand.

Even setting aside theological considerations this seems, ab initio, to be a rather tall order, and it is remarkable that Hawking (1984c) and Hartle & Hawking (1983) have succeeded in developing such a scheme. These ideas have generated intense interest over the last few years and I will devote the rest of this section to a brief overview of the programme (the topic was reviewed comprehensively by Don Page at the GRG11 quantum gravity workshop). Hartle and Hawking start by defining (cf. eq.(4.3))

$$\Psi[g] := \int d\mu[g^4] \exp\text{-}S_{Eucl}[g^4] \tag{4.4}$$

in which the functional integral is over Riemannian metrics g^4 on a 4-manifold M with the property that: (i) the spatial 3-manifold S is the boundary ∂M of M, and (ii) the 3-metric g is the restriction of g^4 to S. This suggestion is striking for two reasons. Firstly, the integral (4.4) is over *Riemannian* metrics, but the functional $\Psi[g]$ satisfies the Wheeler-DeWitt equation (4.1), and this is usually derived from a canonical analysis of the *Lorentzian* theory. To my mind, this is by far the most elegant synthesis yet of these two complementary perspectives on quantum gravity. The second feature of (4.4) is that it solves the problem of fixing boundary conditions by specifying that there are none: the universe is created *Ex Nihilio* since the 4-manifold M has only the connected space Σ as its boundary.

Any 3-manifold Σ is cobordant to the empty set and so an "interpolating" manifold M can always be found, even if Σ is topologically quite complex (eg. Gonchorov & Bytsenko 1985). The scheme can be extended to include manifolds Σ that are disconnected, and for a Kaluza-Klein theory this could be obligatory since not every higher-dimensional manifold is topologically cobordant to the empty set. This is discussed by Halliwell (1986) in the context of six-dimensional Einstein-Maxwell theory. Hawking (1986) has suggested that this extension should also be made when Σ is three-dimensional but with the proviso that, since a "physical" observer can only live in a single connected component of Σ, one should sum over all parts of the state vector associated with the "unphysical" components. This is similar to the philosophy that underlies the Hawking radiation from a black hole where one sums over that part of the state associated with the quantum field on the event horizon. And as far as the observer is concerned, the consequence is the same: the pure state given by eq.(4.4) becomes a mixed, density matrix, state.

Many aspects of the Hartle-Hawking scheme have been investigated over the last few years. One can view eq.(4.4) as giving a particular prescription for fixing the boundary data for the functional derivative Wheeler-DeWitt equation (4.1), and this aspect is prominent in the work of Schleich (1985), who studies the behaviour of $\Psi[g]$ near the singular limit $g \to 0$, and of Wada (1986), who discusses the general consistency of these implicit boundary conditons on $\Psi[g]$. Hajicek (1985) has emphasised that, in comparing the functional integral in eq.(4.4) with the expression (4.3) from normal quantum mechanics, it is important to note that the former is more like a projection operator onto an energy eigenstate than the solution of an evolution equation. Hartle (1984) has shown that the ground state for linearized quantum gravity is reproduced correctly in the weak field limit, and Horowitz (1985) has extended the method to include (Riemann curvature)^2 terms in the action.

Armed with the formalism, we may now ask "Can it be applied to the study of the early universe?". A key tool in the older, operator approach to quantum cosmology was the use of "mini-superspace" models. These were obtained by freezing all but a finite number of the modes of the 3-geometry and then applying the methods of canonical quantization to those that remained. The analogue in the present case is to approximate the functional integral with just a few modes of the 4-metric g^4, and models of this type have been studied extensively. The simplest choice is a single gravitational mode a(t) representing the radius of a Robertson-Walker universe augmented with a single homogeneous mode $\phi(t)$ of a scalar matter field supporting the cosmological expansion (the scheme is readily extended to include such fields). For example, Hawking (1984a) and Page (1986) study a free, massive scalar field; Hawking & Wu (1985) and Hawking & Luttrell (1984a) add (Riemann curvature)^2 terms; and Moss & Wright (1985) and Gonzalez-Diaz (1985) add an "inflationary" potential Log ϕ. All these models tend to predict an inflationary stage for the universe whose state vector is given by eq.(4.4).

Other features of the physical universe that have been studied in this way include the isotropy problem (Hawking & Luttrell 1984b; Amsterdamski 1985) and the fluctuation spectrum (Halliwell & Hawking 1985). The theory has been extended to incoporate spinors by D'Eath & Halliwell (1986), and applied to the inside of a black hole by Laflamme (1986).

In practice, the functional integral (4.4) is used in these models as a means of extracting the appropriate boundary conditions for the Wheeler-DeWitt equation, which is then solved as a partial differential equation. In many cases, these boundary conditions involve setting data on a hypersurface in the finite-dimensional approximation to the configuration space Riem Σ/Diff Σ that is null with respect to a natural hyperbolic metric defined on this space. But the general situation is unclear and an important task for the future is to establish the exact connection between the functional integral prescription and the corresponding boundary conditions for the Wheeler-DeWitt equation.

Another important question is the extent to which the fine details of these results can be believed. The non-renormalizability of the perturbatively quantized theory cautions against reading too much into these finite-mode models unless it can be shown that the theory represented by eq.(4.4) really does exist in some meaningful sense. For this reason, a fascinating and important challenge is to see if the Hartle-Hawking ideas can be applied to superstring theories . However, viewed from all angles this is an exciting and active branch of quantum gravity, and one in which interest is likely to continue for some time to come.

4.2 Conceptual problems

But what does it mean? The canonical approach to quantum gravity is well-known for the profundity of the conceptual issues which it raises, and the Hartle-Hawking prescription does not alleviate these. Since the early work of DeWitt (1967) there have been endless debates about the nature of the "intrinsic" time that is supposedly encoded into the "g" in $\Psi[g]$, and the argument continues. A critical role is played by the concept of a "quantum clock" and of the way in which the classical notion of time emerges from the quantum formalism in a suitable limit. Recent papers dealing with this include Banks (1985), Hajicek (1986), Hosoya (1985), Elitzur et al (1985), Kazama & Nakayama (1985), Page & Wooters (1985), and Zeh (1986).

Another, and closely related, question concerns the status of probability concepts as applied to the "wave function of the universe". It is a moot point whether one should expect to find anything representing a probability density in regimes where the classical concept of time is inapplicable. And if one did have such an object, would it have an operational meaning? Are we to consider an ensemble of universes with some universal external observer who reduces the state vector? Or must we invoke the ontological mysteries of the Everett-Wheeler interpretation of quantum theory? These questions have been with us for a long time and raise deep issues that transcend the specific scheme of Hartle and Hawking. A collection of articles dealing with some of these matters is contained in the book edited recently by Penrose & Isham (1986). However, the Hartle-Hawking approach possesses peculiar features of its own which raise additional questions of principle. The most striking perhaps is that the wave functional generated by the functional integral (4.4) is necessarily *real*. An immediate consequence is that the familiar "Klein-Gordon" type of inner product on the solutions of the Wheeler-DeWitt equation is no longer applicable (it vanishes identically!). The validity of this inner product has always been a problematical feature of canonical quantum gravity and a traditional alternative is to define $< \Psi, \Phi > := \int \Psi^*[g]\, \Phi[g]\, d\nu[g]$ for some measure $d\nu[g]$ on Riem Σ/Diff Σ. This is usually criticized on the grounds that it includes integrating over that part of the metric which corresponds to intrinsic time, and this is not a standard feature of inner products in quantum theory. In the present case there is no obvious alternative; but the question is closely related to the probabilistic interpretation of the theory, and the reader is referred to the articles cited above for further discussion on the matter.

Another peculiar feature stemming from the reality of $\Psi[g]$ can be seen by considering the semi-classical approximation to the functional integral (4.4). For some choices of the 3-metric g on the boundary $\Sigma \approx \partial M$, there will exist a solution to the classical, Riemannian equations of motion, and the semi-classical state vector will be of the form $\Psi \approx \exp -S$, where the action S is evaluated on this solution. However, for other choices of the 3-metric the saddle points will be complex, leading to a wave functional of the form $\Psi \approx \Sigma\, c_n \exp iS_n$ where S_n is evaluated at a solution to the Hamilton-Jacobi equation (and the sum is over all such solutions). Roughly speaking, these correspond to classical cosmologies that are either expanding or collapsing with respect to proper time. But then the enforced reality of Ψ means that an expanding configuration is always accompanied by one that collapses; a rather peculiar situation in a theory that purports to describe the *creation* of the universe! This raises a number of questions concerning the "arrow of time" and the CTP invariance of the theory, and these have been debated carefully by Hawking (1985) and Page (1985).

There is also a striking discrepancy between this picture of quantum creation and the one developed by Vilenkin (1982,1983,1984,1985). Vilenkin's work contains many interesting ideas (especially in relation to the inflationary picture of cosmology) but, in the present context, the crucial point is the way in which he sets his boundary conditions by requiring only expanding modes to be present: producing a wave functional that is necessarily complex. This striking difference between the specification of boundary conditions is partially related to the role assigned to classical spacetime singularities in the quantum theory. For example, Vilenkin connects his boundary conditions with Penrose's suggestion that the Weyl curvature tensor should always vanish at an initial singularity; raising again the issue of the arrow of time. It is therefore most important to understand fully the essential nature of the difference between these two schemes. But the problem is not trivial and touches on many general questions concerning the status of quantization schemes in which, as in eqs.(4.1,2), classical gauge constraints are imposed as operator constraints on allowed state vectors. This has been a subject of active research in its own right and forms the topic of the next, and final, section.

5 CONSTRAINT QUANTIZATION

5.1 Phase space tunnelling

The canonical analysis of a gauge theory involves a set $C_a(q,p)$ of constraint functions whose Poisson brackets close in the form,

$$\{C_a, C_b\}(q,p) = \lambda_{ab}{}^c(q,p)\, C_c(q,p) \tag{5.1}$$

In a Yang-Mills theory with internal symmetry group G, the functions.C_a, a = 1....Dim G, are the canonical generators of the group of gauge transformations and the "structure functions" $\lambda_{ab}{}^c$ are in fact independent of the canonical variables (q,p). In general relativity, the constraints also close in the sense of eq.(5.1), but with the complication that the Poisson bracket of two longitudinal constraints H_L (at different points in the 3-space Σ) involves a genuine function of the 3-metric canonical variable.

One approach to quantizing a structure of this type is to reduce it to a constraint-free canonical system by choosing a gauge and solving the constraints classically. The remaining (physical) degrees of freedom can then be quantized in the usual way. However, the big disadvantage of such a scheme is the loss of the covariance carried by the original constraints, and this led Dirac to suggest an alternative approach in which all degrees of freedom are quantized and the operator constraints are used to project out the allowed state vectors (cf. eqs(4.1,2)),

$$C_a(q,p)\,\Psi = 0\,. \tag{5.2}$$

(A more modern version is the Fradkin-Vilkovisky, BRS-extension of (5.2), but for our purposes it suffices to consider the original method). To ensure consistency it is necessary to preserve the "first class" nature of the classical constraints , ie. the commutators of the operator constraints should preserve the algebraic structure of the classical Poisson brackets (5.1). This problem is particularly difficult if the $\lambda_{ab}{}^c$ are functions when it generates the well-known operator ordering problems.

The justification for imposing (5.2) is clear enough when the constraints are *linear* functions of the momenta p (as in Yang-Mills theory). Indeed, working on the assumption that the state vectors are functions $\Psi(q)$ on the full (ie. unconstrained) configuration space Q, the constraint (5.2) merely affirms that such a function is independent of the coordinates that point along the gauge group orbits. However, the crucial longitudinal constraint in general relativity is *quadratic* in the momenta and, in this case, the status of imposing (5.2) (ie. the Wheeler-DeWitt equation) is more debatable. The problem hinges arounds "phase space tunnelling"; a phenomenon first explicitly identified in general relativity by Kaup & Vittello (1974). As an example,consider a one-dimensional system with the constraint $x^2 + p^2 = 0$. Viewed classically, this implies that the system is confined to the single point in phase space x=p=0. But, quantum mechanically, we could certainly make the substitution p = -id/dx and find solutions to the differential equation

$$x^2\psi(x) - d^2\psi(x)/dx^2 = 0 \tag{5.3}$$

and the resulting function $\psi(x)$ is not concentrated at the point $x = 0$! There are some dubious aspects to this argument but, taken at face value, it shows how it may be possible in quantum mechanics to reach regions that are classically inconsistent with the constraints. This, roughly speaking, is the phenomenon of phase space tunnelling. Viewed from the perspective of geometric quantization one can say (Ashtekar & Stillerman 1985; Gotay 1986b) that the polarization of the unconstrained phase space T^*Q is incompatible with the constraints. Within the group theoretical framework developed in Isham (1984) the problem will appear as an incompatibility of the constraints with the natural choice of the canonical group acting globally on T^*Q.

The phenomenon of phase space tunnelling means that "constraining then quantizing" may not yield the same result as "quantizing then constraining". Is this acceptable ? This is an important question since, for example, Vilenkin's (1985) programme for studying the creation of the universe via the solutions of the Wheeler-DeWitt equation is clearly dependent on this mechanism. On the other hand, Ashtekar & Horowitz (1982) showed how it can produce a negative ADM mass, and Gotay (1986a) has produced a variety of examples illustrating the pathological nature of the issues involved.

To examine the origin of this behaviour we must look more carefully at the operator constraint (5.2). Note that this equation could mean one of three quite different things:

(i) It could be that C_a is a genuine self-adjoint operator on the Hilbert space V carrying the representation of the unconstrained canonical variables, and that zero is a discrete eigenvalue of this operator. Then Ψ will be normalizable and the set V_{phys} of all such physical states is a genuine subspace of V.

(ii) Or zero could be part of the *continuous* spectrum of V. Then Ψ will not be normalizable and V_{phys} is not a subspace of V but must be discussed within the rigged Hilbert space framework associated with the continuous spectrum. A failure to appreciate this point has led to a number of errors in the past.

(iii) The third possibility, and the one that is responsible for phase space tunnelling, is that zero might not be a genuine eigenvalue at all!

This final option may sound like a contradiction in terms; but that it is otherwise can be seen with the aid of the example in eq.(5.3). This equation can certainly be solved *as a differential equation* , and yet we know that the eigenvalues of the harmonic oscillator Hamiltonian do not include zero. In fact we could replace the right-hand side of eq.(5.3) with any number (even negative) multiplied by $\psi(x)$, and still be able to solve the differential equation! Note that, in general, the "eigenfunctions" for both of the options (ii) and (iii) will not be square-integrable, but the implications of this in the two cases are quite different and it is important to distinguish between them carefully.

In solving (5.2) "just as a differential equation", we have dropped the condition that C_a be a self-adjoint operator on the original Hilbert space V. Does this matter? It has often been argued that self-adjointness is an inappropriate requirement for the constraint operators and/or the "gauge" parts of the canonical variables; perhaps the time has come to take this idea seriously? There are certainly some situations that suggest this quite strongly. For example, an incomplete Hamiltonian vector field on a classical phase space will produce such an operator. This applies to the spacetime diffeomorphism group in canonical quantum gravity (Kuchar & Isham 1985a,b); it would also apply to the first-quantized string if the string were required to be spacelike as it propagates through Lorentzian spacetime.

Much could be said about the pros and cons of permitting non self- adjointness. Perhaps the main objection is the considerable difficulty that can attend attempts to control operators of this type. This is particularly true in the case of a quantum field theory where an uncontrolled functional differential equation can have almost no meaning. But there are benefits too. For example, Karel Kuchar spoke at the GRG11 quantum gravity workshop about his recent work on the operator ordering problem associated with classical Poisson brackets like eq.(5.1). He has shown (Kuchar 1986a,b) how this may be solved by requiring the various operators in the theory to be self-adjoint *only* on the subspace of solutions to the formal differential equations. Kuchar's work is also aimed at ensuring the commutativity of the acts of constraining and quantizing and, as such, it is distinctly "anti" phase space tunnelling, even though he too employs operators that are not self-adjoint.

5.2 New canonical variables

Another awkward problem in the constraint quantization of gravity is knowing how to select the set of unconstrained variables that are to be used as the basic canonical operators in the theory. An obvious choice for the configuration variable is of course the 3-metric $g_{ij}(x)$. This raises the question of the status of the det g > 0 condition since Dirac's scheme is not designed to handle constraints in the form of inequalities. However, the probem can be solved using the general group theoretic methods discussed in Isham (1984) and leads to the replacement of the usual Weyl-Heisenberg group by a new canonical group that is consistent with the positivity of det g (Isham & Kakas 1984a,b).

But obvious choices are not necessarily correct, and many workers in quantum gravity have entertained the hope of finding some cunning selection of variables that would substantially simplify the entire quantization programme. Ashtekar (1985) has made a significant advance in this direction based on earlier work proving the Witten positive energy theorems from within a canonical framework. The starting point is the triadic version of canonical general relativity plus a 2-spinor λ_A and a new spinorial connection $A^{\pm}$ defined via the covariant derivative (Sen 1982),

$$D^{\pm}_j\lambda_A := D_j\lambda_A \pm i\,\pi_{jA}{}^B\,\lambda_B \quad , \; j = 1,2,3 \tag{5.4}$$

$$= \partial_j\lambda_A \pm i\,A^{\pm}{}_{jA}{}^B\,\lambda_B$$

where $\pi_{jA}{}^B$ is the usual conjugate momentum with one of the indices written in spinorial form. Ashtekar's proposal is that the set of fields $A^{+}{}_{jA}{}^B$ should be used as the new configuration variables (the set $A^{-}{}_{jA}{}^B$ would have served just as well), and this has some striking implications:

(i) Both longitudinal and transverse constraints become relatively simple expressions when written in terms of the field-strength constructed from the connection A^+ in the usual way.

(ii) A^+ (resp. A^-) are canonical data for self-dual (resp. anti self-dual) solutions of the Einstein equations. Thus, with the aid of these new variables, it may become possible to relate the canonical quantization of gravity to the ideas of Penrose and Newman on non-linear gravitons, H-space etc.

(iii) The choice of $A^{+}{}_{jA}{}^B$ for the canonical variables can be written as

$\delta\Psi/\delta A^{-}{}_{jA}{}^B = 0$ which, it has been argued, could be interpreted as saying that Ψ is an "infinite-dimensional" holomorphic function. This in turn suggests that the Bargmann-Segal representation of the canonical commutation relations may be particularly appropriate for this theory.

(iv) Ashtekar has managed to generate a solution to the operator ordering problem with the aid of these variables.

These ideas have caused much interest and will certainly be developed further in the future. Wald (1986) has emphasised the difference between the canonical and covariant approaches *vis-à-vis* these new variables, and Jacobsen & Smolin (private communication) have been investigating the use of Wilson line integrals over the connection A^+. They believe that by these means they may have a real chance of solving the Wheeler-DeWitt equation in its new form. This would of course be a major breakthrough in the quantization of gravity!

6 CONCLUSIONS

At the end of my review talk at the GRG meeting, two good friends and colleagues spoke to me expressing approval of what I had said. The first because "You made it clear that superstring theory is the only viable way forward in quantum gravity"; and the second because "You made it clear that there is so much more to quantum gravity than superstring theory"! It is nice to keep one's friends, but I doubt very much if the quantum gravity reviewer at the next GRG meeting will be able to maintain such a schizoid profile.

Of the many approaches to quantum gravity currently on offer, only the superstring programme promises any immediate prospect of finding a theory that is genuinely finite. And the rate of progress in this subject is such that, by the summer of 1989, the issue will surely have been settled one way or the other. But it is a "secondary" theory of quantum gravity in which the much-hallowed principles of classical general relativity play a correspondingly minor role; which explains no doubt why many classical relativists seem to view the subject as something of an anathema. It is worth emphasising therefore that, in the first instance, the "finiteness" refers to a particular perturbative analysis of strings moving in a fixed, ten-dimensional Minkowski space. This obviously poses the problem of how to recover the four-dimensional spacetime of phenomenological, macroscopic physics; and hence the current investigations into the compactifications/ground-states of the theory. But in other respects the situation resembles the one that would have arisen if, for some miraculous reason, the Poincaré group invariant, weak-field perturbative approach to conventional primary quantum gravity had turned out to be finite. We would then have had to face the task of reconstructing the "physical" light-cone/causal structure from this graviton-graviton scattering theory; which would almost certainly have entailed a study of non-perturbative features of the theory, including the possibility of different "ground-states" associated with a quantum foam picture of spacetime.

An analogous situation arose in the early 1970s in relation to the study of non-abelian gauge theories. It was only *after* t'Hooft had proved perturbative renormalizability that the subject really got under way; but thereafter, much of the discussion was focussed on non-perturbative properties. General experience in quantum field theory suggests that one should first make sense of the system in some appropriate perturbative way; only then can one proceed with any degree of confidence to study the more speculative, non-perturbative features of the theory. It is in this sense that superstring theory is such a potentially promising approach to quantum gravity; and only time will tell if these expectations are justified. If not, we will have to work even harder at confronting the profound problems that arise when attempting to construct a primary quantization and/or redemption of the classical theory of relativity.

REFERENCES

Amsterdamski, P. (1985). Wave function of an anisotropic universe. Phys. Rev., **D31**, 3073.

Anderson, A. & DeWitt, B. (1985). Does the topology of space fluctuate? University of Texas at Austin preprint.

Antoniadis, I. & Tomboulis, E.T. (1986). Gauge invariance and unitarity in higher derivative quantum gravity. Phys. Rev., **D33**, 2756.

Alvarez-Gaume, L. & Gomez, C. (1983). Remarks on quantum gravity. Comm. Math. Phys., **89**, 235.

Ashtekar, A. (1985). New variables for classical and quantum gravity. University of Syracuse preprint.

Ashtekar, A. & Horowitz, G. (1982). On the canonical approach to quantum gravity. Phys. Rev., **D26**, 3342.

Ashtekar, A. & Stillerman, M. (1985). Geometric quantization and constrained systems. University of Syracuse preprint.

Avramidy, I. & Barvinsky, A. (1985). Asymptotic freedom in higher-derivative quantum gravity. Phys. Letts., **159B**, 269.

Banks, T. (1985). TCP, quantum gravity, the cosmological constant and all that. Nucl. Phys., **B249**, 332.

Banks, T., Susskind, L. & Peskin, M.E. (1984). Difficulties for the evolution of pure states into mixed states. Nucl. Phys., **B244**, 125.

Banks, T., Fischler, W. & Susskind, L. (1985). Quantum cosmology in 2+1 and 3+1 dimensions. Nucl. Phys., **B262**, 159.

Barrett, J.W. (1986). The Einstein tensor in Regge's discrete gravity theory. Class. Quan. Grav., **3**, 203.

Barth, N. & Christensen, S.M. (1983). Quantizing fourth-order gravity theories: The functional integral. Phys. Rev., **159B**, 269.

Barvinsky, A.O. & Ponomariov, V.N. (1986). Quantum geometrodynamics: The path integral and the initial value problem for the wave function of the universe. Phys. Letts., **167B**, 289.

Berg, B. (1985). Exploratory numerical study of discrete quantum gravity. Phys. Rev. Letts., **55**, 904.

Berg, B. (1986). Entropy versus energy on a fluctuating four-dimensional Regge skeleton. Phys. Letts. **176B**, 39.

Berger, B.K. (1985). Path integral quantum cosmology II. Bianchi type I with volume dependent source. Phys. Rev., **D32**, 2485.

Berger. B.K. & Vogeli, C. (1985). Path integral quantum cosmology I. Vacuum Bianchi type I. Phys. Rev., **D32**, 2477.

Boulware, D.G. (1984). *In* Quantum theory of gravity, ed. S. Christensen, pp 267-294. Cambridge: Cambridge University Press.

Boulware, D.G. & Deser, S. (1984). Canonical analysis of the fermion sector in higher-derivative supergravity. Phys. Rev., **D30**, 707.

Boulware, D.G., Deser, S. & Stelle, K.S. (1986). Energy and supercharge in higher-derivative gravity. Phys. Letts., **168B**, 336.

Boulware, D.G. & Gross, D.J. (1984). Lee-Wick indefinite metric quantization: a functional integral approach. Nucl. Phys., **B233**, 1.

Boulware, D.G., Horowitz, G.T. & Strominger, A. (1983). Zero-energy theorem for scale invariant gravity. Phys. Rev. Letts., **50**, 1726.

Cheeger, J., Müller, W. & Schrader, R. (1984). On the curvature of piecewise flat spaces. Comm. Math. Phys., **92**, 405.

Crane, L. & Smolin, L. (1986). Renormalizability of general relativity on a background of spacetime foam. Nucl. Phys., **B266**, 714.

David, F. (1984). A comment on induced gravity. Phys. Letts., **138B**, 383.

David, F. & Strominger, A. (1984). On the calculability of Newton's constant and the renormalization of scale invariant quantum gravity. Phys. Letts., **143B**, 125.

D'Eath, P. & Halliwell, J.J. (1986). Fermions in quantum cosmology. University of Cambridge preprint.

Deser, S. & Jackiw, R. (1984). Three-dimensional cosmological gravity: Dynamics of constant curvature. Ann. Phys., **153**, 405.

Deser, S., Jackiw, R. & t'Hooft, G. (1984). Three-dimensional Einstein gravity: Dynamics of flat space. Ann. Phys., **152**, 220.

DeWitt, B. (1967). Quantum theory of gravity III. Application of the covariant theory. Phys. Rev., **162**, 1239.

Dimock, J. (1985). Scattering for the wave equation on the Schwarzschild metric. Gen. Rel. Grav., **17**, 353.

Dimock, J. & Kay, B.S. (1986). Scattering for massive scalar fields on Coulomb potentials and Schwarzschild metrics. Class. Quan. Grav., **3**, 71.

Elitzur, S., Forge, A. & Rabinovici, E. (1985). The wave functional of a superclock. CERN preprint TH.4265.

Ellis, J., Hagelin, J.S., Nanopoulos, D.V. & Srednik, M. (1984). Search for violation of quantum mechanics. Nucl. Phys., **B241**, 381.

Englert, F., Frere, J.-M., Rooman, M. & Spindel, Ph. (1986). Metric spacetime as fixed point of the renormalization group equations on fractal structure. Université Libré de Bruxelles preprint.

Feinberg, G., Freidberg, R., Lee, T.D. & Ren, H.C. (1984). Lattice gravity near the continuum limit. Nucl. Phys., **B245**, 343.

Fradkin, E.S. & Tseytlin, A.A. (1981). Renormalizable asymptotically free quantum theory of gravity. Phys. Letts., **104B**, 377.

Fradkin, E.S. & Tseytlin, A.A. (1982). Renormalizable asymptotically free quantum theory of gravity. Nucl. Phys., **B201**, 469.

Freese, K., Hill, C. & Müeller, M. (1985). Covariant functional Schrödinger formalism and application to the Hawking effect. Nucl. Phys., **B255**, 693.

Friedan, D. & Shenker, S. (1986). The integrable analytic geometry of quantum string. Enrico Fermi Institute preprint.

Freidberg, R. & Lee, T.D. (1984). Derivation of Regge's action from Einstein's theory of general relativity. Nucl. Phys., **B242**, 145.

Friedman, J.L. & Jack, I. (1986). 3+1 Regge calculus with conserved momentum and angular momentum constraints. University of Wisconsin preprint.

Friedman, J.L. & Witt D.M. (1983). Internal symmetry groups of quantum geons. Phys. Letts., **120**, 324.

Fröhlich, J. (1985). Survey of random surface theory. *In* Recent developments in quantum field theory. eds. J. Ambjørn, B.J. Durhuus & J.L. Peterson, pp 67-93. Elsevier Science Publishers.

Fröhlich, J. (1981). Regge calculus and discretized gravitational integrals. ETH preprint, Zurich.

Gibbons, G.W. (1986). The elliptic interpretation of black holes and quantum mechanics. Nucl. Phys., **B271**, 497.

Gonchorov, Yu.P. & Bytsenko, A.A. (1985). The supersymmetric Casimir effect and quantum creation of the universe with nontrivial topology. Phys. Letts., **160B**, 385 ; **169B**, 171.

Gonzalez-Diaz, P.F. (1985). On the wave function of the universe. Phys.Letts.,**159B**, 19.

Goroff, M.H. & Sagnotti, A.(1985).Quantum gravity at two loops. Phys.Letts.,**160B,** 81.

Goroff, M.H. & Sagnotti, A. (1986). The ultraviolet behaviour of Einstein gravity. Nucl. Phys., **B266**, 709.

Gotay, M. (1986a). Negative energy states in quantum gravity. Class. Quan. Grav., **3**, 487.

Gotay, M. (1986b). Constraints, reduction and quantization. Jour. Math. Phys., (In press).

Gross, D.J. (1984). Is quantum gravity unpredictable? Nucl. Phys., **B236**, 349.

Haag, R., Namhofer, H. & Stein, U. (1984). On quantum field theory in gravitational backgrounds. Comm. Math. Phys., **94**, 219.

Hajicek, P. (1985). Elementary properties of a new kind of path integral. University of Bern preprint.

Hajicek, P. (1986). Origin of non-unitarity in quantum gravity. University of Bern preprint.

Halliday, I.G. (1984). Lattice field theories. Rep. Prog. Phys., **47**, 987.

Halliwell, J.J. (1986). The quantum cosmology of Einstein-Maxwell theory in six dimensions. Nucl. Phys., **B266**, 228.

Halliwell, J.J. & Hawking, S.W. (1985). Origin of structure in the universe. Phys. Rev., **D31**, 1777.

Hamber, H.W. (1984). Simplicial quantum gravity. Les Houches lecture notes.

Hamber, H.W. & Williams, R.M. (1984). Higher derivative quantum gravity on a simplicial lattice. Nucl. Phys., **B248**, 392.

Hamber, H.W. & Williams, R.M. (1985). Non-perturbative simplicial quantum gravity. Phys. Letts., **157B**, 368.
Hamber, H.W. & Williams, R.M. (1986a). Two-dimensional simplicial quantum gravity. Nucl. Phys., **B267**, 482.
Hamber, H.W. & Williams, R.M. (1986b). Simplicial quantum gravity with higher derivative terms: formalism and numerical results in four dimensions. Nucl. Phys., **B269**, 712.
Hartle, J.B. (1984). Ground state function of linearized gravity. Phys. Rev., **D29**, 2730.
Hartle, J.B. (1985a). Simplicial minisuperspace I. General discussion. Jour. Math. Phys., **26**, 804.
Hartle, J.B. (1985b). Unruly topologies in two-dimensional quantum gravity. Class. Quan. Grav., **2**, 707.
Hartle, J.B. & Hawking, S.W. (1983). Wave function of the universe. Phys. Rev., **D28**, 2960.
Hawking, S.W. (1982). The unpredictability of quantum gravity. Comm. Math. Phys., **87**, 395.
Hawking, S.W. (1984a). The quantum state of the universe. Nucl. Phys., **B239**, 257.
Hawking, S.W. (1984b). Non-trivial topologies in quantum gravity. Nucl. Phys., **B244**, 135.
Hawking, S.W. (1984c). Quantum cosmology. *In* Relativity, groups and topology II. eds. B.S. DeWitt & R.Stora, pp. 336-79. Amsterdam: Elsevier Science Publishers.
Hawking, S.W. (1985). Arrow of time in cosmology. Phys. Rev., **D32**, 2489.
Hawking, S.W. (1986). The density matrix of the universe. University of Cambridge preprint.
Hawking, S.W. & Luttrell, P.C. (1984a). Higher derivatives in quantum cosmology. (I) The isotropic case. Nucl. Phys., **B247**, 250.
Hawking, S.W. & Luttrell, P.C. (1984b). The isotropy of the universe. Phys. Letts., **143B**, 83.
Hawking, S.W. & Page, D.N. (1986). Operator ordering and the flatness of the universe. Nucl. Phys., **B264**, 185.
Hawking, S.W. & Wu, Z.C. (1985). Numerical calculations of minisuperspace cosmological models. Phys. Letts., **151B**, 15.
Henneaux, M. (1984). Coordinate invariant states for quantum gravity in 2+1 dimensions. Phys. Letts., **134B**, 184.
Henneaux, M. (1985). Quantum gravity in two dimensions. Exact solution of the Jackiw model. Phys. Rev. Letts., **54**, 959.
t'Hooft, G. (1985). On the quantum structure of a black hole. Nucl. Phys., **B256**, 727.
Horowitz, G.T. (1985). Quantum cosmology with a positive definite action. Phys. Rev., **D31**, 1169.

Hosoya, A. (1985). Synchronous time gauge condition in quantum cosmology. Osaka University preprint.
Isham, C.J. (1984). Topological and global aspects of quantum theory. *In* Relativity, groups and topology II. eds. B.S. DeWitt & R. Stora, pp. 1061-1290, Amsterdam: Elsevier Science Publishers.
Isham, C.J. & Kakas, A.C. (1984a). A group theoretical approach to the canonical quantization of gravity: I. Construction of the canonical group. Class. Quan. Grav., **1**, 621.
Isham, C.J. & Kakas, A.C. (1984b). A group theoretical approach to the canonical quantization of gravity: II. Unitary representations of the canonical group. Class. Quan. Grav., **1**, 633.
Isham, C.J. & Kuchar, K. (1985a). Representations of spacetime diffeomorphisms. I. Canonical parametrized theories. Ann. Phys., **164**, 288.
Isham, C.J. & Kuchar, K. (1985b). Representations of spacetime diffeomorphisms. II. Canonical geometrodynamics. Ann. Phys., **164**, 316.
Jack, I. & Parker, L. (1985). Proof of summed form of proper-time expansion for propagator in curved spacetime. Phys. Rev., **D31**, 2439.
Jackiw, R. (1985). Lower dimensional gravity. Nucl. Phys., **B252**, 343.
Jevicki, A. & Ninomiya, M. (1986). Functional formulation of Regge gravity. Phys. Rev., **D33**, 1634.
Kaku, M. (1986). Generally covariant lattices, the random calculus and the strong coupling approach to the renormalization of gravity. *In* Quantum field theory and statistical physics, eds. I. Batelin, G. Vilkovisky & C.J. Isham. Bristol: Adam Hilger (In press).
Károlyházy, F., Frenkel, A. & Lukács, B. (1986). On the possible role of gravity in the reduction of the wave function. *In* Quantum concepts in space and time, eds. R. Penrose & C.J. Isham, pp. 109-128, Oxford: Oxford University Press.
Kaup, D. & Vittello, A. (1974). Solvable cosmological models and the importance of quantizing in a special canonical frame. Phys. Rev., **D9**, 1648.
Kay, B.S. (1985). The double-wedge algebra for quantum fields on Schwarzschild and Minkowski space. Comm. Math. Phys., **100**, 57.
Kazama, Y. & Nakayama, R. (1985). Wave packet in quantum cosmology. Phys. Rev., **D32**, 2500.
Kondo, Kei-ichi. (1984). Euclidean quantum gravity on a flat lattice. Prog. Theor. Phys., **72**, 841.
Kuchar, K. (1986a). Hamiltonian dynamics of gauge systems. University of Utah preprint.
Kuchar, K. (1986b). Covariant factor ordering of gauge systems. University of Utah preprint.
Laflamme, R. (1986). The wave function of a black hole interior. University of Cambridge preprint.

Mannion, C.L. & Taylor, J.G. (1981). General relativity on a flat lattice. Phys. Letts., **100B**, 261.
Martinec, E. (1984). Soluble systems in quantum gravity. Phys. Rev., **D30**, 1198.
Menotti, P. & Pelissetto, A. (1985). Reflection positivity and graviton doubling in euclidean lattice gravity. University of Pisa preprint.
Menotti, P. & Pelissetto, A. (1986). Conformal, Poincaré and deSitter gravity on the lattice. University of Pisa preprint.
Moss, I.G. & Wright, W.A. (1984). Wave function of the inflationary universe. Phys. Rev., **D29**, 1069.
Müller, B. & Schäfer, A. (1986). Improved bounds on the dimension of spacetime. Phys. Rev. Letts., **56**, 1215.
Page, D.N. (1984). A fractal set of perpetually bouncing universes. Class. Quan. Grav., **1**, 417.
Page, D.N. (1985). Will entropy decrease if the universe recollapses? Phys. Rev., **D32**, 2496.
Page, D.N. (1986). Hawking's wave function for the universe. *In* Quantum concepts in space and time, eds. R. Penrose & C.J. Isham, pp. 274-285. Oxford: Oxford University Press.
Page, D.N. & Wooters, W. (1985). Evolution without evolution: dynamics described by stationary observers. University of Texas at Austin preprint.
Parker, L. & Toms, D. (1985). New form for the coincidence limit of the Feynman propagator, or heat kernel, in curved spacetime. Phys. Rev., **D31**, 953.
Penrose, R. (1986), Gravity and state vector reduction. *In* Quantum concepts in space and time, eds. R. Penrose & C.J. Isham, pp.129-146. Oxford: Oxford University Press.
Penrose, R. & Isham, C.J. (1986). Quantum concepts in space and time. Oxford: Oxford University Press.
Piran, T. & Strominger, A. (1986). Solutions of the Regge equations. Class. Quan. Grav., **3**, 97.
Piran, T. & Williams, R. (1986). Three-plus-one formulation of Regge's calculus. Phys. Rev., **D33**, 1622.
Rocek, M. & Williams, R. (1984). The quantization of Regge calculus. Zeit. Phys. C., **21**, 371.
Römer, H. & Zähringer, M. (1986). Functional integration and the diffeomorphism group in euclidean lattice quantum gravity. University of Freiburg preprint.
Rooman, M. (1986). Stability of the large-scale metric in fractal structures. Université Libré de Bruxelles preprint.
Sánchez, N. & Whiting, B. (1986). Quantum field theory and the antipodal identification of black holes. University of Meudon preprint.

Sánchez, N. & Folacci, A. (1986). Quantum field theory and the "elliptic interpretation" of deSitter space. University of Meudon preprint.
Schleich, K. (1985). The semiclassical wave function of the universe at small three-geometries. University of California at Santa Barbara preprint.
Sen, A. (1982). Gravity as a spin system. Phys. Letts., **119B**, 89.
Sorkin, R.D. (1985a). Introduction to topological geons. Erice lecture notes.
Sorkin, R.D. (1985b). On topology change and monopole creation. University of Syracuse preprint.
Stelle, K. (1977). Renormalization of higher derivative gravity. Phys. Rev., **D16**, 953.
Strominger, A. (1984). Positive energy theorem for $R + R^2$ gravity. Phys. Rev., **D30**, 2251.
Svözil, K. & Zeilinger, A. (1985). Measuring the dimensions of spacetime. Phys. Rev. Letts., **54**, 253.
Tomboulis, E.T. (1984). Unitarity in higher derivative quantum gravity. Phys. Rev. Letts., **52**, 1173.
Toms, D.J. (1986). The functional measure for quantum field theory in curved spacetime. University of Newcastle preprint.
Vilenkin, A. (1982). Creation of universe from nothing. Phys. Letts., **117B**, 25.
Vilenkin, A. (1983). Birth of inflationary universes. Phys. Rev., **D27**, 2848.
Vilenkin, A. (1984). Quantum creation of universes. Phys. Rev., **D30**, 509.
Vilenkin, A. (1985). Boundary conditions in quantum cosmology. Tufts University preprint.
Wada, S. (1986). Consistency of canonical quantization of gravity and boundary conditions for the wave function of the universe. University of Tokyo preprint.
Wald, R. (1986). Non existence of dynamical perturbations of Schwarschild with vanishing self-dual part. Class. Quan. Grav., **3**, 55.
Whitt, B. (1984). Fourth-order gravity as general relativity plus matter. Phys. Letts., **145B**, 176.
Williams, R. (1986). Quantum Regge calculus in the Lorentzian domain and its Hamiltonian formulation. Class. Quan. Grav., (In press).
Witt, D.M. (1986). Symmetry groups of state vectors in canonical quantum gravity. Jour. Math. Phys., **27**, 573.
York, J.W. (1985). Black hole thermodynamics and the euclidean Einstein action. University of North Carolina preprint.
Zeh, H.D. (1986). Emergence of classical time from a universal wave function. University of Heidelberg preprint.

Black Hole Uniqueness Theorems

Pawel Mazur
Department of Physics
Syracuse University
Syracuse, New York 13244-1130 USA

and

Department of Theoretical Physics
Jagellonian University
Krakow, Poland

Introduction

It is well known that spherical gravitational collapse produces a black hole (Oppenheimer and Snyder, 1939). There are reasons to believe that this is true for collapse with small deviations from spherical symmetry (Doroshkevich, Zeldovich and Novikov, 1966) and it was conjectured by Penrose (1969) that no 'naked singularities' can occur during complete gravitational collapse (Cosmic Censor Conjecture). The Cosmic Censor Conjecture is also the fundamental and not yet proved assumption we make when studying black hole equilibrium states. The considerable body of work on perturbations of stationary black holes brought us the presently accepted picture of gravitational collapse (Regge and Wheeler 1957, Vishveshwara 1970, Ipser 1971, Price 1972, Press and Teukolsky 1973, Wald 1973, and others). One would expect that a black hole formed by gravitational collapse would settle down to a stationary equilibrium state. Properties of black hole equilibrium states were extensively studied in the late sixties and early seventies (Carter, 1971, 1973; Bardeen, Carter and Hawking, 1973; Hawking and Ellis, 1973; Hawking, 1973). Theorems of Israel, Carter, Hawking and Robinson obtained between 1967 and 1975 gave proof of the remarkable result that Kerr (1963) black holes are the only possible stationary vacuum black holes. The first black hole uniqueness theorem came as a surprise when Israel (1967, 1968) proved that a static, topologically spherical black hole is described by the Schwarzschild or the Reissner-Nordstrom solutions. No hair and uniqueness theorems for a stationary axisymmetric, topologically spherical black hole were obtained by Carter (1971, 1973), Robinson (1974, 1975), Mazur (1982) and Bunting (1983) using the Ernst (1968) and Geroch (1971) formulation of the Einstein equations for a stationary and axisymmetric gravitational field. The idea of a proof of uniqueness of the charged black hole solution (Newman et al. 1965) is quite simple. The uniqueness proof makes use of the fact that one can construct divergence identities for two solutions of the field equations with a nice positivity property.

The basic assumptions made in a proof of uniqueness theorems were proved (justified) by Hawking (1973) who demonstrated that a stationary black hole must be static or axisymmetric and the event horizon has a spherical topology. Also, Hajicek (1973) has shown that the outer boundary of the ergosphere must always intersect the event horizon. All these results depend on the validity of the Cosmic Censorship Hypothesis. Causality of a part of spacetime outside the event horizon is also assumed in a proof of uniqueness theorems.

In this lecture I describe recent work on black hole uniqueness theorems, reformulating the problem of uniqueness of the Kerr-Newman black hole solutions as a problem in the harmonic map theory. I would like to refer the reader to the excellent review article by Carter (1979) for a more detailed account of basic assumptions made in a proof of uniqueness theorems.

1. The Einstein-Maxwell Equations for Spacetimes with One Killing Vector

The Ernst-Geroch formulation of the Einstein-Maxwell equations for the stationary axisymmetric case can be carried out with respect to one of the two existing Killing vectors. It is crucial for the proof of black hole uniqueness theorems to have such a formulation which leads to the harmonic maps between two Riemannian spaces, because only then may one hope to have a global divergence identity with the required positivity property. One can define, therefore, the Ernst potential with respect to the stationary Killing vector $\partial/\partial t$ or with respect to the axial symmetry Killing vector $\partial/\partial\phi$. However, as we will see below, it is the second choice which leads to a Riemannian metric on the image space of the Ernst potential because the norm $X = (\partial/\partial\phi, \partial/\partial\phi)$ is everywhere positive and vanishes only on the symmetry axis. X is positive because we assume that the black hole spacetimes are causal in the domain of outer communications (Carter, 1979). What happens when we use the locally timelike Killing vector $\partial/\partial t$? First of all, we notice that the norm $- V = (\partial/\partial t, \partial/\partial t)$ of $\partial/\partial t$ changes sign on the ergosurface of rotating black holes. This means that the signature of the metric on the image space N of the Ernst potentials is changing as we cross the ergosurface of the black hole. The metric on N is pseudoriemannian inside the ergosphere. For this reason the global divergence identity is losing its nice positivity property and besides the difficulties with deriving boundary conditions for black holes, in this case, we cannot apply it in the proof of the black hole uniqueness conjecture!

It may be useful to give a short presentation of a reduction of E-M equations from 4-dim to 3-dim using a covariant Geroch formulation for the case with one Killing vector $\xi = \xi^a\partial_a$. Let (M, g_{ab}), F_{ab} be solutions to the Einstein-Maxwell equations

$$R_{ab} - {}^1/_2 \, R \, g_{ab} = 8\pi \, T_{ab} \; , \tag{1}$$

$$D_b \, F^{ab} = 0 \; , \tag{2}$$

$$D_{[a}F_{bc]} = 0 \; , \tag{3}$$

where

$$T_{ab} = (4\pi)^{-1}(F_a{}^cF_{bc} - {}^1/_4 \, g_{ab} \, F_{cd} \, F^{cd}) \; , \tag{4}$$

and F_{ab} is the electromagnetic strength tensor.

Assume that (g_{ab}, F_{ab}) are invariant with respect to a one-parameter isometry group generated by the Killing vector field $\xi = \xi^a\partial_a$, i.e, $L_\xi g = L_\xi F = 0$. Our convention for the signature of g_{ab} is (−+++). The projection of the E-M equation on the space of orbits of ξ is straightforward (we follow Geroch formulation and notation (Geroch 1971)). The metric on the space of orbits M is $h_{ab} = g_{ab} - \lambda^{-1}\xi_a\xi_b$, where $\lambda = \xi_a\xi^a$. At this point it is convenient to notice that when ξ is timelike, then h_{ab} is Riemannian; on the other hand, if ξ is spacelike, as is the case with $\partial/\partial t$ Killing vector inside the ergosphere of a black hole, then h_{ab} is pseudoriemannian and the reduced field equations are hyperbolic in this region! One can also define a twist vector $\tilde{\omega}_a$ which is non-vanishing if ξ is not a hypersurface orthogonal Killing vector, $\tilde{\omega}_a = \epsilon_{abcd}\xi^{b\%\%Dc}\xi^d$. In the vacuum case knowing $(M, h_{ab}, \lambda, \tilde{\omega}_a)$ one can reconstruct (M, g_{ab}) completely. In the electrovacuum case one needs more information, i.e. we have to know the 'electric' and 'magnetic' fields defined as follows: $E_a = -F_{ab}\xi^b$, $B_a = -{}^*F_{ab}\xi^b$, which are automatically defined on M, i.e. $E_a\xi^a = B_a\xi^a = 0$. Knowing E_a, B_a one can reconstruct F_{ab}:

$$F_{ab} = \lambda^{-1}(2\xi_{[a}E_{b]} - \epsilon_{abcd}\xi^c B^d) \quad . \tag{5}$$

It can be easily seen (Carter, 1972) that (2), (3) and the Killing equations on (g,F) imply the existence of 'electric' and 'magnetic' potentials E and B in a simply connected region of M, i.e. $E_a = \nabla_a E$, $B_a = \nabla_a B$. The basic equations for spacetimes with one Killing vector ξ which we apply in derivation of the reduced E-M equations are the following (we present them here for the sake of completeness; it can be seen that using them and the E-M equations, one arrives easily at the final Ernst form of the field equations. In the following, we will only indicate the main steps in the derivation of the reduced equations.)

$$D_a\xi_b = {}^1/_2\ \lambda^{-1}\epsilon_{abcd}\xi^c\tilde{\omega}^d + \lambda^{-1}\xi_{[b}\nabla_{a]}\lambda \ , \tag{6}$$

$$D_a D_b \xi_c = \xi^d R_{dabc} \quad , \tag{7}$$

$$\nabla_{[a}\tilde{\omega}_{b]} = -\epsilon_{abcd}\xi^c R^d{}_e \xi^e \ , \tag{8}$$

$$\nabla_a\tilde{\omega}^a = {}^3/_2\ \lambda^{-1}\tilde{\omega}^a\nabla_a\lambda \quad , \tag{9}$$

$$\nabla_a\nabla^a\lambda = {}^1/_2\ \lambda^{-1}\nabla_a\lambda\nabla^a\lambda - \lambda^{-1}\tilde{\omega}_a\tilde{\omega}^a - 2R_{ab}\xi^a\xi^b \tag{10}$$

$$R_{ab} = {}^1/_2\ \lambda^{-2}(\tilde{\omega}_a\tilde{\omega}_b - h_{ab}\tilde{\omega}_c\tilde{\omega}^c) + {}^1/_2\ \lambda^{-1}\nabla_a\nabla_b\lambda - {}^1/_4\ \lambda^{-2}\nabla_a\lambda\nabla_b\lambda + h_a{}^c h_b{}^d R_{cd} \tag{11}$$

where ${}^{(3)}R_{ab}$ is the Ricci tensor of h_{ab}.

Using the field equation (1), (4) and equations (7), (10), we arrive at

$$\nabla_{[a}\tilde{\omega}_{b]} = -4\ E_{[a}B_{b]} \tag{12}$$

It is easy to see that we can define a curl-less 'twist' vector

$$\omega_a = -\tilde{\omega}_a + 2(BE_a - EB_a) \qquad (13)$$

i.e. $\nabla_{[a}\omega_{b]} = 0$. In a simply connected region one can define therefore a potential ω: $\omega_a = \nabla_a\omega$. It is easy to check that if we conformally rescale the metric h_{ab} using a scale factor depending only on λ we can cancel the second derivatives of λ in the expression for the Ricci tensor on the manifold of orbits. The required conformal transformation has the form: $h_{ab} = \lambda^{-1}\gamma_{ab}$.

Introducing the complex Ernst potentials ω and ψ as follows

$$\epsilon = -\lambda - \psi\bar{\psi} + i\omega, \quad \psi = E + iB, \qquad (14)$$

and taking all covariant derivatives ∇_a with respect to γ_{ab}, we can write equations (11), (12) and (13) in a nice symmetric form of 3-dim Einstein equations coupled to the Ernst equations:

$$-\lambda\nabla_a\nabla^a\epsilon = (\nabla_a\epsilon + 2\bar{\psi}\nabla_a\psi)\nabla^a\epsilon \quad , \qquad (15)$$

$$-\lambda\nabla_a\nabla^a\psi = (\nabla_a\epsilon + 2\bar{\psi}\nabla_a\psi)\nabla^a\psi \quad , \qquad (16)$$

$$R_{ab}(\gamma) = {}^1/_2\ \lambda^{-2}\ [\nabla_{(a}\epsilon\nabla_{b)}\bar{\epsilon} + 2\psi\nabla_{(a}\epsilon\nabla_{b)}\bar{\psi} + 2\bar{\psi}\nabla_{(a}\bar{\epsilon}\nabla_{b)}\psi + 4\lambda\nabla_{(a}\psi\nabla_{b)}\bar{\psi}] \qquad (17)$$

Now it can be easily seen that equations (15), (16) and (17) can be derived from the action principle of the 3-dim gravity coupled to a harmonic map model

$$S = \int |\gamma|^{1/2}\ d^3x\ (R - L) \qquad (18)$$

where R is the Ricci scalar of the metric γ_{ab} and L is given by

$$L = {}^1/_2\ \lambda^{-2}[|\nabla\epsilon + 2\ \bar{\psi}\nabla\psi|^2 + 4\lambda|\nabla\psi|^2] \tag{19}$$

L defines an 'energy' integral for a harmonic map between two spaces (M,γ_{ab}) and (N,H_{AB}), i.e. a map between the space of orbits M of ξ with its conformal metric γ_{ab} and the space N of Ernst's potentials $\phi^A = (\epsilon,\psi)$ with the metric H_{AB} as defined by L. At this point let me recall the definition of the *harmonic map*. I will give here a definition using coordinate systems on two spaces M and N, but one can easily see that it is a coordinate invariant statement.

Definition

Let (M,γ) and (N,H) be two (pseudo)Riemannian manifolds with metrics γ and H respectively. A map g: $M \to N$ which in local coordinate systems X^a and ϕ^A on M and N, respectively, can be written as: $\phi^A = g^A(x^a)$ is called a *harmonic map* if the energy functional

$$I[g] = {}^1/_2 \int_M |\gamma|^{1/2}dx\ \gamma^{ab}H_{AB}(\phi\ )\nabla_a\phi^A\nabla_b\phi^B \quad , \tag{20}$$

is stable under small deformations of g, i.e. if the Lagrange-Euler equations for the functional I[g] are satisfied (Bochner, 1940; Fuller, 1954, Eells and Sampson, 1964). N is called the image space of the harmonic map.

The Lagrangian (19) defines the hermitian metric on the image space N of the Ernst potentials ϵ and ψ

$$ds^2 = {}^1/_2\ (\mathrm{Re}\,\epsilon + \psi\bar{\psi})^{-2}[|d\epsilon + 2\bar{\psi}\ d\psi|^2 - 4(\mathrm{Re}\,\epsilon + \psi\bar{\psi})d\psi d\bar{\psi}] \tag{21}$$

It is quite interesting to study properties of this metric. The metric (21) is Riemannian only when $\lambda = -(\mathrm{Re}\,\epsilon + \psi\bar{\psi}) > 0$; otherwise it is a pseudoriemannian metric. It will be demonstrated later that the image space N is in fact the symmetric homogeneous space $N = G/H = SU(2,1)/S(U(2) \times U(1))$ when $\lambda > 0$, and $N = SU(2,1)/S(U(1,1) \times U(1))$ when $\lambda < 0$ (Mazur, 1982a,b). Harmonic maps onto homogeneous spaces G/H are known in physics as nonlinear sigma models. This peculiar sigma model property of the E-M field equations with one Killing vector proved to be crucial in proving black hole uniqueness theorems for stationary and axisymmetric black holes. Because we are interested in stationary and axisymmetric solutions we have to consider further reduction of the E-M field equations. We assume that there is another Killing vector $\eta = \eta^a \partial_a$ commuting with the previous Killing vector $\xi = \xi^a \partial_a$, i.e. $[\xi, \eta] = 0$. Subsequent $3 \rightarrow 2$ reduction from the space of orbits of ξ to the space of orbits of ξ and η is straightforward. The Ernst potentials ϵ and ψ defined with respect to ξ will be well defined on the space of orbits $M' = M/G_2$, where G_2 is the abelian isometry group generated by ξ and η, only if the action of G_2 on M has the orthogonal transitivity property. It means that there exist a family of 2-surfaces orthogonal to 2-surfaces of transitivity of G_2. The dimensional $4 \rightarrow 2$ reduction of the E-M equations from 4 to 2 dimensions can be carried out if the reduction does not depend on the order of the reduction in two steps. We first consider $4 \rightarrow 3$ reduction with respect to ξ and later $3 \rightarrow 2$ reduction with respect to η. The result of this reduction should be the same if we change the order of subsequent steps, i.e. we first reduce with respect to η and later with respect to ξ. The consistency conditions for the $4 \rightarrow 2$ reduction are, of course, the vanishing of Lie derivatives of ϵ and ψ with respect to the second Killing vector, i.e. $L_\eta \epsilon = L_\eta \psi = 0$, where ϵ and ψ are defined with respect to ξ. Analogous conditions should be satisfied for the reversed order of the two steps of reduction; they are equivalent to the Frobenius integrability condition for the orthogonal transitivity of G_2. As a result of reduction from 4 to 2 dimensions, we will obtain a 2-dimensional gravity. Because any 2-dimensional manifold is conformally flat and a nonlinear sigma model is conformally invariant in two dimensions, we observe that nonlinear sigma model equations will decouple from the two dimensional gravity. It means that we can solve first nonlinear sigma model equations on the 2-dimensional manifold of orbits M' choosing the flat metric on M'. The remaining 2-dim Einstein equations will determine the metric $g_{\mu\nu} = e^{2\Gamma}\delta_{\mu\nu}$, $\mu = 1,2$, i.e. its conformal scale Γ on M' (we have chosen a 'conformal gauge' on M'). Once the solution to nonlinear sigma model equations is known we can determine the metric $g_{\mu\nu}$ on M' and we can reconstruct the four dimensional spacetime (M, g_{ab}) completely. One can see, therefore, that the problem of uniqueness of black hole solutions of the Einstein-Maxwell equations can be studied as a two-dimensional boundary value problem for the nonlinear sigma model equations.

2. Black Hole Boundary Conditions

To this end let me present the black hole boundary value problem. We refer the reader to the more detailed derivation of these conditions discussed by Carter (Carter, 1973, 1979). We will follow Carter's notation. Thus if $\xi = \partial/\partial\phi$ we denote $\lambda = X$ and $\omega = Y$. The determinant of the matrix of scalar products of two Killing vectors $\xi = \partial/\partial\phi$, $\eta = \partial/\partial t$ defines the real scalar field ρ: $-\rho^2 = (\xi,\xi)(\eta,\eta) - (\xi,\eta)^2$. The scalar field ρ vanishes on the symmetry axis A, which is a set of fixed points of $\xi = \partial/\partial\phi$ and on the event horizon H. ρ is also a harmonic function on the space of orbits M' of ξ and η. After reductions of the E-M equations for stationary and axisymmetric fields, we obtain a set of equations on the space of orbits M'

$$\nabla_\mu \nabla^\mu \rho = 0 \quad , \tag{22}$$

$$-X\nabla_\mu(\rho\nabla^\mu \epsilon) = \rho(\nabla_\mu \epsilon + 2\bar{\psi}\nabla_\mu \psi)\nabla^\mu \epsilon \quad , \tag{23a}$$

$$-X\nabla_\mu(\rho\nabla^\mu \psi) = \rho(\nabla_\mu \epsilon + 2\ \bar{\psi}\nabla_\mu \psi)\nabla^\mu \psi \quad , \tag{23b}$$

where ∇_μ is the covariant derivative with respect to the metric $g_{\mu\nu}$ on the space of orbits M'.

The black hole solutions are solutions to (22) and (23) for the Carter boundary conditions (Carter, 1973, 1979), which can be written explicitly once we fix a coordinate system on $(M', g_{\mu\nu})$. Once we know that the topology of the event horizon of stationary black holes is $R^1 \times S^2$ (Hawking, 1972, 1973) and the topology of the domain of outer communications is $R^2 \times S^2$ we can show, using the Morse theory, that the harmonic function ρ does not have *critical* points on $\bar{M}'$. It means that a gradient of ρ and its harmonic conjugate function z is non-zero on $\bar{M}'$ (Carter, 1973). From the asymptotic flatness condition, we have that ρ at large distances behaves as the Weyl canonical cylindrical coordinate. The size of the event horizon fixes the overall scale of ρ. We can take ρ and z, its harmonic conjugate function, as the globally well behaved coordinates on $\bar{M}'$

$$\rho^2 = c^2(x^2 - 1)(1 - y^2), \ z = cxy \quad , \tag{24}$$

where

$$c^2 = M^2 - (J/M)^2 - Q^2 \quad .$$

$x = 1$ defines the location of the event horizon H, $y = \pm 1$ are the two branches of the symmetry axis A, and we reach the spatial infinity as $x \to +\infty$. The black hole's boundary conditions are parametrized by three parameters: the angular momentum J, the electric charge Q and the parameter c or the total mass M of a black hole. The black hole solution is described by the asymptotically flat spacetime which is regular on the axis A and on the event horizon H. These conditions can be translated into the formalism discussed above. The asymptotic boundary conditions are:

$$
\begin{aligned}
E &= -Qy + O(x^{-1}) \; , \\
B &= O(x^{-1}) \; , \\
Y &= 2Jy(3 - y^2) + O(x^{-1}) \; , \\
X &= c^2x^2(1 - y^2) + O(x) \; ,
\end{aligned}
\tag{25}
$$

as $x \to \infty$. The symmetry axis boundary conditions are: E, B, X and Y, should be well behaved functions of x and y which satisfy

$$
\begin{aligned}
&\partial_x E = O(1 - y^2) \; , \quad \partial_y E = O(1) \; , \\
&\partial_x B = O(1 - y^2) \; , \quad \partial_y B = O(1) \; , \\
&\partial_x Y = O(1 - y^2) \; , \quad \partial_y Y + 2(E\partial_y B - B\partial_y E) = O(1 - y^2) \\
&X = O(1 - y^2), \; X^{-1}\partial_y X = -2y(1 - y^2)^{-1} + O(1 - y^2)
\end{aligned}
\tag{26}
$$

as $y \to \pm 1$. The event horizon boundary conditions demand only that E, B, X and Y are well behaved functions of x and y as $x \to 1$.

$$
\begin{aligned}
&\partial_x E = O(1) \; , \quad \partial_y E = O(1) \; , \\
&\partial_x B = O(1) \; , \quad \partial_y B = O(1) \; , \\
&\partial_x Y = O(1) \; , \quad \partial_y Y = O(1) \; ,
\end{aligned}
\tag{27}
$$

$$X = O(1) \;, \quad X^{-1} = O(1) \;.$$

It was conjectured that the only stationary and axisymmetric black hole solutions satisfying these boundary conditions are described by the Kerr-Newman et al. (Kerr, 1963; Newman, et al., 1965) family of solutions satisfying the condition of $c^2 = M^2 - (J/M)^2 - Q^2 > 0$. Great progress in proving correctness of this conjecture was achieved initially in the early seventies (Carter, 1971, 1973; Robinson, 1974) and then it culminated in a remarkable black hole uniqueness theorem for vacuum black holes (Robinson, 1975).

The proof of no hair theorems of Carter and Robinson and Robinson's uniqueness theorem for the Kerr black hole made use of remarkable divergence identities. The common ancestor of these mysterious identities is the Green identity for the Laplace equation, which is usually applied to show that the Dirchlet or Neuman boundary value problem is well posed. The no hair theorems of Carter and Robinson are basically statements about the absence of nontrivial bifurcations for solutions of the harmonic map equations (22) and (23) with the black hole boundary conditions (25), (26) and (27). It means that if we consider the linearized harmonic map equations for small perturbations of a harmonic map around a black hole solution with fixed values of M, J and Q then there are no perturbations satisfying linearized black hole boundary conditions. The conclusion one could draw from these results (Carter, 1973) was that if there were other solutions than the Kerr-Newman black hole solutions, they should form a disjoint family of solutions parametrized also by three conserved quantities: the total mass M, the total angular momentum J and electric charge Q.

What remained to be done was to exclude the possiblity of other than the Kerr-Newman families of black hole solutions, proving, therefore, the black hole uniqueness theorem. For the vacuum black holes, this goal was achieved by D. C. Robinson in 1975 who found a nonlinear version of Carter's divergence identity and applied it to the proof of the black hole uniqueness theorem for the Kerr black hole. The most general form of the black hole uniqueness theorem was proven only recently.

Black Hole Uniqueness Theorem (Mazur, 1982a,b; Bunting, 1983)

The only possible stationary and axisymmetric black hole solutions of the Einstein-Maxwell equations satisfying boundary conditions (25), (26) and (27) are the Kerr-Newman solutions subject to the constraint $M^2 - (J/M)^2 - Q^2 > 0$. Black holes are *completely* characterized by three parameters only: the total mass M, the angular momentum J and the electric charge Q. If there is a magnetic charge P in Nature, then black holes will be described completely by the four parameter family of Kerr-Newman solutions with $M^2 - (J/M)^2 - Q^2 - P^2 > 0$.

Before proceeding with the proof of this uniqueness theorem for black holes, I would like first to introduce very useful techniques of the harmonic map theory. The remarkable divergence identities of Carter and Robinson will emerge as a natural consequence of the harmonic map property of the Einstein-Maxwell equations for stationary and axisymmetric fields.

3. Divergence Identities for Harmonic Maps and No Hair Theorems

Harmonic maps enjoy a lot of nice properties; especially we can say this for harmonic maps between two Riemannian manifolds where the image manifold has non-positive sectional curvature (Yau, 1982). It is well known that if N has non-positive curvature, then every smooth map from M to N is homotopic to a harmonic map with minimal 'energy'. This existence theorem for harmonic maps between two compact Riemann manifolds was established by Eells and Sampson (1964). Moreover, there exists one and only one harmonic map in each homotopy class of maps between two compact Riemann manifolds M and N, when N has negative curvature and if the image of the first homotopy group $\pi_1(M)$ of M in $\pi_1(N)$ is not a cyclic group (Hartman, 1967). Mathematically, the black hole uniqueness theorem is an example of a result which shows that under certain boundary conditions the same uniqueness theorem for harmonic maps can be extended to the non-compact case. I will present here results discussed first in (Mazur, 1984a) and independently by Bunting (1983). Because the no hair theorems are concerned with small perturbations of black hole solutions, we will be concerned here with the question of 'stability' of harmonic maps to a negatively curved Riemann manifold N. The Euler-Lagrange equations for harmonic maps can be derived from the variational principle (20)

$$\nabla_a \nabla^a \phi^A + \Gamma_{BC}{}^A \nabla_a \phi^B \nabla^a \phi^C = 0 \ . \tag{28}$$

For stationary and axisymmetric E-M fields eq. (28) is modified in the way that a covariant Laplacian in (28) is replaced by $\rho^{-1}\nabla(\rho\nabla)$, where ρ is a non-negative harmonic function. This modification does not change conclusions we will draw about properties of harmonic maps, because all divergence identities presented below will be multiplied by a non-negative function ρ.

In order to study the problem of bifurcations off a family of black hole solutions we need the linearized form of (28). Let us consider a one-parameter family of maps f_τ: $M \to N$, represented locally by functions $\phi^A(\tau,x)$. A tangent vector $\partial/\partial\tau$ to a curve f_τ with components, $X^A = \partial\phi^A/\partial\tau$ satisfies the 'small perturbations' equation (the Jacobi 'geodesic deviation' equation; see e.g. Misner, 1978)

$$D_a D^a X^A + R^A{}_{BCD}\nabla_a\phi^B\nabla^a\phi^C X^D = 0 \ , \tag{29}$$

where

$$D_a X^A = \nabla_a X^A + \Gamma^A{}_{BC}\nabla_a\phi^B X^C \ . \tag{30}$$

It is a general property of nonlinear systems that the nonexistence of linearized perturbations for a given boundary value problem is sufficient to exclude the possiblity of a corresponding bifurcating family of exact solutions. The following divergence identity is useful to determine the possiblity of bifurcation of solutions to (28)

$$\nabla^a(H_{AB}X^A D_a X^B) = H_{AB}D_a X^A D^a X^B + R_{ABCD}X^A\nabla_a\phi^B X^C\nabla^a\phi^D. \tag{31}$$

The divergence term on the l.h.s. of (31), when integrated over M gives the surface term contribution which vanishes for a large class of boundary conditions (when M has a boundary ∂M; for M compact it vanishes identically). The first term on the r.h.s. of (31) is non-negative when M and N are Riemannian. The second term is also non-negative only when the sectional curvature of N is non-positive. If the surface term vanishes, as happens for black hole solutions, then it follows that each term on the r.h.s of (31) vanishes separately i.e. $D_a X^A = 0$ and $X^{[A}\nabla_a\phi^{B]} = 0$. Then from (31) it follows that $\Lambda^A{}_B X^B = 0$, where $\Lambda^A{}_B = R^A{}_{CDB}\nabla_a\phi^C\nabla^a\phi^D$. There can exist a non-zero solution to the zero eigenvalue problem for the matrix Λ if its determinant is zero. In general, a given solution of (28) for which we are seeking bifurcations has the property that Λ is a non-degenerate matrix. One can also see that the number of bifurcations is equal to dim N - rank Λ. We conclude that, in general, there are no bifurcations for harmonic maps when the sectional curvature of N is non-positive (unless we have a degenerate case). The Carter and Robinson no hair theorems are simply very special cases of the result I have described briefly above. The positive definite divergence identity (31) when applied to electrovacuum black holes, i.e. to solutions of (22) and (23) satisfying black hole boundary conditions (25), (26) and (27) gives as a result the black hole no hair theorem (Carter, 1971, 1973; Robinson, 1974; Mazur, 1984a). This is so because the image space of the Ernst potentials ϵ and ψ is a Riemannian space with negative sectional curvature (Mazur, 1982 a,b). It is quite interesting that the complicated divergence identities of Carter and Robinson are just special cases of the general divergence identity (31) for perturbations of harmonic maps to negatively curved Riemann manifold N. It is obvious that using a particular Ernst's coordinate system on (N,H_{AB}), as given by (21), one can obtain a rather unenlightening form of the identity (31). It is the advantage of the geometrical approach of harmonic map theory which could bring us a better understanding of global black hole solutions. We have, therefore, a nice theorem.

No Hair Theorem (Carter, 1971; Robinson, 1974)

There do not exist regular, small perturbations of the stationary and axisymmetric black hole solutions of the E-M equations preserving boundary conditons (25), (26) and (27) for fixed values of the mass M, the angular momentum J and the electric charge Q of a black hole. The only deformations of black holes that do exist are those obtained by a change of M, J and Q.

It is also easy to see that starting with the

divergence identity for small perturbations of harmonic maps one can obtain the *global* divergence identity with the required positivity property when N has non-positive sectional curvature. Consider a curve of harmonic maps f_τ, $\tau \in [0,1]$. Then f_0 and f_1 are homotopically equivalent. For each τ the r.h.s. of (31) is non-negative. If we integrate both sides of the divergence identity (31) along a geodesic curve f_τ the right hand side (r.h.s.) will be non-negative. The left hand side (l.h.s.) will have a form

$$\int_0^1 d\tau \nabla_a \nabla^a (H_{AB} X^A X^B) = \nabla_a \nabla^a \int_0^1 d\tau (\partial/\partial\tau,\ \partial/\partial\tau) \geq 0\ . \qquad (32)$$

Taking advantage of the fact that there always exists a unique geodesic joining two points f_0 and f_1 on a simply connected Riemannian manifold N with a negative sectional curvature one can evaluate the integral (32). Because f_τ is a geodesic, it means that the norm $(\partial/\partial\tau, \partial/\partial\tau)$ is a constant along f_τ, i.e. $(\partial/\partial\tau,\ \partial/\partial\tau) = c^2$ = constant. One can easily see that the geodesic distance between f_0 and f_1 is:

$$S(f_0, f_1) = \int_0^1 d\tau\ \sqrt{(\partial/\partial\tau,\ \partial/\partial\tau)} = c \int_0^1 d\tau = c$$

It means that the τ integral in (32) is equal to $S^2(f_0, f_1$ In this way we have arrived at the Bunting divergence inequality (Bunting, 1983)

$$\nabla_a \nabla^a S^2(f_0, f_1) \geq 0 \qquad (33)$$

A similar identity where S^2 was replaced by an arbitrary function of S^2 with positive first and second derivatives was proposed also by the present author (Mazur, 1984a). One can apply the divergence identity (33) to the space N of Ernst's potentials is Riemannian and has negative sectional curvature. The 'distance' function $S(f_0,f_1)$ for two solutions of the harmonic map equations (28) (and in the special case of Ernst equations (23)) satisfying the same boundary conditions is non-negative and vanishes only on the boundary. For the case of Ernst's equations for the stationary and axisymmetric fields the divergence identity is modified by the presence of the harmonic, positive function ρ: $\nabla_\mu(\rho\nabla^\mu S^2(f_0,f_1) \geq 0$. When we integrate this identity over the manifold of orbits M' we obtain surface terms which vanish. The contribution from the symmetry axis A and the event horizon H vanishes in virtue of the boundary conditions (26) and (27) and because ρ is vanishing there. The asymptotic boundary conditions (25) imply vanishing of a surface integral at infinity where $\rho \to \infty$. The r.h.s. of the identity (33) vanishes only when $S(f_0, f_1)$ is constant everywhere. But $S(f_0,f_1)$ vanishes on the boundary, so it must vanish everywhere. If we take f_0 to be the Kerr-Newman black hole solution and for f_1 we take another possible black hole solution, then vanishing of $S(f_0,f_1)$ implies $f_0 \equiv f_1$, i.e. the black hole uniqueness theorem (Mazur, 1982a,b; Bunting, 1983). We have seen that the no hair theorems and the black hole uniqueness theorem can be proved applying a generalization of the Green divergence identity to the case of harmonic maps to Riemannian manifolds with non-positive sectional curvature. As it was mentioned above, one can easily generalize the inequality (33) by taking an even function of S with positive first and second derivatives with respect to S. Consider a function of $\sigma = S^2$, $h(\sigma)$ and evaluate its Laplacian $\nabla^2 h(\sigma) = h'(\sigma)\nabla^2\sigma + h'(\sigma)(\nabla\sigma)^2$. $\nabla^2(\sigma)$

≥ 0 if $h' \geq 0$ and (33) is satisfied, i.e. if N has non-positive sectional curvature.

The method the present author used in his proof of the black hole uniqueness theorem is based on a divergence identity for the Ernst equations (23), whose derivation was based on the observation that the image manifold of Ernst's potentials is a homogeneous, symmetric Kähler manifold $N = SU(2,1)/S(U(2) \times U(1))$. This method seems to be much simpler in application because for the electrovacuum black holes the Riemannian space N of Ernst's potentials ϵ and ψ is a symmetric manifold with a SU(2,1) isometry group. Because of this large symmetry, one can use group theoretical methods to construct an SU(2,1) invariant two-point function $h(f_0,f_1)$ which is a certain function of the geodesic distance between two harmonic maps f_0 and f_1. The algebraic explicit construction of the global divergence identity offers a simple understanding of the otherwise mysterious Robinson identity (Mazur, 1982a,b; Mazur, 1984a,b), because the identity produced this way reproduces in the vacuum case the Robinson identity (Robinson, 1975). The construction I am going to describe here has a much broader range of applications than solely the black hole uniqueness theorems. It can be applied to any nonlinear elliptic sigma model on symmetric Riemannian spaces with noncompact isometry groups or, which is equivalent, with negative sectional curvature (Mazur, 1984a; Mazur and Richter, 1985; Breitenlohner and Maison, 1986).

4. A Global Identity for Nonlinear Sigma Models and its Applications (Black Hole Uniqueness Theorem)

Before discussing the most general case of sigma models on symmetric Riemannian and hyperbolic spaces and the divergence identity associated with them, let me first demonstrate explicitly that the metric (21) related to the electrovacuum Ernst equations is the left invariant metric on the homogeneous, symmetric and Kähler space N = SU(2,1)/S(U(2) x U(1)). The detailed construction is described in (Mazur, 1982a,b, 1983). There are many ways to see this. The metric (21) is an example of a hermitian metric on an (almost) complex manifold N with complex coordinates z^α (in the case of electrovacuum Ernst equations, $\alpha = 1,2$, $z^1 = \epsilon$, $z^2 = \psi$): $ds^2 = \kappa_{\alpha\bar{\beta}} dz^\alpha d\bar{z}^\beta$. A complex hermitian manifold N is a Kähler manifold if it admits a closed nondegenerate (1,1) form $\omega = -{}^i/_2\ \kappa_{\alpha\bar{\beta}} dz^\alpha d\bar{z}^\beta$, $d\omega = (\partial + \bar{\partial})\omega = 0$. One can easily see that the metric (21) is Kähler (Mazur 1983), i.e. $\partial_\gamma \kappa_{\alpha\bar{\beta}} = \partial_{\bar{\alpha}} \kappa_{\gamma\bar{\beta}}$, $\partial_{\bar{\gamma}} \kappa_{\alpha\bar{\beta}} = \partial_{\bar{\beta}} \kappa_{\alpha\bar{\gamma}}$, where $\partial_\alpha = \partial/\partial z^\alpha$ and $\partial_{\bar{\alpha}} = \partial/\partial \bar{z}^\alpha$. Moreover, the metric (21) has 8 holomorphic Killing vectors which generate the Lie algebra of the pseudounitary group SU(2,1). The group G = SU(2,1) acts on N nonlinearly by holomorphic (homographic) maps. One can see that by introducing new Ernst coordinates w^α on N: $z^1 = (w^1 - 1)/(w^1 + 1)$, $z^2 = w^2/(w^1 + 1)$ and showing that the following holomorphic transformations of w^α are invariant transformations of the metric (21)

$$w^{1\prime} = (u^1{}_1 w^1 + u^1{}_2 w_2 + u^1{}_3)/(u^3{}_1 w^1 + u^3{}_2 w^2 + u^3{}_3),$$

$$w^{2\prime} = (u^2{}_1 w^1 + u^2{}_2 w^2 + u^2{}_3)/(u^3{}_1 w^1 + u^3{}_2 w^2 + u^3{}_3),$$

where the matrix $u^\alpha{}_\beta$, $\alpha,\beta = 1,2,3$ satisfies the pseudo unitarity condition

$$u^+ \eta u = \eta, \quad \eta = \mathrm{diag}(1,1,-1)$$

The group G = SU(2,1) acts on N simply transitively. This means that N is a homogeneous Kähler manifold. Every homogeneous manifold of G is diffeomorphic to the left coset space G/H for some subgroup H of G. In our case, it can be easily seen that $H = S(U(2) \times U(1))$. Fix a point $p_0 \in N$, such that $w^a(p_0) = 0$. Every point $p \in N$ can be reached from p_0 by a map u: $p = up_0$, $u \in G$. Two maps u, u' $\in G$ represent the same point if $u \sim u' = uh$, $h \in H$. H is the isotropy subgroup of p_0 which is isomorphic to $S(U(2) \times U(1))$. The equivalence relation $\sim$ defines the coset space G/H. The homogeneous space $N = SU(2,1)/S(U(2) \times U(1))$ is also a symmetric space. The last property helps to construct a unique left G invariant metric on N in terms of a coset representative g.

We would like to reformulate the Ernst equations (23) associated with the metric (21) in terms of group theoretical objects, like a coset representative $g(uH) = g(u)$, making their SU(2,1) covariance explicit. This will help us to construct the generalized Robinson identity in the obvious way. Exploiting the SU(2,1) covariance of the field equations (23) one can obtain an SU(2,1) invariant divergence identity for two solutions of (23). The metric (21) is a left SU(2,1)-invariant metric on $N = SU(2,1)/S(U(2) \times U(1))$. We can write it as a standard left G-invariant metric on a symmetric space $N = G/H$. This leads to a nonlinear sigma model form of the Ernst equations (23), once we go from the Ernst parametrization of N to a coset representation of N. We give here a short derivation of a global divergence identity for nonlinear sigma models on symmetric spaces and apply it to the electrovacuum Ernst equations.

We define the symmetric space as a triple (G,H,μ), where G is a connected Lie group, and H is a closed subgroup of G defined by an involutive automorphism μ of G such that $(G_\mu)_0 \subset H \subset G_\mu$ with G_μ and $(G_\mu)_0$ being the set of fixed points of μ and its identity component, respectively. An involutive automorphism μ defines also a smooth mapping g of a coset space G/H into G: g(u) and $u\mu(u)^{-1}$, $g(uH) = g(u)$.

The G-valued field g satisfies a constraint: $g\mu(g) = I$. In terms of g one can naturally introduce the left G-invariant metric on G/H: $dS^2 = {}^1/_2\, Tr(dgg^{-1})^2$. If G is a noncompact group and H is its maximally compact subgroup, then the metric $Tr(dgg^{-1})^2$ is Riemannian. A harmonic map from a Riemannian space M to a Riemannian coset space N = G/H is called a nonlinear sigma model with the Lagrangian

$$L = {}^1/_2\, Tr(J_a J^a),\ J_a = \nabla_a g g^{-1}\ , \tag{35}$$

and field equations

$$\nabla_a J^a = 0\ . \tag{36}$$

The field equation (36) is covariant under the left G translations on G/H. A global divergence identity for the nonlinear sigma model (35) can be obtained exactly in the same way as the Green identity for the Laplace equation. To this end, consider two fields g_0 and g_1, not necessarily solutions of (36). Define a field $\phi = g_0 g_1{}^{-1}$, which tranforms under the rigid left G-translations on N, $u_0 \to uu_0$, $u_1 \to uu_1$, $u \in G$, in a simple way: $\phi \to u\phi u^{-1}$. One can easily see that the trace of ϕ is invariant under the rigid G-translations of two points on N. It means that $Tr\phi$ is a function of the geodesic distance between g_0 and g_1: $Tr\phi = h(s)$, $s = s(g_0, g_1)$. Evaluating the covariant Laplacian of $Tr\phi$ and using (35) we arrive at the generalized Green identity for nonlinear sigma models (Mazur, 1982a,b, 1984a)

$$\nabla_a \nabla^a h = Tr[\phi \nabla_a (J^{(0)a} - J^{(1)a})] + Tr[\phi (J^{(0)}{}_a J^{(0)a} + J^{(1)}{}_a J^{(1)a} - 2J^{(0)}{}_a J^{(1)a})]\ , \tag{37}$$

where $J^{(i)}{}_a = \nabla_a g_i g_i{}^{-1}$, $i = 0,1$.

Lemma (Mazur, 1982b, 1984a)

A global divergence identity (37) for Riemannian nonlinear sigma models has a positive definite right hand side if g_0 and g_1 are solutions to (36) and the following equivalent conditions on N are satisfied:

a) N = G/H is a Riemannian symmetric space with noncompact isometry group G;

b) N has a non-positive sectional curvature.

We can see now a direct correspondence of this result to the basic properties of harmonic mappings to a Riemannian space N with non-positive sectional curvature discussed before. For the electrovacuum Ernst equations, the identity (37) is a generalization of the Robinson identity and also it coincides with Robinson's identity for the vacuum case. This form of the divergence identity (37) proves to be very useful in evaluating surface terms for black hole boundary conditions because the function $h = \mathrm{Tr}\Phi$ turns out to be only a rational function of Ernst's potentials. The symmetric space $N = SU(2,1)/S(U(2) \times U(1))$ is singled out by the existence of an inner involutive automorphism of $G = SU(2,1)$: $u \to \mu(u) = \eta u \eta^{-1}$, $\eta = (1,1,-1)$, because $\mu(H) = H$, $H = S(U(2) \times U(1))$. A coset representative $g = u\mu(u)^{-1}$ which is an element of $G = SU(2,1)$, i.e, $g^+\eta g = \eta$, satisfies also a constraint $I = g\mu(g) = g\eta g\eta^{-1}$. This implies that g is a hermitian matrix $g^+ = g$. Writing $g = \eta + 2P$, one can see that P satisfies the following constraints $P^+ = P$, $(P\eta)^2 = -P\eta$. We will recover the Ernst parametrization of $SU(2,1)/S(U(2) \times U(1))$ and satisfy constraints on P if we write $P^{\alpha\beta} = v^\alpha \bar{v}^\beta$, where $v^\alpha \eta_{\alpha\beta} \bar{v}_\beta = -1$, $\alpha, \beta = 1,2,3$, and define $w^1 = v^1/v^3$, $w^2 = v^2/v^3$. w^1 and w^2 can be written in terms of Ernst's potentials ϵ and ψ as follows: $w^1 = (1 +$

$\epsilon)/(1 - \epsilon)$, $w^2 = 2\psi/(1 + \epsilon)$. Using the Ernst parametrization of the coset space element g which was discussed above one can calculate a function $h(s) = \mathrm{Tr}\Phi$ in terms of the Ernst potential X_i, Y_i, E_i and B_i, $i = 0,1$

$$h(s) = \mathrm{Tr}\Phi = 3 + X_0^{-1}X_1^{-1}\{(X_0 - X_1)^2 + 2(X_0 + X_1)[(E_0 - E_1)^2 + (B_0 - B_1)^2] + [(E_0 - E_1)^2 + (B_0 - B_1)^2]^2 + [Y_0 - Y_1 + 2(E_1B_0 - E_0B_1)]^2\} \tag{38}$$

The black hole uniqueness theorem for electrically charged black holes can be obtained by a straightforward application of the divergence identity (37). When we integrate the l.h.s. of (37) we obtain a surface term which can be shown, after using (38), to be vanishing for the black hole boundary conditions (25), (26) and (27). Because the r.h.s. of (37) is positive definite, it means that h = const. This is consistent with the boundary conditions and (38) only when $X_0 = X_1$, $Y_0 = Y_1$, $E_0 = E_1$ and $B_0 = B_1$. It means that the Kerr-Newman family of solutions with $M^2 - (J/M)^2 - Q^2 >0$ characterizes completely the stationary equilibrium black hole states in the Einstein-Maxwell theory. Concluding this talk, I would like to point out that it was only possible to prove black hole uniqueness theorems for the stationary black holes which are also axisymmetric. It is fortunate that the Einstein-Maxwell equations reduce to harmonic mapping equations which are moreover conformally invariant in the case of stationary and axisymmetric black holes. One may hope that the reasonable extension of arguments presented here may lead to the solution of the uniqueness problem for stationary black holes, without the assumption of axial symmetry.

This work was supported in part by the National Science Foundation through Grant No. PHY 83-18350 with Syracuse University and a travel grant, also. I would like to thank the General Relativity Society for giving me the opportunity to present this material.

References

Bardeen, J. M., Carter, B., and Hawking, S. W., (1973). "The four laws of black hole mechanics", Commun. Math Phys., 31, 161.

Bochner, S., (1940). "Harmonic surfaces in Riemann metric", Trans. Amer. Math. Soc. 47, 146.

Breitenlohner, P. and Maison, D., (1986). "Solitons in Kaluza-Klein theories", in Solitons in General Relativity, eds. H. Morris and R. Dodd, (New York: Plenum).

Bunting, G., (1983). "Proof of the uniqueness conjecture for black holes", Ph.D. thesis (unpublished) University of New England, Armidale, N.S.W.

Carter, B., (1971). "Axisymmetric black hole has only two degrees of freedom", Phys. Rev. Lett. 26, 331.

Carter, B., (1973). "Black hole equilibrium states", in Black Holes, eds. C. DeWitt and B. S. DeWitt, pp. 57-214, New York: Gordon and Breach.

Carter, B., (1979). "The general theory of the mechanical, eletromagnetic and thermodynamic properties of black holes", in An Einstein Centenary Survey, eds. S. W. Hawking and W. Israel, pp. 294-369, Cambridge: Cambridge University Press.

Doroshkevich, A. G., Zeldovich, Ya.B., and Novikov, I.D., (1966). "Gravitational collapse of non-symmetric and rotating bodies", Sov. Phys. J.E.T.P. 22, 122.

Eells, J. and Sampson, J. H. (1964). "Harmonic mappings of Riemannian manifolds", Amer. J. Math. 86, 109.

Ernst, F. J., (1968). "New formulation of the axially symmetric gravitational field problem I, II", Phys. Rev. 167, 1175; 168, 1415.

Fuller, F. B., (1954). "Harmonic mappings", Proc. Nat. Acad. Sci. USA 40, 987.

Geroch, R. P., (1971). "A method for generating solutions of Einstein's equations", J. Math. Phys. 12, 918.

Hajicek, P., (1973). "General Theory of Vacuum Ergospheres", Phys. Rev. D7, 2311 (1973).

Hartle, J. B., and Hawking, S. W., (1972). "Solutions of the Einstein-Maxwell equations with many black holes", Commun. Math. Phys. 26, 87.

Hartman, P., (1967). "On homotopic harmonic maps", Canada J. Math. 19, 673.

Hawking, S. W., (1972). "Black holes in general relativity", Commun. Math. Phys. 25, 152.

Hawking, S. W., (1973). "The event horizon", in Black Holes, eds. C. DeWitt and B. S. DeWitt, pp. 5 - 54, New York: Gordon and Breach.

Hawking, S. W. and Ellis, G. F. R., (1973). "The large scale structure of spacetime", (Cambridge: Cambridge University Press).

Ipser, J., (1971). "Electromagnetic test fields around a Kerr-Metric black hole", Phys. Rev. Lett. 27, 529.

Israel, W., (1967). "Event horizons in static vacuum spacetimes", Phys. Rev. 164, 1776.

Israel, W., (1968). "Event horizons in static electrovac spacetimes", Commun. Math. Phys. 8, 245.

Kerr, R. P., (1963). "Gravitational field of a spinning mass a an example of algebraically special metrics", Phys. Rev. Lett. 11, 237.

Mazur, P. O., (1982a). "Proof of uniqueness of the Kerr-Newman black hole solution", J. Phys. A15, 3173.

Mazur, P. O., (1982b). "Properties and integrability of the Ernst equations", Ph.D. thesis, Jagellonian University, Krakow, Poland (in Polish).

Mazur, P. O., (1983). "A relationship between the electrovacu Ernst equations and nonlinear sigma model", Acta. Phys. Polon. B14, 219.

Mazur, P. O., (1984a). "A global identity for nonlinear sigma models", Phys. Lett. 100A, 341.

Mazur, P. O., (1984b). "Black hole uniqueness from a hidden symmetry of Einstein's gravity", GRG16, 211.

Mazur, P. O. and Richter, E., (1985). "Harmonic maps and uniqueness of axisymmetric monopole solutions", Phys. Lett 109A, 429.

Misner, C. W., (1978). "Harmonic maps as models for physical theories", Phys. Rev. D18, 4510.

Newman, E. T., Couch, E., Chinnapared, K., Exton, A., Prakash, A., and Torrence, R., (1965). "Metric of a rotating charged mass", J. Math. Phys. 6, 918.

Oppenheimer, J. R. and Snyder, H., (1939). "On continued gravitational contraction", Phys. Rev. 56, 455.

Penrose, R., (1969). "Gravitational collapse: the role of general relativity", Riv. del Nuovo Cimento 1, 252.

Price, R. H., (1972). "Nonspherical perturbations of relativistic gravitational collapse. I. Scalar and gravitational perturbations", Phys. Rev. D5, 2419.

Press, W. H. and Teukolsky, S. A., (1973). "Perturbations of rotating black hole. II. Dynamical stability of the Kerr metric", Phys. Rev. D5, 2439.

Regge, T. and Wheeler, J. A., (1975). "Stability of a Schwarzschild singularity", Phys. Rev. 108, 1063.

Robinson, D. C., (1974). "Classification of black holes with electromagnetic fields", Phys. Rev. D10, 458.

Robinson, D. C., (1975). "Uniqueness of the Kerr black hole", Phys. Lett. 34, 905.

Wald, R. M., (1972). "Nonspherical gravitational collapse and black hole uniqueness", Ph.D. thesis, Princeton University (Unpublished).

Wald, R. M., (1973). "On perturbations of a Kerr black hole", J. Math. Phys. 14, 1453.

Vishveshwara, C. V., (1970). "Stability of the Schwarzschild metric", Phys. Rev. D1, 3870.

Yau, S. T., (1982). "Survey on partial differential equations in differential geometry", in Seminar on Differential Geometry, ed. Shing-Tung Yau, (Princeton: Princeton University Press).

Twistors in General Relativity

by
Roger Penrose
Mathematical Institute
24-29 St. Giles
Oxford

Abstract

The two most significant definitions of a twistor in (conformally) flat space-time are that given by solutions of the twistor equation and that given by α-planes in the complexification of the space-time. It is shown that these concepts are naturally dual to one another, and that they give rise to various different twistor concepts in curved space-time when referred to submanifolds of differing dimension. One of these concepts, that of 2-surface twistors, gives rise to a quasi-local definition of mass (and angular momentum). The status of this definition is explored in relation to various examples.

1. Introduction

There are many different ways of introducing the ideas of twistor theory. Fortunately there are now two reasonably comprehensive books available on the subject (Penrose and Rindler 1986, Huggett and Tod 1985) and it will not be necessary for me to be in any way complete in my descriptions of twistors.

The most primitive idea of a twistor can be taken to be a (geometrical) light ray in space-time, i.e. a null geodesic. In twistor theory one transfers one's attention from the space-time to twistor space (the space of twistors); so according to this primitive picture, we think of a space-time point in terms of the family - a (Riemann) sphere's worth - of light rays which pass through that point. Thus, a space-time point is interpreted, twistorially, as a Riemann sphere in twistor space, whereas a light ray in space-time is interpreted as a single point in twistor space.

There are various motivations which underlie this procedure. Quantum gravity could allow the space-time points actually to become "fuzzy" - rather than merely the light cones in space-time becoming fuzzy - this being allowed for if it is the twistor space, rather than the space-time which is taken as primary. Moreover, there is the possibility that one might be able to find an alternative to a space-time description at curvature singularities, so that some kind of coherent physical description could still persist at the big bang, or at black-hole cores, even when our conventional notions of space-time give out. Twistor descriptions are essentially non-local, which could fit in with the known non-local features of quantum theory (as manifested, for example, in the EPR-Bohm type of experiment of Freedman and Clauser 1972, Aspect, Grangier and Roger 1982). Twistors have an essential chirality, which fits in with the known chiral behaviour of Nature as manifested in weak interactions. Massless fields and conformally invariant interactions seem to have a primitive role to play in Nature, these being very naturally treated within twistor theory. Perhaps most importantly of all, twistors are fundamentally complex (holomorphic) objects and serve to unify the basic complex numbers of quantum theory (probability amplitudes) with the very structure of space-time.

2. Twistors in Flat Space-time

I shall concentrate on just two of the various different more complete ways of regarding a twistor in standard (four-dimensional, signature +---) Minkowski space-time $\mathbb{M}$. The first is as a solution of the twistor equation

$$\nabla_{A'}^{(A}\omega^{B)} = 0.$$

(See Penrose and Rindler 1984 for the conventions concerning the 2-spinor notation used here.) The solution space of this equation is four-complex-dimensional, and we have

$$\omega^A = \mathring{\omega}^A - ix^{AA'}\mathring{\pi}_{A'}$$

$$\pi_{A'} = \mathring{\pi}_{A'}$$

where $x^a(= x^{AA'})$ is the position vector relative to some fixed origin O, and where the symbols with the small circle above denote constant spinors, whose values coincide with that of the corresponding spinor field (denoted without the circle above) at the origin O. The pair of spinor fields ω^A and $\pi_{a'}$ are the spinor parts of the twistor Z^α (where α is a four-dimensional abstract index), and we may write

$$Z^\alpha = (\omega^A, \pi_{A'}).$$

The second way of viewing a twistor that I am considering here, is in terms of an α-plane. For this we need the complexification $\mathbb{C}\mathbb{M}$ of Minkowski space $\mathbb{M}$ - and , moreover, the conformal compactification $\mathbb{C}\mathbb{M}^\#$ of this. An α-plane is a complex 2-plane which is totally null, in the sense that the (complexification of the) space-time metric g_{ab} vanishes identically over the entire plane. In fact there are <u>two</u> disjoint families of such totally null planes, namely the α-planes and the β-planes. Any complex null vector has the spinor form

$$\lambda^A \pi^{A'},$$

and the tangents to an α-plane are the vectors of this form for which $\pi^{A'}$ is held fixed and λ^A varies. Correspondingly the tangents to a β-plane have this form where λ^A is held fixed and $\pi^{A'}$ varies. Clearly all the tangent vectors to an α-plane (or β-plane) are orthogonal to one another (e.g. $(\lambda^A\pi^{A'})(\lambda_A\pi_{A'}) = 0$, since $\pi^{A'}\pi_{A'} = 0$), whence the induced metric vanishes identically. An α-plane (or a β-plane) which contains a real point, contains precisely a single light ray of real points (given by $\lambda^A = k\bar{\pi}^A$, $k \in \mathbb{R}$), and this regains our original <u>light-ray</u> way of viewing certain twistors - these being the <u>null</u> twistors.

The fact that α-planes in $\mathbb{CM}$ have totally null tangent 2-planes of this type immediately implies that they are flat (totally geodesic) - by a Lie bracket argument. In fact we can take $\pi_{A'}$ to be <u>constant</u> over the α-plane (the "covariant" version $\pi_{A'}$ being preferred over the "contravariant" version $\pi^{A'}$ for the reason that this constancy is thereby rendered conformally invariant). This gives us an alternative definition of a (non-zero) twistor, as an α-plane with constant $\pi_{A'}$ associated with it. The freedom in the choice of $\pi_{A'}$ is simply that of a rescaling

$$\pi_{A'} \mapsto \chi\, \pi_{A'} \qquad (0 \neq \chi \in \mathbb{C}).$$

The choice of scaling for $\pi_{A'}$ fixes the <u>scaling for the twistor</u>.

The equivalence (in $\mathbb{CM}^{\#}$) between this definition and that provided by solutions of the twistor equation is seen from the fact that the <u>zero-locus</u> of any (non-zero) solution is an α-plane. The points with position vectors x^a, at which $\omega^A = 0$, are given by

$$\overset{\circ}{\omega}{}^A = ix^{AA'}\overset{\circ}{\pi}_{A'}.$$

If we start from one such point at which this equation holds, the directions along which we can move, while maintaining satisfaction of this equation, clearly have the form

$$\lambda^A \overset{\circ}{\pi}{}^{A'}$$

for arbitrary λ^A (and recall that $\overset{\circ}{\pi}{}^{A'} = \pi^{A'}$), showing that this locus is indeed an α-plane and the scaling is provided by $\pi_{A'}$ as before. We should note that the <u>constant</u> solutions of the twistor equation evade this description, since now $\pi_{A'} = 0$. For these, we invoke the conformal invariance (that we shall examine shortly) of the twistor equation. By transforming null infinity to a finite place we observe that there is indeed a zero-locus of ω^A, which is an α-plane lying entirely at null infinity.

3. Duality Between the Two Definitions

I want to point out an essential duality that underlies the relation between these two definitions of a twistor. In fact dual twistors are identified with complex conjugate twistors, so it is the twistor concept associated with a β-plane which must be seen to be dual to that associated with a solution of the twistor equation. Indeed, we can use this duality to see what the significance of that particular equation that we have called the twistor equation actually is!

Suppose we simply have some spinor field ω^A which is taken to define some object that is naturally dual to the β-plane type of twistor, i.e. to totally null complex 2-planes (β-planes) in $\mathbb{C}\mathcal{M}^\#$ with constant spinors λ_A attached to them, the tangent vectors to the 2-plane having the form

$$\lambda^A\xi^{A'}$$

for varying $\xi^{A'}$. What equation should ω^A satisfy? Now the constancy of λ_A over the β-plane asserts

$$\lambda^A\xi^{A'}\nabla_{AA'}\lambda_B = 0$$

for all $\xi^{A'}$, i.e.

$$\lambda^A\nabla_{AA'}\lambda_B = 0 \quad .$$

The duality with the ω^A-field is expressed in the existence of a scalar product between the ω^A-field and the λ_A-scaled β-plane, which must be an element of $\mathbb{C}$. We take this scalar product to be

$$\omega^A\lambda_A,$$

and the fact that it is a single element of $\mathbb{C}$ requires that this be constant over the β-plane:

$$\lambda^A\xi^{A'}\nabla_{AA'}(\omega^B\lambda_B) = 0$$

for all $\xi^{a'}$. Using the above relation we obtain $\lambda^A\lambda_B\nabla_{AA'}\omega^B = 0$, i.e.

$$\lambda_A\lambda_B(\nabla^{(A}_{A'}\omega^{B)}) = 0.$$

This must hold for all β-planes and thus for all λ_A, so the bracketed expression $\nabla^{(A}_{A'}\omega^{B)}$ must indeed vanish, which is the required twistor equation.

4. The Geometry and Physics of Twistors

The α-plane and twistor equation concepts of a twistor lead naturally in somewhat different directions. First, the α-plane concept leads us directly into various geometrical considerations. We have seen that in the real space $\mathbb{M}$, the α-plane definition provides the picture of a (null) twistor as a (scaled) light ray. Also, in complex terms, the space $\mathbb{CM}^{\#}$ is seen to be the Klein representation of projective lines in projective twistor space $\mathbb{PT}$ (a complex projective 3-space). Points in $\mathbb{PT}$ correspond to β-planes in $\mathbb{CM}^{\#}$. Lines in $\mathbb{PT}$ correspond to points in $\mathbb{CM}^{\#}$. Pairs consisting of planes in $\mathbb{PT}$ with single points marked on them correspond to null geodesics in $\mathbb{CM}^{\#}$ - the intersections of α-planes with β-planes. The complex conjugate of an α-plane is a β-plane. Thus, in $\mathbb{PT}$, the complex conjugate of a point is a plane. For a null twistor, the point lies on the plane, and there is a corresponding real null geodesic in $\mathbb{M}^{\#}$.

The definition in terms of the twistor equation leads more directly to connections with physics. In the first place, any solution ω^A of the twistor equation acts as a <u>spin-lowering</u> operator for massless fields. If $\varphi_{AB\ldots L} = \varphi_{(AB\ldots L)}$, with n indices (i.e. spin n/2), satisfies the massless free-field equation

$$\nabla^A_{A'}\varphi_{AB\ldots L} = 0$$

then so does

$$\omega^L\varphi_{AB\ldots KL},$$

with n-1 indices (i.e. spin (n-1)/2), since

$$\nabla^A_{A'}(\omega^L\varphi_{AB\ldots KL}) = \varphi_{AB\ldots KL}\nabla^{(A}_{A'}\omega^{L)} = 0.$$

Similarly ω^A can be used as a <u>spin-raising</u> operator since, by a direct calculation

$$\bar{\omega}^{M'}\nabla_{MM'}\varphi_{AB\ldots L} - i(n+1)\bar{\pi}_{(M}\varphi_{AB\ldots L)}$$

also satisfies the massless free-field equation.

The twistor equation plays yet another role: a twistor has an intepretation as providing the <u>angular momentum</u> $M^{ab}(= -M^{ba})$ and <u>momentum</u> p_a of a (classical) massless particle. The conditions that p_a be future-null and proportional to the Pauli-Lubański spin vector

$$S_a = 1/2\ e_{abcd}p^bM^{cd},$$

according to

$$S_a = sp_a,$$

can be encoded in the fact that, for some ω^A and $\pi_{A'}$,

$$p_a = \bar{\pi}_A \pi_{A'}$$

and

$$M^{ab} = i\,\omega^{(A}\bar{\pi}^{B)}\epsilon^{A'B'} - i\,\epsilon^{AB}\,\bar{\omega}^{(A'}\pi^{B')} \quad .$$

We regard these as field quantities, the field-point being the origin about which the angular momentum is being taken (p_a being, of course, constant). Taking $\pi_{A'}$ as likewise constant (and non-zero) we find that the required position dependence

$$M^{ab} = \mathring{M}^{ab} - x^a p^b + x^b p^a$$

is precisely equivalent to the position-dependence of ω^A as a solution of the twistor equation, as given above, with the same $\pi_{A'}$, i.e. to

$$\nabla_{AA'}\omega^B = -i\epsilon_A{}^B\pi_{A'} .$$

It may be remarked that the principal null directions of M^{ab} are the flagpole directions of ω^A and $\pi_{A'}$. For $\pi_{A'}$, the flagpole direction is constant, being that of the 4-momentum, while for ω^A the directions are those of a Robinson congruence - whose projections into a Euclidean 3-space of constant time are tangent to the family of circles of a stereographically projected system of Clifford parallels on S^4.

For a general physical system with 4-momentum p_a and angular momentum

$$M^{ab} = \bar{\mu}^{AB}\epsilon^{A'B'} + \epsilon^{AB}\mu^{A'B'} ,$$

the dependence on position (i.e. on choice of origin) is still twistorial, being defined by that of the spinor parts of the angular momentum twistor

$$A_{\alpha\beta} = \begin{pmatrix} 0 & p_A{}^{B'} \\ p^{A'}{}_B & -i\mu^{A'B'} \end{pmatrix}$$

This is subject to the Hermiticity condition

$$A_{\alpha\beta} I^{\beta\gamma} = \bar{A}^{\gamma\beta} I_{\beta\alpha}$$

where the infinity twistors $I^{\alpha\beta}$, $I_{\alpha\beta}$ are defined by

$$I^{\alpha\beta} = \begin{pmatrix} \epsilon^{AB} & 0 \\ 0 & 0 \end{pmatrix} , \qquad I_{\alpha\beta} = \begin{pmatrix} 0 & 0 \\ 0 & \epsilon^{A'B'} \end{pmatrix}$$

and where the rules for forming the complex conjugate of a higher-valence twistor follow from that for a twistor of valence one, whereby complex conjugation sends

$$Z^\alpha = (\omega^A, \pi_{A'})$$

to the dual twistor

$$\bar{Z}_\alpha = (\bar{\pi}_A, \bar{\omega}^{A'}).$$

The corresponding α-plane is sent to its complex conjugate β-plane. We note that the rest-mass m satisfies

$$m^2 = -1/2\ A_{\alpha\beta}\bar{A}^{\alpha\beta}.$$

The Hermitian form $Z^\alpha\bar{Z}_\alpha$ has signature (++--) and in fact has the physical interpretation

$$Z^\alpha\bar{Z}_\alpha = 2s,$$

where the helicity s is the proportionality factor in $S_a = sp_a$. When $s = 0$, Z^α is a null twistor and, as we have seen, the α-plane determined by Z^α contains a real null straight line: the world-line of the massless particle whose momentum and angular momentum are p_a and M^{ab}. When $s \neq 0$ no such unique world-line exists. When $\pi_{A'} = 0$ (and $\omega^A \neq 0$) we get a limiting world-line which is a generator of $\mathscr{I}$.

5. Twistor Concepts for Curved Space-Time

It has been remarked that the twistor equation is conformally invariant, i.e. under conformal rescaling

$$g_{ab} \mapsto \hat{g}_{ab} = \Omega^2 g_{ab}$$

solutions ω^A of the twistor equation remain solutions when the Christoffel-Levi Civita connection ∇_a is replaced by the corresponding $\hat{\nabla}_a$. The corresponding $\pi_{A'}$ is modified, however, by the addition of a term involving ω^A, and we have

$$\hat{\omega}^A = \omega^A, \quad \hat{\pi}_{A'} = \pi_{A'} + i\ \omega^A\Omega^{-1}\nabla_{AA'}\Omega.$$

This shows us how to extend the flat-space twistor concept to any conformally flat curved space-time - at least in any simply-connected region.

For a general curved space-time $\mathcal{M}$, the situation is much less clear. Several different concepts of twistor have arisen, e.g. the local twistor, 2-surface twistor, 3-surface twistor, hypersurface twistor and asymptotic twistor concepts. None of these is universally satisfactory, but each is of value within the context in which it applies. In order to systematize this somewhat bewildering multiplicity, I have classified these various concepts in Table 1. In each case the concept arises as an adaptation of either the "α-plane definition" or else the "twistor equation definition" of a twistor, the two columns distinguishing these two different approaches.

Table 1

Submanifold referred to	α-plane definition	twistor equation definition
point	—	local twistor: $(\omega^A, \pi_{A'})$ at point $\hat{\omega}^A = \omega^A,\ \hat{\pi}_{A'} = \pi_{A'} + i\,\omega^A \Omega^{-1} \nabla_{AA'}\Omega$
(open) curve (tangent t^a)	—	local twistor transport: $t^a \nabla_a \omega^B = -i\, t^{BA'} \pi_{A'}$ $t^a \nabla_a \pi_{B'} = -i\, t^a P_{aBB'} \omega^B$ $(P_{ab} = \frac{1}{12} R\, g_{ab} - \frac{1}{2} R_{ab})$
2-surface $\mathcal{S}$ (spacelike, $\cong S^2$)	Spinor $\pi_{A'}$ at a point of $\mathbb{C}\mathcal{S}$	2-surface twistor: solution of $ð'\omega^0 = \sigma'\omega^1,\ ð\omega^1 = \sigma\omega^0$
hypersurface $\mathcal{H}$ (normal n^a)	hypersurface twistor: α-curve, with $\pi_{A'}$-scaling, on $\mathbb{C}\mathcal{H}$ $\pi_{A'} n^{AA'} \pi_{B'} \nabla_A{}^{B'} \pi_{C'} = 0$ asymptotic twistor: special case when $\mathcal{H} = \mathscr{I}$.	3-surface twistor: solution of $n^{A'(A} \nabla^B{}_{A'} \omega^{C)} = 0$ (needs magnetic part of Weyl tensor and its normal derivative both to vanish at $\mathcal{H}$)
whole space-time (complex)	"non-linear graviton construction" requires $\tilde{\Psi}_{A'B'C'D'} = 0$ $(C_{abcd} = \Psi_{ABCD}\varepsilon_{A'B'}\varepsilon_{C'D'} + \varepsilon_{AB}\varepsilon_{CD}\tilde{\Psi}_{A'B'C'D'})$	googly??

The various rows of the table refer to the different dimensions of submanifolds with respect to which the twistor definition refers. With respect to a point, the twistor equation definition provides the concept of a local twistor $Z^\alpha = (\omega^A, \pi_{A'})$ which is replaced, under conformal rescaling of the space-time metric, by $Z^\alpha = (\hat{\omega}^A, \hat{\pi}_{A'})$ as given by the above expressions. With respect to an (open) curve, the twistor equation reduces to local twistor transport of local twistors along the curve, which is a conformally invariant propagation. The definition is such as to agree with the behaviour of the ω^A and $\pi_{A'}$ fields in the case when $\mathcal{M}$ is conformally flat. Without conformal flatness, local twistor transport is not integrable, so the twistor concept does not globalize throughout $\mathcal{M}$. The local twistor curvature involves both the Weyl tensor and its (contracted) covariant derivative. Local twistors can be used to provide a Yang-Mills theory for conformal gravity, the Yang-Mills current being the Bach tensor (see Merkulov 1984). They also lead to a systematic procedure for generating conformally invariant objects and operations in four-dimensional space-time manifolds.

Neither with respect to a point nor a curve do we get an α-plane concept of a twistor, since the dimensions are such that an α-plane would, in general, not intersect the given submanifold. However, with respect to a submanifold $\mathcal{S}$ of dimension two, the α-plane concept of a twistor becomes relevant. For an analytic $\mathcal{S}$ in $\mathbb{M}$, its complexification $\mathbb{C}\mathcal{S}$ will (generally) meet some open family of α-planes lying in $\mathbb{C}\mathbb{M}$. (The symbol "$\mathbb{C}$" used in this way always means "complexified". Here $\mathbb{C}\mathcal{S}$ would be some complex "thickening" of $\mathcal{S}$ in $\mathbb{C}\mathbb{M}$, i.e. a complex two-dimensional submanifold of $\mathbb{C}\mathbb{M}$ whose intersection with $\mathbb{M}$ is $\mathcal{S}$.) At any point of intersection of $\mathbb{C}\mathcal{S}$ with an α-plane, the spinor $\pi_{A'}$ would determine the α-plane, so we can think of a twistor (belonging to the given open family) as being defined by a point of $\mathbb{C}\mathcal{S}$ together with a spinor $\pi_{A'}$ at that point. This definition carries over to an (analytic) surface $\mathcal{S}$ in a general (analytic) space-time $\mathcal{M}$. (Some subtleties arise when the flagpole direction of $\pi_{A'}$ becomes close to being either normal or tangent to $\mathbb{C}\mathcal{S}$, but I shall not go into these here). For a spacelike 2-surface $\mathcal{S}$ (with S^2 topology) the twistor-equation definition will also apply, and it leads us to the definition of a 2-surface twistor. This important case, and its significance for general relativity, will be discussed in the next two sections.

For a three-dimensional analytic hypersurface $\mathcal{H}$ we have a concept of twistor arising both from the α-plane definition and from the twistor-equation definition, providing hypersurface twistors and Tod's (1983,1984) 3-surface twistors, respectively. In complexified (conformally) flat space-time, α-planes exist and these intersect $\mathbb{C}\mathcal{H}$ in complex curves called α-curves defined by the equation

$$\pi^{A'} n^{A'B'} \pi_{B'} \nabla_{AA'} \pi_{C'} = 0,$$

where n^a is normal to $\mathbb{C}\mathcal{H}$ and $\pi^{A'} n^{A'B'} \pi_{B'}$ is tangent to the curve. The α-curve defines the α-plane, and the twistor scaling is fixed by $\pi_{A'}$. In general there are no α-planes in $\mathbb{C}\mathcal{M}$, but we can still use α-curves (with $\pi_{A'}$-scaling) to provide a twistor concept - in fact, the hypersurface twistor concept (Penrose 1975, Penrose and Rindler 1986). The space $\mathcal{T}^\bullet(\mathcal{H})$ of hypersurface twistors for $\mathcal{H}$ is a complex manifold, but it does not generally have a linear structure. This is a general feature of the α-plane definition. The twistor-equation definition, on the other hand, always provides a linear structure, but only applies to certain special 3-surfaces, namely those in which the magnetic part of the Weyl tensor vanishes and the normal derivative of this magnetic part also vanishes (this latter being equivalent, in the case of a vacuum space-time, to the condition that the curl of the electric part vanishes). These are the integrability conditions for the 3-surface twistor equation

$$n^{A'(A} \nabla^{B}_{A'} \omega^{C)} = 0,$$

this equation being the restriction of the twistor equation to directions lying within $\mathcal{H}$ (see Tod 1983). When a complete 4-parameter family of solutions (3-surface twistors) exists, we can identify the hypersurface twistors in terms of the zero-loci of 3-surface twistors. More directly, we can see that the complex conjugates of hypersurface twistors are naturally dual to solutions of the 3-surface twistor equation, and this leads to a linear structure to the space of hypersurface twistors in this case.

Despite the lack of linear structure, in the general case, hypersurface twistors are still of value. As the hypersurface $\mathcal{H}$ is moved within $\mathcal{M}$, the structure of $\mathcal{T}^\bullet(\mathcal{H})$ evolves. This can be related to the initial value problem and Hamiltonian evolution for Einstein's equations (Mason 1986 cf. also LeBrun 1984). The definition of hypersurface twistors is conformally invariant, and so it can be applied to $\mathscr{I}$ in the case when $\mathcal{M}$ is (appropriately) asymptotically flat. The twistors so obtained are called <u>asymptotic twistors</u> (Penrose and MacCallum 1972). In this case the α-curves become null geodesics on $\mathbb{C}\mathscr{I}$.

When applied to the whole space-time $\mathcal{M}$, neither the α-plane definition nor the twistor-equation definition directly applies in general. Defining an α-plane (or "α-surface") in $\mathbb{C}\mathcal{M}$ to be a complex 2-surface whose tangent vectors at any one point all have the form $\lambda^A\pi^{A'}$ for fixed $\pi^{A'}$ and varying λ^A, we find by a Lie bracket argument, that

$$\tilde{\Psi}_{A'B'C'D'}\pi^{A'}\pi^{B'}\pi^{C'}\pi^{D'} = 0.$$

Thus, if there is an α-plane corresponding to each $\pi_{A'}$ through each point of $\mathbb{C}\mathcal{M}$ (giving a complex three-parameter family of α-planes in all) then we must have

$$\tilde{\Psi}_{A'B'C'D'} = 0$$

throughout $\mathbb{C}\mathcal{M}$, which is to say that $\mathbb{C}\mathcal{M}$ is right-conformally flat. (The complex Weyl curvature is given by $C_{abcd} = \Psi_{ABCD}\epsilon_{A'B'}\epsilon_{C'D'} + \epsilon_{AB}\epsilon_{CD}\tilde{\Psi}_{A'B'C'D'}$ with Ψ_{ABCD} and $\tilde{\Psi}_{A'B'C'D'}$ both symmetric, and describing the anti-self-dual and self-dual parts of the conformal curvature, respetively.) For $\mathbb{C}\mathcal{M}$ to be actually the complexification of a Lorentzian space-time $\mathcal{M}$, this would require $\mathcal{M}$ to be conformally flat. I shall abuse the notation, somewhat, here and simply use $\mathbb{C}\mathcal{M}$ to denote a complex space-time.

For any right-conformally flat $\mathbb{C}\mathcal{M}$, the projective twistor space $\mathbb{P}\mathcal{T}$, whose points represent α-planes in $\mathbb{C}\mathcal{M}$ (considered only locally in $\mathbb{C}\mathcal{M}$), and the non-projective twistor space $\mathcal{T}$, for which the scaling of the $\pi_{A'}$-spinors is also included, are well defined and can be used to reconstruct the (complex) conformal metric of $\mathbb{C}\mathcal{M}$. When $\mathbb{C}\mathcal{M}$ is also Ricci-flat - so $\mathbb{C}\mathcal{M}$ is <u>right flat</u>, i.e. a general <u>anti-self-dual</u> solution of the Einstein fluid equations - the metric itself emerges from the structure of $\mathcal{T}$, which is now strengthened by the existence of a holomorphic fibration

$$\mathcal{T} \to \mathbb{C}^2 - \{0\},$$

(where $\mathbb{C}^2$ stands for the space of $\pi_{A'}$) whose fibres are complex symplectic 2-manifolds (i.e. with a complex area form) with appropriate homogeneity properties. The construction of $\mathcal{T}$ requires the solution of <u>no algebraic or differential equations whatever</u>, but the construction of the points of $\mathbb{C}\mathcal{M}$ from $\mathcal{T}$ requires finding the family of holomorphic cross-sections of the above fibration, each cross-section giving a point of $\mathbb{C}\mathcal{M}$, which may be hard to do explicitly. This is the "non-linear graviton" (or "leg-break") construction (Penrose 1976; see also Ward 1978, Tod and Ward 1979, Hitchin 1979, 1982, for explicit examples, and Ward 1980 for the inclusion of a cosmological constant).

There are now some good indications that a corresponding construction for the self-dual Einstein equations can be found in terms of a twistor space (as opposed to a dual twistor space - which could be trivially obtained by the previous construction merely by a change of conventions). Such a construction is referred to as a "googly" (in view of a cricketing analogy). The googly construction appears to have features in common with a twistor-equation definition rather than an α-plane definition of a twistor, but the matter is not yet wholely clear. The idea is that both the googly and leg-break structures of $\mathcal{T}$ can be imposed simultaneously, so that there is then the possibility of constructing general solutions of the full (i.e. neither anti-self-dual nor self-dual) Einstein equations. All this is work in progress.

6. Quasi-Local Mass and Angular Momentum

Let us return to the case of a 2-surface $\mathcal{S}$, which we take to be spacelike, with S^2 topology and we consider the twistor-equation definition of a twistor. Choosing a spin-frame o^A, ι^A whose flagpole directions are normal to $\mathcal{S}$, the required tangential part of the twistor equation becomes

$$\eth'\omega^0 = \sigma'\omega^1, \quad \eth\omega^1 = \sigma\omega^0.$$

It turns out (by an argument involving the Atiyah-Singer index theorem) that there are always four independent solutions to this pair of equations and, in general, no more than four. (Jeffryes has constructed examples where there are more than four, but it is not known whether this can ever happen in physically realistic situations cf. also Jeffryes 1984.) When four-dimensional, the solution space of these equations is referred to as the 2-surface twistor space $\mathbb{T}^\alpha(\mathcal{S})$, and a solution ω^A is referred to as a 2-surface twistor on $\mathcal{S}$, denoted $Z^\alpha \in \mathbb{T}^\alpha(\mathcal{S})$.

The main significance of 2-surface twistors is that they can be used as spin-lowering operators on $\mathcal{S}$ which lower the "spin 2" curvature tensor, in two steps, to a "spin 1" Maxwell-type field on $\mathcal{S}$. This leads to a definition of the total mass-momentum and angular momentum within $\mathcal{S}$

in terms of "charge integrals" for the Maxwell-type field (Penrose 1982, 1984, Penrose and Rindler 1986). This complex of quantities constitutes the <u>2-surface angular momentum twistor</u> $A_{\alpha\beta} \in \mathbb{T}_{(\alpha\beta)}(\mathscr{S})$ given by

$$A_{\alpha\beta} \underset{1}{Z}^{\alpha} \underset{2}{Z}^{\beta} = \frac{-1}{16\pi} \oint \eta \, R_{abCC'DD'} \underset{1}{\omega}^{C} \underset{2}{\omega}^{D} \epsilon^{C'D'} dx^{a} \wedge dx^{b}$$

integrated over $\mathscr{S}$, where η varies over $\mathscr{S}$ in direct proportion to the determinant

$$\begin{vmatrix} \underset{1}{\omega}^{0} & \underset{2}{\omega}^{0} & \underset{3}{\omega}^{0} & \underset{4}{\omega}^{0} \\ \underset{1}{\omega}^{1} & \underset{2}{\omega}^{1} & \underset{3}{\omega}^{1} & \underset{4}{\omega}^{1} \\ \eth\underset{1}{\omega}^{0} & \eth\underset{2}{\omega}^{0} & \eth\underset{3}{\omega}^{0} & \eth\underset{4}{\omega}^{0} \\ \eth'\underset{1}{\omega}^{1} & \eth'\underset{2}{\omega}^{1} & \eth'\underset{3}{\omega}^{1} & \eth'\underset{4}{\omega}^{1} \end{vmatrix}$$

The numbers written beneath the symbols Z and ω refer to (corresponding) different arbitrarily chosen 2-surface twistors. The η-factor can be taken to be unity in many circumstances (i.e. wherever the determinant is constant over $\mathscr{S}$), but in the general case there is an uncertainty as to the appropriate way in which η should be normalized (corresponding to the most appropriate choice of 2-surface Levi-Civita twistor $\epsilon_{\alpha\beta\gamma\delta}$). There is an uncertainty, also, as to the most appropriate definition for a complex conjugation map

$$Z^{\alpha} \mapsto \bar{Z}_{\alpha}$$

and for an infinity twistor $I^{\alpha\beta} = -I^{\beta\alpha}$. (Such an infinity twistor would have to be of a "de Sitter" rather than a "Poincaré" type, in general, i.e. it would have to have matrix rank four rather than two.) Modulo such uncertainties we can define the total rest-mass within $\mathscr{S}$ using the formula $m^2 = -1/2\, A_{\alpha\beta}\bar{A}^{\alpha\beta}$ as given above (or else using $m^4 = 1/6\, A_{\alpha\beta}A_{\gamma\delta}A_{\lambda\mu}A_{\rho\sigma}\epsilon^{\alpha\gamma\lambda\rho}\epsilon^{\beta\delta\mu\sigma}$, an alternative suggested by Tod).

For an asymptotically flat space-time, the quasi-local 4-momentum reduces, at space-like infinity, to that of Arnowitt, Deser and Misner, and the quasi-local angular momentum reduces to that of Ashtekar and Hansen (Shaw 1983, 1984). At null infinity the quasi-local 4-momentum reduces to that of Bondi and Scahs, whereas the quasi-local angular momentum provides a new and rather more satisfactory definition than had been obtainable previously (Penrose 1982, Penrose and Rindler 1986). it has still not proved possible to find a completely satisfactory flux law for angular momentum at null infinity, however (Shaw 1984).

7. Examples of Quasi-Local Mass

The quasi-local mass has been calculated for a number of particular examples, mainly by Tod (1983, 1985) who has employed the device of embedding $\mathscr{S}$ in a conformally flat space-time in such a way that the induced metric on $\mathscr{S}$ remains unchanged and so do the extrinsic curvature quantities at $\mathscr{S}$. (These extrinsic quantities are simply the spin-coefficients $\sigma, \sigma', \rho, \rho'$ at $\mathscr{S}$.) A surface $\mathscr{S}$ for which it is possible to do this is called uncontorted; otherwise $\mathscr{S}$ is called contorted. Uncontorted surfaces have the advantage that we can choose

$\eta = 1$, since the relevant determinant is always constant over $\mathscr{S}$. Moreover, the complex conjugate map $Z^{\alpha} \mapsto \bar{Z}_{\alpha}$ is naturally defined in these cases, so that m^2 aquires an unambiguous definition. Calculations are facilitated by the fact that the 2-surface twistors are all induced from solutions of the twistor equation in the conformally flat embedding space.

In the Schwarzschild space-time, any 2-surface $\mathscr{S}$ lying in a hypersurface of revolution

$$t = f(r)$$

is uncontorted (this hypersurface admitting 3-surface twistors). It turns out $m = 0$ if such an $\mathscr{S}$ does not link the source, while m becomes precisely

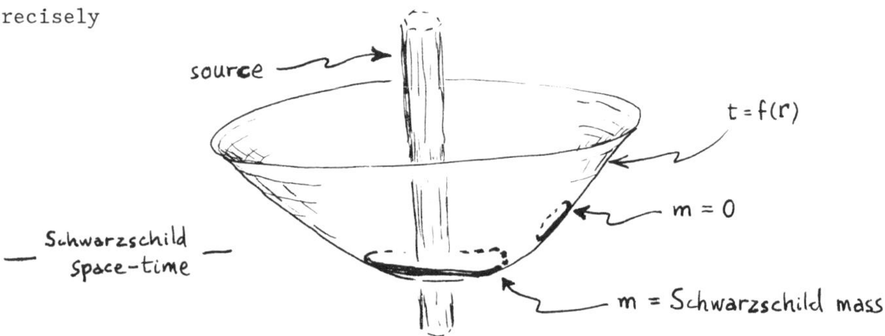

the Schwarzschild mass whenever $\mathscr{S}$ does link the source (once). For the Reissner-Nordstrom space-time a similar result holds except that there is a correction to the mass (as expected) arising from the electromagnetic energy density (Tod 1983).

Another of Tod's examples arises when $\mathscr{S}$ lies in a spacelike hypersurface $\mathscr{H}$ which is a spatially conformally flat moment of time-symmetry. Suppose that on $\mathscr{H}$, there are localized regions of matter $\mathscr{R}_1$, $\mathscr{R}_2$, ... (or else wormholes connecting $\mathscr{H}$ to a second copy of itself). Again $\mathscr{S}$ is uncontorted (and $\mathscr{H}$ admits 3-surface twistors) and, for $\mathcal{S}$ entirely in the vacuum region, m turns out to be a function only of the topological linking that $\mathcal{S}$ has with the various source regions. Moreover

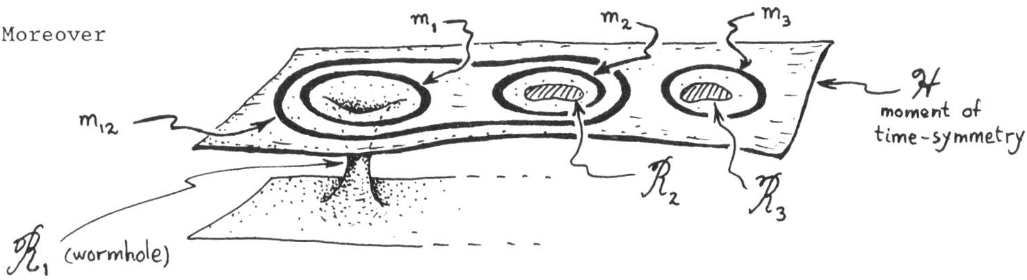

if m_1, m_2 and m_{12} are the respective masses obtained when the regions $\mathcal{R}_1$, $\mathcal{R}_2$ and $\mathcal{R}_1 \cup \mathcal{R}_2$ are surrounded, then the physically reasonable inequality

$$m_{12} < m_1 + m_2$$

is obtained. In the limit of large separations the difference

$$m_{12} - m_1 - m_2$$

agrees with the expected negative contribution from the Newtonian potential energy. This is very gratifying, and it is further reassuring that (as Jeffryes 1986 has recently shown) in the Newtonian limit of a general system, the quasi-local mass expression correctly incorporates the mass due to Newtonian potential energy.

Tod has also shown that for a static "black hole in a cage" - where the cage is to distort the black hole out of spherical shape - with $\mathcal{S}$ chosen at the hole's horizon, of area A, there is an inequality $m \leqslant (A/16\pi)^{1/2}$, equality holding only in the spherically symmetrical case. This also seems physically reasonable, there being an energy due to the distortion of the hole out of spherical shape.

Recently, Tod (1986) (and also Jeffryes and Reula) have examined cylindrically symmetrical space-times, where $\mathcal{S}$ is a cylinder with flat ends. A dependence of the mass on the length ℓ of the cylinder is found which falls into three different categories:

Hyperbolic: exponential dependence on ℓ

Elliptic: sinusoidal dependence on ℓ

Parabolic: linear dependence on ℓ.

At first, such dependence seems puzzling, particularly in the elliptic case, since the mass would seem to go negative after a critical length is reached. However, these results may be compared with earlier calculations by Tod concerning the Friedman-Robertson-Walker models (Tod 1983). There a similar behaviour was found and taken to be "physically reasonable". The puzzling elliptic case now occurs with k = +1 models, but here the cylinder $\mathcal{S}$ becomes an "equatorial" torus (cut at one place) just at the

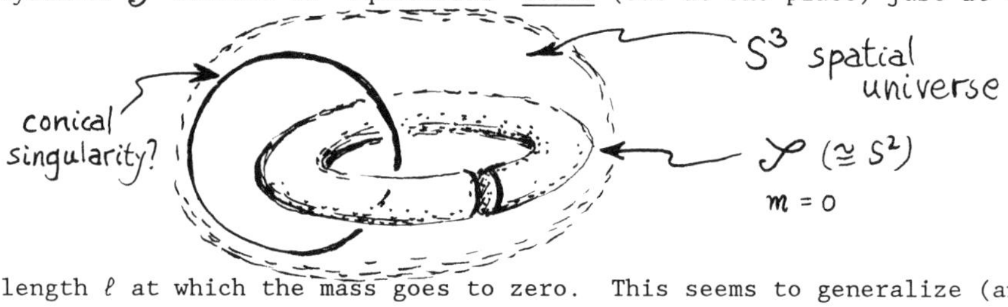

length ℓ at which the mass goes to zero. This seems to generalize (and be compatible with) the result that the total mass of a closed universe should be zero.

One might envisage a similar situation but where the mass is allowed to become negative, by introducing a conical singularity linking the torus, so that the mass goes negative before the torus closes. However, this introduces negative mass outside $\mathcal{S}$, which must be regarded as physically unreasonable. There is, however, a moral to this which is a warning to anyone attempting to prove a positivity theorem for quasi-local amss: the space must be physically reasonable both inside and outside of $\mathscr{S}$.

Finally, the calculations of Kelly, Tod and Woodhouse (1986) concerning m for small spheres and ellipsoids may be mentioned. If $\mathscr{S}$ is such a small ellipsoid, its linear size being characterized by a parameter r, they find that at order r^{-3} there is the expected contribution from the energy-momentum tensor times volume. In vacuum the first possibility of a non-zero contribution occurs at order r^{-5}. Without the η-factor in the quasi-local mass definition anomalies arise ($m^2 < 0$) when $\mathscr{S}$ is contorted. But these miraculously vanish when η is included. Further work is in progress.

REFERENCES

Aspect, A., Grangier, P. & Roger, G. (1981). Phys. Rev. Lett. 47, 460

Freedman, S.J. & Clauser, J.F. (1972). Phys. Rev. Lett. 28, 938

Hitchin, N.J. (1979). Polygons and gravitons, *Math. Proc. Camb. Phil. Soc.* 85, 465-76

Hitchin, N.J. (1982). Complex manifolds and Einstein's equations, in *Twistor Geometry and Non-Linear Systems* ed. H.D. Doebner and T.D. Palev (Springer-Verlag, Berlin), pp. 73-99

Huggett, S. & Tod, K.P. (1985). Introduction to Twistor Theory, London Math. Soc. Lecture Note Series (Cambridge University Press, Cambridge)

Jeffryes, B.P. (1984b). Two-surface twistors and conformal embedding in *Asymptotic Behaviour of Mass and Spacetime Geometry* , Proc. Corvalis, Oregon 1983, ed. F.J. Flaherty (Springer-Verlag, Berlin)

Kelly, R., Tod, K.P. & Woodhouse, N.M.J. (1986). Classical and Quantum Gravity (to appear)

LeBrun, C.R. (1984). Twistor CR-manifolds and three-dimensional conformal geometry, *Trans. Amer. Math. Soc. 278, 209-31*

Mason, L. (1986). Twistors in curved space time, D.Phil. thesis, Oxford

Merkulov, S.A. (1984). A conformally invariant theory of gravitation and electromagnetism, Class. Quantum Grav. , 1, 349-54

Penrose, R. (1975). Twistor theory: its aims and achievements, in *Quantum Gravity, an Oxford Symposium* , eds. C.J. Isham, R. Penrose and D.W. Sciama (Oxford University Press, Oxford)

Penrose, R. (1976). Nonlinear gravitons and curved twistor theory, *Gen. Rel. Grav.* 7,, 31-52; The non-linear graviton, ibid., 171-6

Penrose, R. (1982). Quasi-local mass and angular momentum in general relativity, *Proc. Roy. Soc. London* A381, 53-63

Penrose, R. (1984). Integrals for general-relativistic sources: a development from Maxwell's electromagnetic theory, in *Maxwell Symposium Volume* , ed. M.S. Berger (North-Holland, Amsterdam)

Penrose, R. & MacCallum, M.A.H. (1972). Twistor theory: an approach to to the quantization of fields and space-time, *Phys. Repts.* 6C, 241-315

Penrose, R. & Rindler, W. (1984). Spinors and Space-Time, Vol. 1, Two-Spinor Calculus and Relativistic Fields (Cambridge University Press, Cambridge)

Penrose, R. & Rindler, W. (1986). Spinors and Space-Time Vol. 2, Spinor and Twistor Methods in Space-Time Geometry (Cambridge University Press, Cambridge)

Shaw, W.T. (1983). Twistor theory and the energy-momentum and angular momentum of the gravitational field at spatial infinity, *Proc. Roy. Soc. London* A390, 191-215

Shaw, W.T. (1984). Twistors, asymptotic symmetries and conservation laws at null and spatial infinity, in *Asymptotic Behaviour of Mass and Spacetime Geometry* , Proc. Corvalis, Oregon 1983, ed. F.J. Flaherty (Springer-Verlag, Berlin)

Tod, K.P. (1983). some examples of Penrose's quasi-local mass construction, *Proc. Roy. Soc. London* A388, 457-77

Tod, K.P. (1984). Three-surface twistors and conformal embedding, *Gen. Rel. Grav.* 16, 435-43

Tod, K.P. and Ward, R.S. (1979). Self-dual metrics with self-dual Killing vectors, *Proc. Roy. Soc. London* A368, 411-27

Ward, R.S. (1978). A class of self-dual solutions of Einstein's equations, *Proc. Roy. Soc.* A363, 289-95

Ward, R.S. (1980). Self-dual space-times with cosmological constant, *Comm. Math. Phys.* 78, 1-17

NUMERICAL RELATIVITY FROM GRAVITATIONAL RADIATION TO COSMOLOGY

Tsvi Piran

The Racah Institute for Physics, The Hebrew University, Jerusalem, Israel.
and The Institute for Advanced Study, Princeton, NJ, USA.

Abstract

I describe six new general relativistic numerical calculations : one dimensional relativistic stellar dynamics, spherical collapse of a massless scalar field, inflationary cosmology using Regge calculus, rotating collapse and emission of gravitational radiation, evolution of three dimensional gravitational wave packet and three dimensional relativistic hydrodynamics around a Kerr black hole. These solutions demonstrate the recent developments in numerical relativity i.e: construction of accurate and well tested codes, adaptation of new numerical techniques, application of the calculations to a variety of unsolved problem in general relativity and the existence of fully relativistic three dimensional codes.

Introduction

Three themes characterize the recent progress in Numerical Relativity: 1) Development of accurate codes, 2) proliferation of new numerical solutions and formalisms and 3) potential for investigating the most general i.e. three dimensional, time dependent general relativistic problems. In this talk I try to demonstrate these developments with six calculations (I refer the reader to the proceedings of the Drexel conference on Numerical Relativity (Centrella, 1986) for discussion of other recent work that is not reviewed here). These solutions represent new techniques and new formalisms, as well as new physical results. I begin with two one dimensional spherical solutions (one space dimension as well as time). I continue with a Regge calculus calculation, which is in some sense three dimensional. Later I discuss a two dimensional solution and I conclude with two three dimensional solutions.

Relativistic Stellar Cluster

The fate of a relativistic stellar cluster has intrigued relativists since the thirties (Einstein, 1939). Interest rose in the mid sixties with the discovery of QSO's and the work of Zel'dovich and Podurets (1965). At first it was speculated that the high redshift observed in Qsos might be the central redshift in a relativistic stellar cluster. Later it became clear that the redshift is cosmological. Still the fate of such a cluster is interesting as it provides an excellent candidate for a progenitor to the massive black holes that are believed to exist in Qsos' cores. Zel'dovich and Podurets (1965) suggested that the combined effects of gravitational collapse of stars to compact objects and star-star collisions and coalescence will drive a stellar cluster towards higher and higher central densities. The cluster becomes unstable at high central density. A central supermassive black hole evolves then, in what is called a gravitational catastrophe, within a few dynamic time scales. This instability was first demonstrated by Ipser and Thorne (1968) using perturbation calculations. However only a full nonlinear calculation could reveal the actual fate of the cluster and in particular could answer questions like what is the final mass of the black hole or what is the duration of the collapse.

Fig. 1: *Comparison of the lapse function in Oppenheimer Snyder homogeneous dust collapse calculated analytically (solid line) and numerically (dots), (from ST, 1985b).*

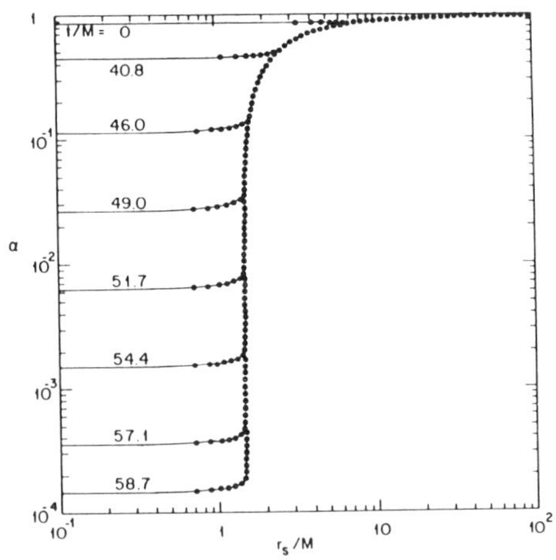

In pioneering work, Shapiro and Teukolsky (1985a,b,c 1986a,b denoted later ST) have combined the Einstein equations with an N body type of calculation to follow the evolution of a relativistic cluster. The Einstein equations were written

Fig. 2: *Binding energy and stability for a one parameter power law sequence of stellar clusters. Open circles mark stability, black circles mark instability, (from ST, 1985c).*

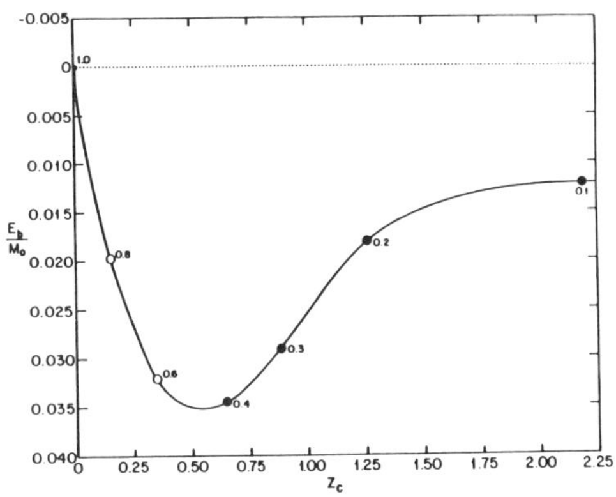

using the 3+1 formalism where several coordinate systems have been tried. The code was subjected to a long series of test bed calculations (ST, 1985b). Fig. 1 shows one of these tests, which compares analytic results of Oppenheimer Snyder homogeneous dust collapse with numerical calculations. As in many other cases a large part of the work on the test was devoted to conversion of the analytic results to a form which could be compared with the numerical ones. In this case Petrich et al, (1985) transformed the analytic results from the Oppenheimer Snyder coordinates to a coordinate system, based on maximal slicing that was used in the numerical calculations.

Ipser's (1980) calculations suggested that the first maximum of the binding energy graph is a sufficient criterion for stability. This means that for a one parameter family of solutions all solutions which are before that maximum are stable. Trial function calculations cast doubt on the conjecture that this is also a necessary condition. The numerical calculations (ST, 1985c) have demonstrated, without any doubt, that this criterion is indeed sufficient. An example is shown in Fig. 2. Within the unstable region, in most of the cases, the instability led to a collapse into a black hole. However, somewhat unexpectedly, in some cases an unstable configuration has expanded and has settled down after a relaxation period as a new stable configuration.

One might expect that when the initial density profile is very centrally concentrated only the inner dense region will collapse into a black hole. In their original paper Zel'dovich and Podurets (1965) conjectured that large black holes will form even in this situation. They have envisioned an "avalanche type contraction", in which stars from the outer regions that move on highly eccentric orbits are captured when they pass near the central core, increasing its mass and leading to capture of other envelope stars. Again the numerical calculations have demonstrated this conjecture and have shown that black holes, that contained a significant fraction of the initial cluster, can form even when the initial core contains only a small fraction of the total mass.

These results, demonstrate the potential importance of relativistic stellar clusters as progenitors of massive black holes. It is interesting to remark that the large number of coalescences of various compact objects during such a collapse might provide an interesting cosmological source of gravitational radiation. However, the present calculations embody spherical symmetry so that gravitational radiation can not yet be calculated.

Characteristic Methods, Scalar Field and Cosmic Censorship

One of the possible approaches to Numerical Relativity is the characteristic method where we solve Einstein's equations along null directions. The advantages of this approach are clear, the partial differential equations become ordinary ones. Furthermore the solution is integrated easily all the way to null infinity (see for example the work of Isaacson, Welling and Winicour (1983, 1986) and of Corkill and Stewart, (1983)). A major unsolved problem with this approach is the potential appearance of caustics. Stewart and Friedrich (1982) Friedrich and Stewart (1983) and Stewart (1984) have classified these singularities and proposed a way to integrate the equations across them. So far this method has not been implemented in an actual solution.

The caustic problem disappears under cylindrical or spherical symmetry. Piran, Safier and Stark (1985), have solved numerically the general cylindrical vacuum problem using a characteristic scheme. The solution revealed a new phenomenon of scattering and rotation of the polarization vector of one gravitational radiation pulse by another.

There are no spherical gravitational waves. Wave propagation can be studied however, under spherical symmetry, for a massless scalar field. This motivated Choptuik (1986a,b) and Choptuik and Unruh (1986) to study, using a 3+1 formalism, the evolution of spherical massless scalar field. Choptuik has developed and tested different second order schemes, demonstrating that the supposed inconsistency of the constraints and evolution equations in numerical relativity can be eliminated. In particular he was able to demonstrate convergence of the constraints to second order while using a second order free evolution scheme.

Christodoulou has studied analytically spherical collapse of a massless scalar field (1986). He has shown that for a weak initial field there exist a regular solution at arbitrary long times, providing the first proof of existence of long time solutions to Einstein equations. A weak initial scalar field converges inwards, bounces and disperses to infinity. One expects a gravitational collapse when the initial field is stronger. However, we do not know what forms a black hole or a naked singularity. This question is, of course, a specific case of the cosmic censorship hypothesis.

In an unpublished work Christodoulou (1983) has shown that if the initial scalar field, ϕ ,specified at $u = -1$, satisfies the equations:

$$r(\frac{1}{2}\bar{g} - r)\frac{d\theta}{dr} = -[\frac{1}{2}(g + \bar{g}) - 2r]\theta + \kappa \quad , \tag{1a}$$

$$g(r) = exp[4\pi \int_0^r \tilde{r}\theta^2 d\tilde{r}] \quad , \tag{1b}$$

$$\bar{g}(r) = \frac{1}{r}\int_0^r g(\tilde{r})d\tilde{r} \quad , \tag{1c}$$

and

$$\phi = -\int_r^\infty \theta(\tilde{r})d\tilde{r} \quad , \tag{1d}$$

its future evolution satisfies:

$$\phi(r, u) = \phi(r/|u|, -1) - \kappa ln(|u|) \quad . \tag{2}$$

A numerical solution of Eqs. 1a-d (with $\kappa = \sqrt{(\pi)}$) is shown in Fig. 3.

Fig. 3: *A numerical solution of Eqs. 1a-d for* $\kappa = \sqrt{(\pi)}$, *the solution ranges from 0 to* $r_0 = 0.5213014$ *(from GP).*

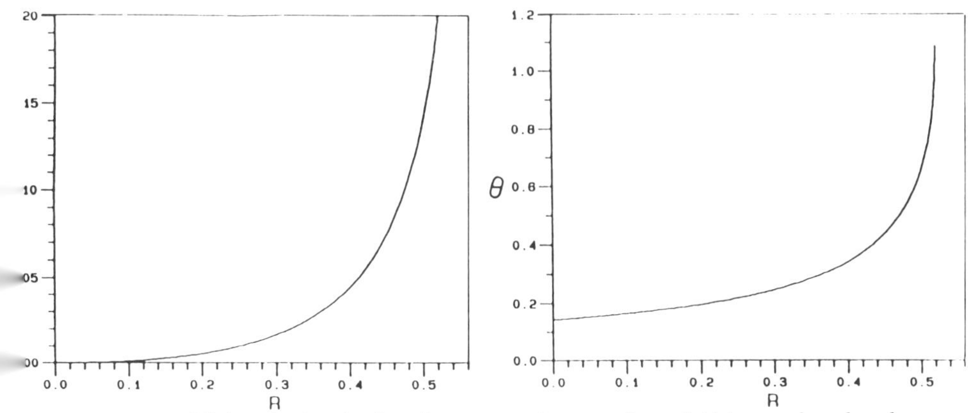

This solution is singular at $r = 0$, $u = 0$, and it is regular elsewhere, so a black hole does not form before $u = 0$. It seems that this solution provides a counter example to the cosmic censorship conjecture. However, Eqs. 1a-d specify the initial data only for $0 \leq r \leq r_0$, with the lhs of Eq. 1a vanishing at r_0. This solution, which we call the interior solution, describes the spacetime only in the domain of dependence of the initial region $\{0 \leq r \leq r_0, u = -1\}$. To obtain a global solution one has to specify initial data on $r > r_0$ and worry whether a black hole forms before $u = 0$ in the exterior region.

It is important to realize that with Christodoulou's construction it is impossible to specify the vacuum as exterior initial data. The initial value equation yields $\phi(r_0) \neq 0$. Specifying , for example, $\phi = 0$ for $r > r_0$ gives rise to a delta function energy distribution just at r_0. One must supplement the solution of Eqs. 1a-d with exterior initial data at $r > r_0$ whose evolution might not be trivial. The global structure of the solution depends on the region exterior to the null geodesic that passes through $(r_0, u = -1)$. In particular if a black hole forms at the exterior before $u = 0$, the singularity at r = 0 and $u = 0$ is not naked.

Goldwirth (1986) and Goldwirth and Piran (1986) (denoted later GP) have constructed a characteristic code for spherical evolution of a scalar field. The evolution equations (see GP for a complete set of the equations) are rewritten as a set of coupled ordinary differential equations (one equation for each null trajectory) and these are solved using standard integration techniques for ODE. The coupling is in fact in the form of integrals, but this is not a significant complication.

The code was first tested with evolution of arbitrary initial data of the scalar field. A typical solution is shown in Fig. 4. A simple way to characterize the evolution is using the scalar $M(r,u)/r$, where $M(r,u)$ is the mass interior to the sphere with radius r at time u. When $2M(r)/r$ remains much smaller than unity the the scalar field converges towards the origin and bounces back to infinity (Fig 4). When $2M(r)/r$ approaches unity a black hole forms (see Fig. 4). In our coordinate system:

$$ds^2 = -g\bar{g}du^2 - 2gdudr + r^2 d\Omega^2 \ . \tag{3}$$

with the normalization $g(0) = \bar{g}(0) = 1$, g and $\bar{g}$ diverge on a black hole's horizon (note that $2M(r)/r = (1 - \bar{g}/g)$). Hence we cannot follow a black hole appearance in these calculations. We can, still, estimate quite well where and when a black hole horizon appears from graphs of $2M(r)/r$, g or $\bar{g}$. A black hole appearance is also marked clearly on the trajectories of null geodesics which converge inwards in the (r,u) plane.

Consistency of the code was checked in many ways (GP). One of these is the remarkable agreement between calculations with different grid sizes and different

Fig. 4: *$2M(r)/r$ for a scalar field bounce (left with a vertical scale of $0 - 10^{-5}$) and for scalar field collapse (right with a vertical scale of $0 - 1$). The formation of a black hole can be seen when $2M(r)/r$ approaches unity (from GP).*

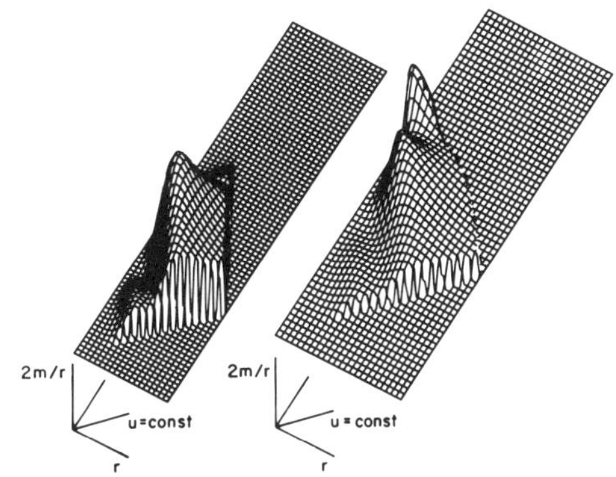

time steps and between calculations performed with different integration techniques for the ODE as well as the integral equations.

To examine Christodoulou's problem we solve Eqs. 1a-d for the interior initial data supplementing it by exterior data. We then integrate the evolution equations to obtain a full solution. Since the initial datum on r_0 determines analytically the data on the null geodesic that passes through $(r_0, -1)$, it is sufficient to integrate the solution only at the exterior. Such a solution is shown in Fig. 5. The graph of $Max[2M(r)/r]$ vs. u (Fig. 6) indicates clearly that a black hole forms before $u = 0$.

We were unable to find initial exterior data that does not lead to a black hole formation before $u = 0$. A unique feature of the initial data satisfying Eq. 1a-d is that on $r_0(u)$, the null geodesic that passes through $(r_0, -1)$, $\bar{g}(r_0(u)) = -2r_0$ which leads to $r_0(u) = -ur_0(-1)$. It is easily realized from this that 2M/r is constant on this null geodesic, i.e.: $2M(r_0(u), u)/r_0(u) = 2M(r_0, -1)/r_0$. Using this we find a lower limit to u_{crit} where $Max[M(r)/r] = 1$:

$$u_{crit} \leq u_0 = min\left[\frac{1/2r - M(r,-1)}{1/4\bar{g}(r,-1) + M(r_0,-1)} - 1\right] . \tag{4}$$

All our solutions satisfy this limit. As u_0 is determined by the initial data one can scan quickly over different sets of initial data looking for $u_0 > 0$. However we were not able to find such initial data. This suggests that without an additional trick, the solution of Eqs. 1a-d is not a counter example for cosmic censorship.

Fig. 5: $2M(r)/r$ *in the exterior region for* $\kappa = \sqrt{(\pi)}$ *and a particular exterior initial datum (from GP).*

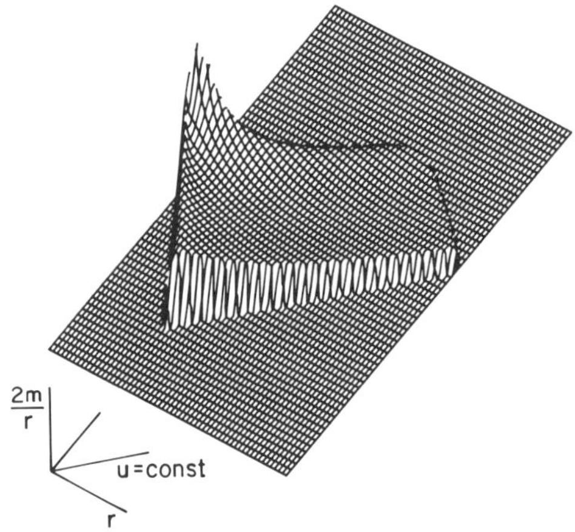

Fig. 6: $Max[2M(r)/r]$ *vs.* u *for* $\kappa = \sqrt{(\pi)}$ *and a particular exterior initial datum (from GP)* .

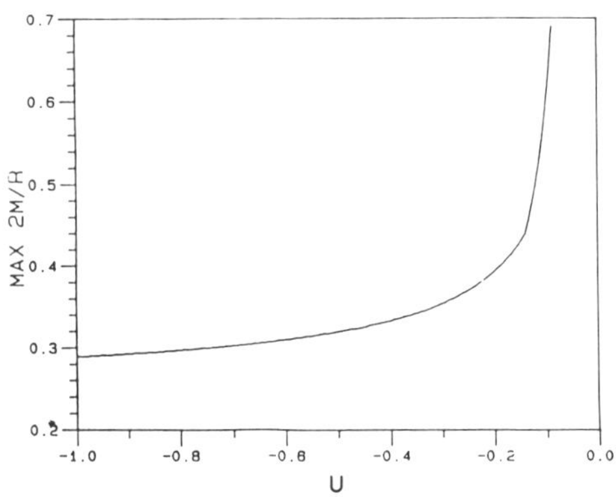

Regge Calculus, Cosmology and Inflation

Regge calculus (Regge, 1961) is the most beautiful way to convert Einstein's equations to a discrete set of equations. Regge calculus is based on constructing a piecewise linear spacetime and writing Hilbert's action in terms of geometrical quantities such as area elements and deficit angles. To obtain the field equations one minimizes the action by varying the length of the skeleton's edges. The recent interest in Regge calculus is expressed in work that ranges from proofs of the convergence of the Regge action to the continuum action (Cheeger et al, 1985; Friedberg and Lee,

1984; Feinberg et al., 1984), to applications to Quantum Gravity (see a review by Isham 1986) and to numerous attempts of solving specific numerical problems (e.g. Dubal, 1986a,b; Brewin, 1986). The beauty of Regge calculus, its coordinate free specification, is however a major obstacle to its practical application. The equations are not much simpler even when high symmetry is imposed, and the bookkeeping demanded from a general solution is formidable. In spite of these obstacles Dubal (1986a,b) has constructed some spherical solutions while Brewin (1986) has constructed a homogeneous cosmological solution. A major achievement of these works is the inclusion of fluid equations in the calculation, a procedure which is done using the velocity potential formalism (Schutz 1970).

One of the problems in application of Regge calculus is the specification of initial data and the selection of evolution procedure. We do not know yet if it is possible to select a set of edges on which the initial data is specified and to determine equations from which the rest of the edge lengths could be evaluated. Lund and Regge (1974) have suggested using a 3+1 version of Regge calculus. The original Lund and Regge scheme dealt only with homogeneous spacetimes. Recently Piran and Williams (1985, 1986) and independently Friedman and Jack (1986a,b) have constructed an inhomogeneous version of 3+1 Regge calculus. In this formalism the spacelike hypersurfaces are three dimensional piecewise linear manifolds. The edge lengths are continuous functions of time, which remains continuous. Four dimensional elements are curved, but they have three dimensional spacelike, flat sections.

Piran and Williams (1986) construct the analogue to the ADM action:

$$S = \int dt \left[\frac{1}{4} \sum_{\alpha} \left(\frac{1}{N} \right)_{\alpha} \mathcal{G}^{\alpha ab} \dot{g}_a \dot{g}_b + \sum_{\mu} N_{\mu} \sum_{a \ni \mu} g_a^{1/2} \epsilon_a - 16\pi \sum_{\alpha} N_{\alpha} V_{\alpha} \mathfrak{L}_{matter}|_{\alpha} \right] , \quad (5)$$

where the basic variables g_a are squares of the edge lengths, ϵ_a is the deficit angle on edge a, N_μ is the lapse function defined on vertex μ, N_α and $(1/N)_\alpha$ are averaged over a tetrahedron α, V_α is the volume of tetrahedron α and $\mathcal{G}^{\alpha ab}$ is the analog of the super-metric G^{ijkl}:

$$\mathcal{G}^{\alpha ab} = -\frac{1}{V_\alpha} \sum_{a,b \in \alpha} \frac{\partial^2 V_\alpha^2}{\partial g_a \partial g_b} \quad . \quad (6)$$

In the above equations $a \in \alpha$ means an edge a that belongs to a tetrahedron α, similarly $a \ni \mu$ means an edge a that touches a vertex μ. From this action one can either derive the second order field equations or construct a conjugate momentum and a Hamiltonian. Piran and Williams have introduced a lapse function, defined on the vertices (Friedmann and Jack have chosen to define it on the tetrahedra). The lapse function is a Lagrange multiplier and as such it introduces a Hamiltonian constraint just like in the continuum case. Unlike the continuum case, but similar to other finite difference schemes, the Hamiltonian constraint is not conserved by the evolution equations of the 3+1 Regge calculus. A possible remedy to this problem is to choose the lapse function such that the Hamiltonian constraint is conserved. It was shown by Friedman and Jack (1986a,b) and in a different context by Brewin (1986) that this is possible. However this seems unnatural and one may proceed with the unconserved equations, as there are arguments suggesting that the deviations of the constraints converge as the skeleton becomes finer.

It is debatable whether one should introduce a shift vector which will generate momentum constraints. Lack of spatial coordinates suggests that a shift vector, whose role in the continuum case is to determine these coordinates, is not needed. However one can imagine a Regge skeleton in a flat space. One can choose the vertices to be anywhere and imagine an evolution in which the vertices move with arbitrary specified velocities, which are just the shift vector. Furthermore, introduction of the shift and with it of momentum constraints provides a potential solution to the inconsistency problem, the full system of constraints might be closed. These arguments led Friedman and Jack (1986a,b) to introduce a shift vector, however, they were not able to find a closed constraint system and it seems that such a closed system does not exist. This is unpleasant and may be a serious blow to the attempts to use the 3+1 Regge calculus as a basis for quantum gravity, as some people have hoped. However this problem is well known in any attempt of finite differencing the Einstein equations (Piran, 1982) and is not a reason to abandon this formalism for numerical calculations.

A major open question in Cosmology is the scope of initial conditions under which a universe undergoes an inflationary phase. One needs a scalar field with a suitable potential for inflation (Turner, 1986, Piran, 1986). However, even if these exist, inflation may not occur if the initial conditions are not favorable. In particular one worries whether inhomogeneities might inhibit the onset of inflation. The best way to explore this question is by numerical calculations based on the 3+1 Regge calculus. Sensible models can be constructed with as few as 15 variables satisfying 15 coupled ordinary differential equation! Furthermore, the lack of a grid pattern makes Regge calculus particularly suitable for cosmological studies where we are interested in evolution of random deviations from homogeneity.

The simplest possible model for a closed universe is a 5 simplex, whose five tetrahedra provide a piecewise linear approximation for a three sphere. This model has a natural homogeneous, but anisotropic configuration (Fig. 7) with identical vertices and a homogeneous scalar field ϕ but with two unequal edge length

Fig. 7: *Five tetrahedra, which compose the surface of a four homogeneous but anisotropic simplex with two different edge lengths (solid lines and dashed lines). (from Piran and Williams 1985).*

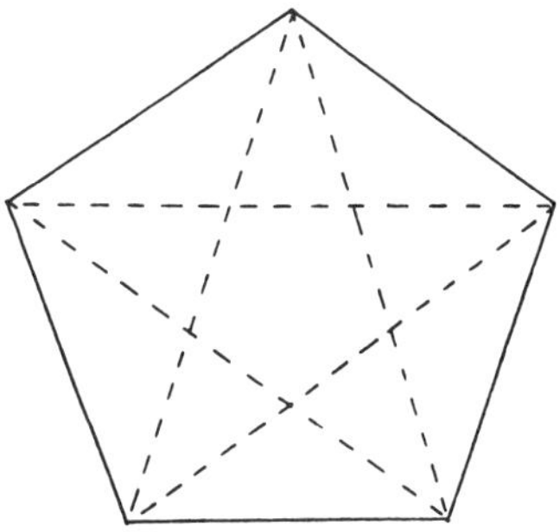

l_1 and l_2. Defining a scale length $R = sqrt(l_1 l_2)$, an expansion rate $H = \dot{R}/R$, and an anisotropy parameter $\omega = l_1/l_2$ we obtain a generalization of the Friedmann equation:

$$H^2 + \frac{2}{3}H\dot{\omega}\frac{\alpha'}{\alpha} - \frac{\dot{\omega}^2\beta}{12\omega^2\alpha} + \frac{4}{R^2\alpha}(\delta(\omega)\sqrt{\omega} + \frac{\delta(1/\omega)}{\sqrt{\omega}}) = \frac{4\pi}{3m_p^2}(\dot{\phi}^2 + m^2\phi^2) \quad . \tag{7a}$$

A scalar field equation:

$$\ddot{\phi} + (3H + \frac{\dot{\omega}\alpha'}{\alpha})\dot{\phi} + m^2\phi = 0 \quad , \tag{7b}$$

and an equation for the anisotropy factor:

$$\ddot{\omega}(\frac{\omega\alpha'}{3\alpha} + \frac{\beta}{4\omega\alpha'}) + \dot{\omega}H(\frac{\omega\alpha'}{\alpha} + \frac{3\beta}{4\omega\alpha'}) + \dot{\omega}^2(\frac{\alpha''\omega}{3\alpha} + \frac{\beta}{8\alpha\omega} + + \frac{\beta'}{8\alpha'\omega^2} - \frac{\beta}{4\alpha'\omega^2})$$
$$= \frac{3}{R^2\alpha'}(\delta(\omega)\sqrt{\omega} - \frac{\delta(1/\omega)}{\sqrt{\omega}}) - \frac{2\omega}{R^2\alpha}(\delta(\omega)\sqrt{\omega} + \frac{\delta(1/\omega)}{\sqrt{\omega}}) \quad , \tag{7c}$$

which becomes an identity when the anisotropy vanishes. The functions α , β and the deficit angles (Williams, 1985) $\delta(\omega)$ and $\delta(1/\omega)$ are functions of ω only:

$$\alpha = \sqrt{(-\omega^3 + 2\omega + 2\omega^{-1} - \omega^{-3})} \;\; ; \;\; \beta = \frac{((\omega)3\omega^3 + 2\omega + 2\omega^{-1} + 3\omega^{-3})}{\alpha} \quad , \tag{7d}$$

$$\delta(\omega) = 2\pi - \cos^{-1}(\frac{2\omega^4 - 2\omega^2 + 1}{4\omega^2 - 1}) - 2\cos^{-1}(\frac{\omega(3 - 2\omega^2)}{\sqrt{(4\omega^2 - 1)}\sqrt{(4 - \omega^2)}}), \tag{7e}$$

and $'$ denotes a ω derivative.

For a homogeneous universe these equations reduce to the Friedmann equations for a closed universe with $4\sqrt{2}(2\pi - 3cos^{-1}(1/3)) = 14.65$ replacing k=

Fig. 8: *Hubble's constant (solid line), the anisotropy parameter (dashed line), and normalized $\dot{\phi}$ (dashed-dotted line) during a typical evolution of an anisotropic but homogeneous toy universe with a scalar field. The initial values were chosen so that the universe will get into (at $t \approx 1$) and out of (at $t \approx 9.5$) an inflationary phase. (from Piran and Williams 1985).*

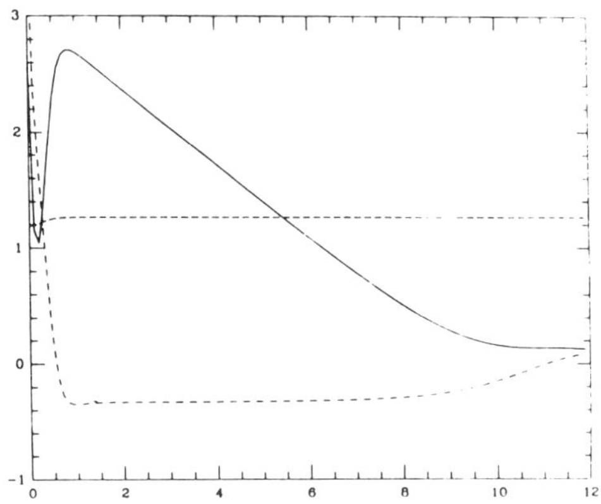

1. The difference between the discrete and the continuum curvature terms is not surprising since the 5 simplex is a relatively poor approximation to S^3. With a better approximation, say 8 vertices one obtains $3\sqrt{2}(2\pi - 4cos^{-1}(1/3)) = 5.76$ and with 120 vertices one obtains $12\sqrt{2}(2\pi - 5cos^{-1}(1/3))/5 = .436$ (see Williams, 1985 for discussion of regular tetrahedral tessellations of S^3).

One interesting feature observed in this model is that when the scalar field is large enough (we are considering mainly chaotic inflation (Linde, 1983, 1985)) the universe evolves into an inflationary phase, in spite of the anisotropy (see Fig. 8). The anisotropy is frozen during and after the inflationary phase (note that $\ddot{\omega} = \dot{\omega} = 0$ is a solution of Eq. 7c). It will be interesting to generalize these calculations. There is no way, that I know, to do so but to use numerical techniques.

Rotating Gravitational collapse and Gravitational Radiation

Prediction of the amount of gravitational radiation emitted in rotating gravitational collapse is one of the current challenges to theoretical relativistic astrophysics. The only hope to solve this nonlinear problem is using numerical techniques. Nakamura (1981) studied numerically the fate of rotating configurations as a function of their dimensionless angular momentum parameter a, $a = J/M^2$, without estimating the gravitational radiation emission. The tiny amount of gravitational radiation emitted constitutes a tough challenge. The code must be extremely accurate to calculate the expected emission efficiencies at a level of 10^{-3} to 10^{-4}. Bardeen and Piran (1983) have designed a computational scheme to solve this problem and Stark and Piran (1986b) have implemented this scheme in an actual computer code. The code, which is second order accurate, follows to a large extent the numerical ideas of Wilson (1979). The code has passed a long series of tests. I discuss two of them here. The complete list as well as the numerical methods are described elsewhere (Stark and Piran, 1986a).

Fig. 9: *Comparison test for the waveform from spheroidal dust collapse (from SP 1985b,1986a).*

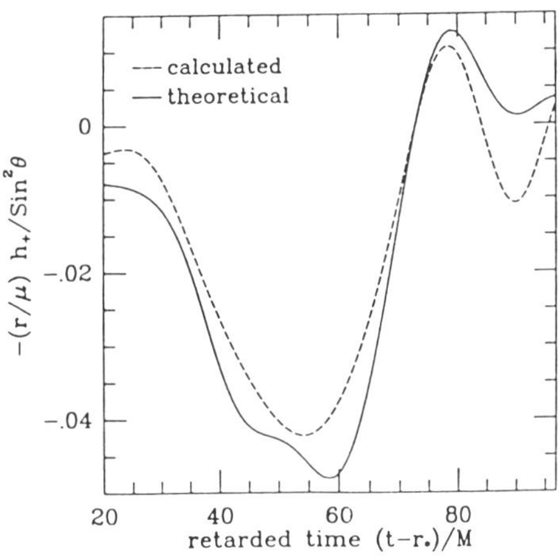

Petrich et al. (1985) have recently calculated the gravitational radiation emission by an ellipsoidal dust shell infalling on a Schwarzschild black hole. They integrate contributions from the shell using the test particle (Davis et al. 1972) formula . With a uniform elliptical shell of mass μ and outer and inner radii of $12M$ and $20M$ the resulting emission efficiency $4.1 \times 10^{-5}\mu/M$ is less than 1/200 of a single particle emission efficiency (due to phase cancellations between different sections of the shell). With $\mu = .19M$ we obtain numerically $\Delta E_{num} = 1.7 \times 10^{-6}M$ which should be compared with the analytic value of $\Delta E_{anal} = 1.48 \times 10^{-6}M$. Both analytic and numerical waveforms of the emitted radiation are shown in Fig. 9. The agreement is excellent, specifically when one considers the small emission efficiency.

In another test Stark and Piran compared numerically and analytically calculated propagation of Teukolsky waves (1982). These are linearized free quadrupolar gravitational waves whose propagation was calculated first by Teukol-

Fig. 10: *Comparison tests for the waveform (left) and energy (right) for Teukolsky waves (from SP 1985b,1986a).*

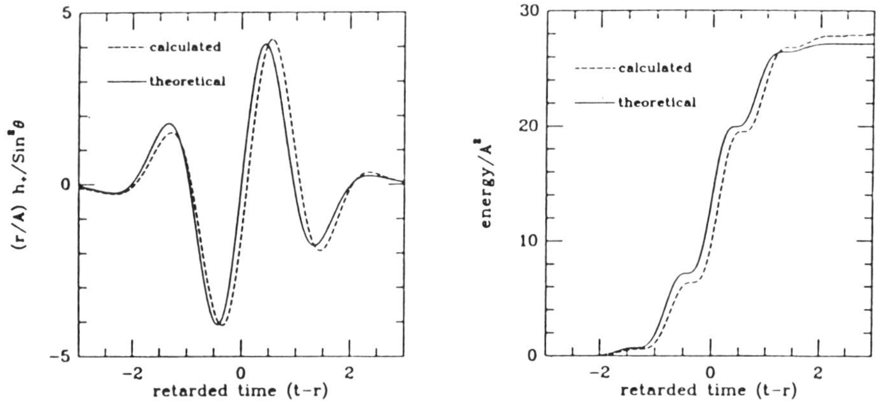

sky in the transverse-traceless gauge and later by Bardeen (1983) in the radial gauge in the numerical calculations. A comparison between the analytic and numerical solutions is shown in Fig. 10. Again, the agreement is excellent.

Calculations of rotating gravitational collapse with this code (Stark and Piran, 1985a,b, 1986a; Piran and Stark, 1986a,b denoted later SP) revealed a few new physical phenomena. The most startling one is the uniqueness of the waveform of the emitted radiation which is practically independent of the details of the collapse (Fig 11). This previously unexpected feature is not so surprising in retrospect. The wave form is that of the lowest quasi-normal modes of a black hole (Detweiler, 1980). Some of the radiation is emitted earlier, in the hydrodynamic phase, and this radiation does depend on details of the collapse. This phase contains, however, only a small fraction of the emitted energy and can be ignored. Most of the radiation is emitted by a deformed black hole horizon which oscillates mainly in its lowest normal modes and emits its energy excess as gravitational radiation. The realization of the uniqueness of the waveform emitted during black hole formation should help us to detect gravitational radiation by providing a waveform one should look for. Furthermore, detection of such waveform will at the same time be a positive identification of black hole formation!

The calculated efficiency scales, for small specific angular momentum, as J^4 (Fig. 12):

$$\Delta E = 1.4 \times 10^{-3} (J/M^2)^4 M \ . \tag{8}$$

This was predicted by perturbation calculations (Cunningham et. al., 1978,1979,1980;

Fig. 11: *h_+ (left) , $h_\times$ (right) waveforms vs. retarded time for various a collapses (from SP 1985a).*

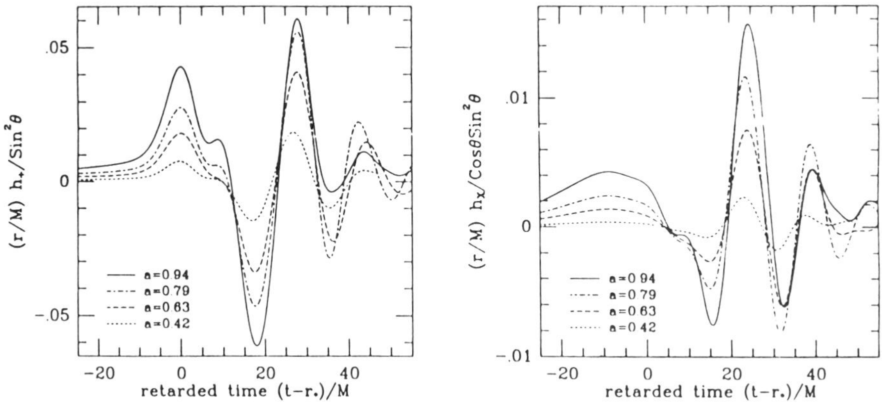

Fig. 12: *Total Gravitational wave energy emitted vs. a for two different types of collapse (from SP 1985a).*

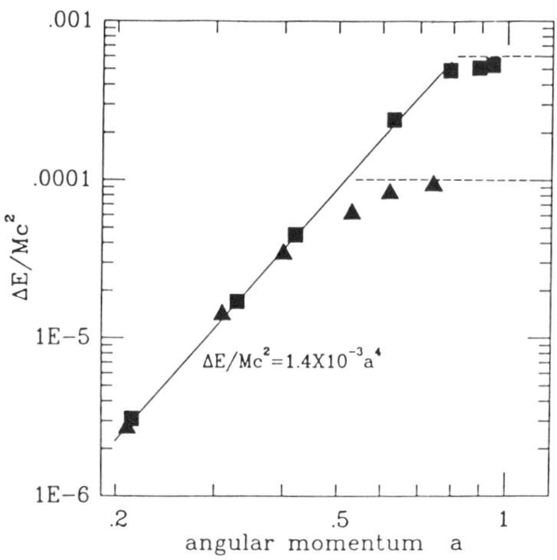

Saenz and Shapiro, 1978). However the efficiency saturates at some level,depending on the details of the collapse, reaching a maximal efficiency of about 0.7×10^{-3}. As expected most of the energy is emitted in the + mode with $(\Delta E/Mc^2)_+ > 10(\Delta E/Mc^2)_\times$.

Translation of these results to maximal strain on earth from an astro-

Fig. 13: *Energy spectra, for even and odd modes and $a = 0.79$ normalized to a collapse of a star of 1 $M_\odot$ at a distance of 10 kpc (from SP 1985a).*

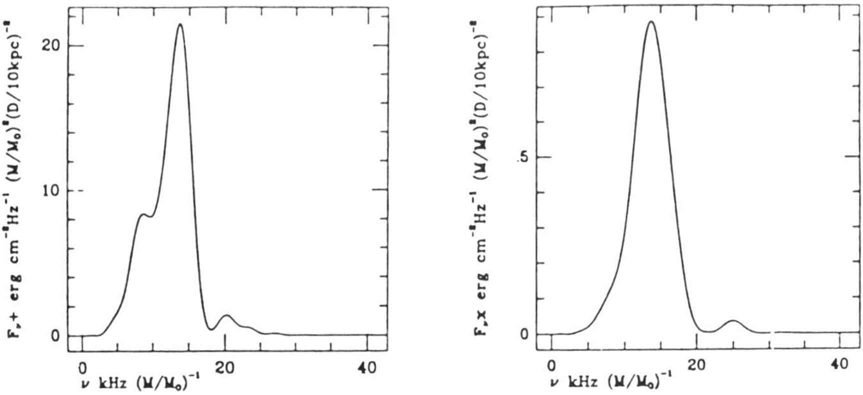

physical event at distance r yields:

$$|\Delta l/l|_{max} = \begin{cases} |h_+|_{max} = 4.0 \times 10^{-19}(M/1.4M_\odot)(r/10\text{kpc})^{-1} \\ \sqrt{G/c^3 F(\nu_{max})} = 3.6 \times 10^{-19}(M/1.4M_\odot)(r/10\text{kpc})^{-1} \end{cases}, \quad (9)$$

where the two measures correspond to resonant (bar) and non- resonant (interferometer) experiments respectively. The maximum spectral energy is:

$$F(\nu_{max}) = 45(M/1.4M_\odot)^2(r/10\text{kpc})^{-2} \ \text{erg cm}^{-2}\text{Hz}^{-1} , \quad (10)$$

which for a galactic event is within the capability of current gravitational radiation detectors. For an extragalactic event within the Virgo cluster this is within the range of current hopes for the next generation of gravitational radiation detectors.

An example of the energy spectrum (i.e. energy/ unit area/ unit frequency vs. frequency) corresponding to the waveforms of Fig. 11 is shown in Fig. 13. Corresponding to the insensitivity of the waveform shape of the gravitational radiation from collapse to a black hole to the initial conditions, the shape of the energy spectra are similarly insensitive. For large values of angular momentum a low frequency component due to the initial flattening of the star appears. The energy spectra peak at frequencies of $0.035 < \nu < 0.08$ $(GM/c^3)^{-1}$ Hz (i.e. $7 < \nu < 16$ $(M/M_\odot)^{-1}$ kHz) corresponding to wavelengths of between 12 and 28 (GM/c^2). The peak frequency and full width half maximum frequency are:

$$\nu_{max} = 10(M/1.4M_\odot)^{-1}\text{kHz} , \quad (11)$$

$$\Delta\nu_{1/2} = 4(M/1.4M_\odot)^{-1} kHz \ . \tag{12}$$

For a $1.4M_\odot$ collapse ν_{max} is significantly higher than the resonance frequency of most current bar detectors.

Fig. 14: $rh_+/\sin^2\theta$ *vs.* t *for different* r *for* $a = 0.79$. *(from SP 1984).*

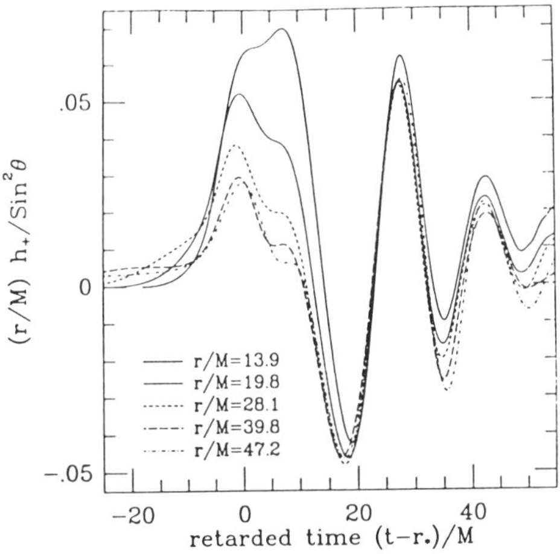

The wave form obtains its asymptotic shape remarkably deep in the inner zone. This unexpected phenomenon is seen clearly in Fig. 14 where the waveforms at different radii are compared. Notice that with a wavelength , $\lambda \approx 20M$, the wave reaches its asymptotic form already at $r \approx \lambda$! This observation was possible since we have used variables, suggested by the Bardeen and Piran scheme, that trace effectively the gravitational radiation. This phenomenon, which might be related to the fact that most of the emission is due to quasi-normal mode oscillations of the black hole, deserves some analytic explanation and may shed some light on the actual process of generation of gravitational radiation in the strong field zone (see however, Schutz, 1986).

Three Dimensional Codes

Some realistic questions cannot be answered without three dimensional non linear, time dependent calculations. The computer power needed for such a calculation can be easily estimated (Piran 1983). With the advance of supercomputers, the necessary computer power is available and the task of solving the Einstein equations in the most general case, i.e. without imposing any symmetry of any sort, is finally feasible.

Figure 15: *Energy density of asymmetric gravitational radiation packet, the initial configuration is on the top right and the time evolution proceeds to the left and to the down (From Nakamura 1986a,b)*

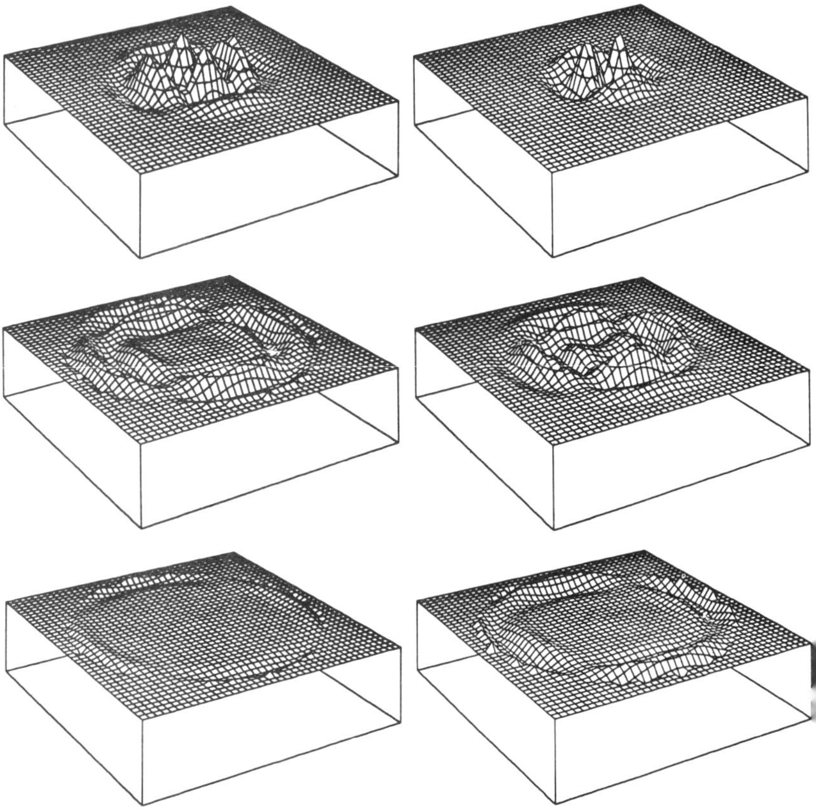

As a first stage towards three dimensional calculations Oohara (1986) has solved the three dimensional initial value problem. Later Nakamura (1986a,b) has constructed the first three dimensional code for vacuum Einstein's equations. Nakamura's code employs a general metric, with a constant lapse and zero shift. This is quite different from the common approach which is based on simplifying coordinate conditions (Wilson 1977; Smarr, 1979; Bardeen and Piran, 1983; Piran, 1983). A suitable coordinate choice can simplify greatly the calculations of the Ricci tensor, avoid potential singularities, and might even isolate the dynamic variables in a way that increases the relevant accuracy of the code by a huge factor. The simplifying coordinate conditions demand, however, a solution of possibly coupled, elliptic (or at least parabolic) equations for the lapse and the shift. With a general metric one disposes of the elliptical equations for the lapse and the shift. The evaluation of the Ricci tensor is, however, substantially more complicated. The technical difficulty involved in first computing all the Ricci tensor components and then programming them is the major obstacle to construction of a general "coordinate free" type of code. The enormous number of terms in these equations makes a common programming procedure hopeless. It is just unlikely that one could code all the necessary terms without introducing errors that would be impossible to trace. In constructing his code, Nakamura has solved this problem by fulfilling the dream of combining algebraic computer calculations with numerical ones. After using an algebraic manipulation language (MacCallum, 1986) to calculate the components of the Ricci tensor Nakamura has converted the results, again using algebraic manipulation language directly into FORTRAN statements. Clearly this is the procedure that should be followed. I am sure that eventually this will be the way that all codes will be written in the future.

Nakamura (1986a,b) has studied evolution of free gravitational waves. An example of propagation of a asymmetric weak gravitational wave is shown in Fig. 15.

Figure 16: Isodensity surface obtained after evolving the asymmetric density distribution $:D(0,\vec{R}) = D_0 \frac{A}{\|\vec{R}-\vec{R}_0\|} + \epsilon$, $\vec{R}_0 = (r = \frac{2}{3} r_b, \theta = \frac{\pi}{3}, \varphi = 0)$ around a Schwarzschild black hole. It can be seen that matter flows onto the black hole following a bridge between $\vec{R}_0$ and the horizon (from Marck and Piran, 1986).

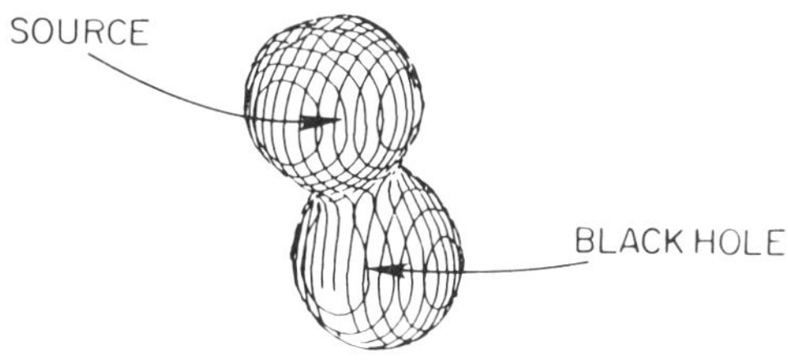

In another three dimensional project, Marck and Piran (1986) have constructed a three dimensional relativistic hydrodynamics code to study matter accretion onto a Kerr black hole. These calculations generalize the pioneering two dimensional calculations of Wilson (1972) and their extended and extremely accurate version in Hawley et al. (1984a,b). However, the numerical techniques are different. Marck and Piran used, following a test by Bonazzola and Marck, (1986) spectral methods (Gotlieb and Orszag, 1977) instead of the common finite differencing methods. The advantages of this approaches are numerous: it is easier to program, easier to vectorize (so to make the numerical algorithm compatible with the basic structure of supercomputers) and it is easier to transform to it from an algebraic computer language. Again I believe that in the future many more computations will be performed using these methods, rather than the traditional finite differencing ones. The code of Marck and Piran is still in a testing phase. In tests with spherical or axial symmetry, the initial symmetry is conserved. Other tests are in progress now. Fig. 16 displays a preliminary three dimensional calculation of evolution of asymmetric density distribution around a Schwarzschild black hole.

Conclusions

It was almost thirty years ago at the Chapel Hill meeting, that Bryce DeWitt suggested the use of computers to solve Einstein's equations and to study the full nonlinear structure of relativity. The first numerical codes appeared in the mid sixties (Hahn and Lindquist, 1964; May and White, 1966) but numerical relativity did not gain momentum until the works of Wilson, Smarr and others in the mid seventies. Today, the applications of numerical relativity extend to different branches of general relativity and relativistic astrophysics and we are not far from a full numerical solution. I hope that I have demonstrated, in this talk, that the dream of solving Einstein's equations numerically has been fulfilled. I am certain that with the rapid development of computers and numerical techniques we are going to learn a lot about general relativity from numerical relativity in the next few years.

Acknowledgment

I am would like to thank J. M. Bardeen, D. Christodoulou, D. Goldwirth, J. A. Marck, R. F. Stark and R. Williams for helpful conversations. This work was supported by NSF grant PHY84-07219.

References

Bardeen, J., M. and Piran, T., (1983) Physics Reports,**96**,20.

Bardeen, J. M., (1983) in *Gravitational Radiation*, eds. Deruelle N. and Piran, T., (North Holland, Amsterdam,1983)

Bonazzola, S. and Marck, J. A., (1986) Astron. and Astrophys., **164** , 300.

Brewin, L., (1986) Paper presented at the GRG 11 meeting, Stockholm.

Centrella J. (ed) (1986) *Dynamical Spacetime and Numerical Relativity*, Cambridge University Press.

Centrella, J., Matzner, R. A., Rothman, T. and Wilson, J., (1986) Nucl. Phys. **B266**, 171.

Christodoulou, D., (1983) private communications.

Christodoulou, D., (1986) Comm. Math. Phys, **105**, 337.

Cheeger, M. J. Muller, W. and Schrader, R., (1984) Comm. Math. Phys, **92**, 405.

Choptuik, M., (1986) Phd. Thesis, University of British Columbia.

Choptuik, M. and Unruh, W. G., (1986) Preprint.

Corkill, R. W. and Stewart, J. M., (1983) Proc. Roy Soc. (London) **386** 373.

Cunningham, C., T., Price R. H. and Moncrief, V., (1978) Astrophys. J. , **224**,643.

Cunningham, C., T., Price R. H. and Moncrief, V., (1979) Astrophys. J. , **230**,870.

Cunningham, C., T., Price R. H. and Moncrief, V., (1980) Astrophys. J. , **263**, 674.

Detweiler, S., (1980) Astrophys. J. ,**239**,292.

Davis, M., Ruffini, R. and Tiomno, J., (1972) Phys.Rev.D,**5**,2932.

Dubal, M. R., (1986a) Preprint .

Dubal, M. R., (1986b) Paper presented at the GRG 11 meeting, Stockholm.

Einstein, A., (1939) Ann. Math. **40**, 922.

Evans, C., (1986) in *Dynamical Spacetime and Numerical Relativity*, ed. Centrella J. Cambridge University Press.

Feinberg, G. R., Friedberg, R. Lee, T. D. and Ren, H. C., (1984) Nucl. Phys. **B245**, 343.

Friedberg, R. and Lee, T. D., (1984) Nucl. Phys. **B252**, 145.

Friedman, J. L. and Jack. I., (1986) J. Math. Phys. in press.

Friedman, J. L. and Jack. I., (1986) Paper presented at the GRG 11 meeting, Stockholm.

Friedrich, H. and Stewart, J. M., (1983) Proc. Roy Soc. (London) **385** 345.

Goldwirth, D., (1986) Msc Thesis, Hebrew University.

Goldwirth, D. and Piran, T., (1986) in preparation. GP.

Gotlieb, D. and Orszag, S., 1977 :*Numerical Analysis of Spectral Methods : Theory and Application*, Regional Conference Series in Applied Mathematics.

Hahn, S. G. and Lindquist, R. W. (1964) Ann. Phys. **29**, 304.

Hartle, J., B., (1985) Classical and Quantum Gravity, **2**, 707.

Hawley, J. F., Smarr, L. L. and Wilson J. R. (1984a) Astrophys. J., **277**, 296.

Hawley, J. F., Smarr, L. L. and Wilson J. R. (1984a) Astrophys. J. Supp., **55**, 211.

Ipser, J. R. and Thorne, K. S., (1968) Astrophys. J. **154**, 251.

Ipser, J. R., (1980) Astrophys. J. **238**, 1101.

Isham, C. J. (1986) in this volume

Isaacson, R. A., Welling, J. S. and Winicour J. (1983) J. Math. Phys. **24**, 1824.

Isaacson, R. A., Welling, J. S. and Winicour J. (1986) in *Dynamical Spacetime and Numerical Relativity*, ed. Centrella J. Cambridge University Press.

Linde, A., (1983) Phys. Lett. **129B**, 177.

Linde, A., (1985) Phys. Lett. **162B**, 281.

Lund, H. and Regge, T., (1974) unpublished.

MacCallum, M. A. H., (1986) in *Dynamical Spacetime and Numerical Relativity*, ed. Centrella J. Cambridge University Press.

Marck, J. A. and Piran, T. (1986) in *Proceedings of the Yamada conference*, eds. Sato, H. and Nakamura, T., in press.

May, M. M. and White, R. H. (1966) Phys. Rev. **141**, 1232.

Nakamura, T., (1981) Prog. Theor. Phys.,**65**,1876.

Nakamura, T., (1986a) in *Proceedings of the Yamada conference*, eds. Sato, H. and Nakamura, T., in press.

Nakamura, T., (1986b) Paper presented at the GRG 11 meeting, Stockholm.

Oohara, K. I., (1986) in *Proceedings of the Yamada conference*, eds. Sato, H. and Nakamura, T., in press.

Pantano, O. and Miller, J. C. (1986) preprint.

Petrich, L. I., Shapiro, S. L. and Wasserman, I., (1985) Astrophys. J. Supp., **58**, 297.

Petrich, L., Shapiro, S. L. and Teukolsky, S. A. (1985) Phys. Rev. , **31**, 2549.

Piran, T. (1982) in *Proceedings of the Second Marcel Grossmann meeting*, ed. R. Ruffini, North Holland.

Piran, T., (1983) in *Gravitational Radiation*, eds. Deruelle N. and Piran, T., (North Holland, Amsterdam,1983)

Piran, T., (1986) Phys. Lett. B in press.

Piran, T., Safier P., D., and Stark R., F., (1985) Phys. Rev **D32**, 3101.

Piran, T. and Stark, R. F., (1985) in *Dynamical Spacetime and Numerical Relativity*, ed. Centrella J. Cambridge University Press, SP.

Piran, T. and Stark, R., F., (1986) in *The 12th. Texas Symposium*, eds. Livio, M. and Shaviv, G. Annals of the N.Y. Acad. Sci., **470**, 247, SP.

Piran, T. and Williams, R. M. (1985) Phys. Lett **163B**, 331.

Piran, T. and Williams, R. M. (1986) Phys. Rev. **D33**, 1622.

Regge,. T. (1961) Nuovo Cimento, **19**, 558.

Rocek, M. and Williams, R. M., (1985) Classical and Quantum Gravity, **2**, 701.

Saenz, R. A. and Shapiro, S., L. (1978) Astrophys. J. ,**221**,286.

Schutz, B. F., (1970) Phys. Rev. **D2**, 2762.

Schutz, B. F., (1986) in *Dynamical Spacetime and Numerical Relativity*, ed. Centrella J. Cambridge University Press.

Shapiro, S. L. and Teukolsky, S. A. (1979) Astrophys. J. Lett, **234**, L177.

Shapiro, S. L. and Teukolsky, S. A. (1980) Astrophys. J. , **235**, 199.

Shapiro, S. L. and Teukolsky, S. A. (1985a) Astrophys. J. Lett, **292**, L 41. ST.

Shapiro, S. L. and Teukolsky, S. A. (1985b) Astrophys. J., **298**, 113. ST.

Shapiro, S. L. and Teukolsky, S. A. (1985c) Astrophys. J., **298**, 58. ST.

Shapiro, S. L. and Teukolsky, S. A. (1986a) Astrophys. J., **307**, 575. ST.

Shapiro, S. L. and Teukolsky, S. A. (1986b) in *Dynamical Spacetime and Numerical Relativity*, ed. Centrella J. Cambridge University Press, ST.

Smarr, L. L., (1979) in *Sources of Gravitational Radiation*, ed. Smarr, L. L., Cambridge University Press, England.

Stark, R. F. and Piran, T., (1985a) Phys. Rev. Lett., 55, 891. SP.

Stark, R. F. and Piran, T. (1985b) in *Proceedings of the Forth Marcel Grossmann meeting*, ed. R. Ruffini, North Holland, in press. SP

Stark, R. F. and Piran, T. (1986a) "A General Relativistic Code for Rotating Axisymmetric Configurations and Gravitational Radiation: Numerical Methods and Tests" to appear in Computer Physics Reports. SP.

Stark, R. F. and Piran,T. (1986b) in *Proceedings of the Yamada conference*, eds. Sato, H. and Nakamura, T., in press. SP.

Stewart, J. M. and Friedrich, H. (1982) Proc. Roy Soc. (London) **384** 427.

Stewart, J. M., (1984) in *Proceedings of the NATO advanced research workshop on Problems of Collapse and Numerical Relativity, Toulouse, France*, Eds. Bancel, D. and Signore, M, Reidel.

Teukolsky, S. A., (1982) Phys.Rev.D,**26**,745.

Turner, M. S., (1986) in this volume.

Williams, R. M., (1985) Gen. Rel. and Grav., **17**, 559.

Williams, R. M. and Piran, T. (1985) in *Proceedings of the Fourth Marcel Grossmann Meeting*, ed. R. Ruffini, North Holland, in Press.

Wilson, J. R. (1972), Astrophys. J., **173** , 431.

Wilson, J. R. (1977) Lecture given at GRG 8 Waterloo, Canada.

Wilson, J. R. (1979) in *Sources of Gravitational Radiation*, ed. Smarr, L., Cambridge University Press, England.

Zel'dovich, Ya. B. and Podurets, M. A., (1965) Astr. Zh., **42**, 693 translation in Soviet Astr. - AJ, **9**, 742.

HOW EINSTEIN DISCOVERED GENERAL RELATIVITY: A HISTORICAL TALE WITH SOME CONTEMPORARY MORALS

J.J. Stachel
Einstein Project, Boston University, Boston, MA 02215, U.S.A.

INTRODUCTION

Like a classical drama, Einstein's attempt to generalize his original 1905 (special) theory of relativity is divided into three acts:

In 1907, Einstein attempted to set up a relativistically invariant theory of gravitation. He realized, however, that this theory failed to obey the equivalence principle: all bodies did not fall with the same acceleration in a gravitational field. He regarded satisfaction of the equivalence principle as the fundamental criterion for an acceptable gravitational theory and decided that the 1905 relativity theory must be generalized in order to include gravitation.

In 1912, he concluded that no scalar generalization of Newtonian gravitation theory would be adequate. He decided that a non-flat generalization of the four-dimensional metric tensor of Minkowski space-time was the correct mathematical representation of the gravitational field.

After several years during which he was convinced that the gravitational field equations could not be generally covariant, in 1915 Einstein adopted (the practically unique) generally covariant field equations for the metric tensor. He found that the spherically symmetric solution to these equations accounted for the well-known anomaly in the precession of Mercury's orbit.

In Stachel (1979), I gave a summary of the historical aspects of this drama; in Stachel (1980) I went into more detail about what happened during the crucial second act. Here, I shall discuss the conceptual difficulties facing Einstein at each step, and how he coped with them. Since physicists usually do not recapitulate the development of a theory when learning it, they are often not aware of such difficulties, nor of how they were overcome. These historical issues are closely connected with some of the unique difficulties encountered in current attempts to quantize the gravitational field, so this story is of more than historical interest.

Finally, I shall discuss some ways in which the currently accepted formulation of the mathematical foundations of the general theory does not do full justice to Einstein's conception of the theory, and suggest some ways in which this formulation might be improved.

THE EQUIVALENCE PRINCIPLE

The main difficulty at this stage was to see that, in contrast to all other forces of nature, gravitation bears a special relationship to inertia. As Einstein put it, gravitation and inertia are `wesensgleich' essentially the same. This insight implies that the class of global inertial frames singled out in the special theory of relativity can have no place in a relativistic theory of gravitation.

> You consider the transition to special relativity as the most essential thought of relativity, not the transition to general relativity: I consider the reverse to be correct. I see the most essential thing in the overcoming of the inertial system, a thing that acts upon all processes but undergoes no reaction. This concept is, in principle, no better that of the center of the universe in Aristotelian physics. (Einstein to Georg Jaffe, 19 January 1954)

Other physicists then working on a relativistic theory of gravitation found this idea difficult if not impossible to accept, as do many physicists to this day. As a consequence of this insight, Einstein was skeptical of all attempts to start with a special-relativistic quantum field theory and make it generally covariant.

> Contemporary physicists do not see that it is hopeless to take a theory that is based on an independent rigid space (Lorentz-invariance) and later hope to make it general relativistic (in some natural way).(ibid.)

Such comments apply *a fortiori* to attempts to quantize the general theory itself by treating it as if were no more than a special-relativistic theory which happens to be non-linear and hence invariant under a particularly nasty gauge group.

THE METRIC TENSOR

The main difficulty at this stage was to grasp the dual nature of the metric tensor: it is both the mathematical object which represents the space-time structure (chronogeometry) and the set of `potentials,' from which representations of the gravitational field (Christoffel symbols) and of the tidal `forces' (Riemann tensor) may be derived. The space-time structure of general relativity is generally non-flat; more important, it is itself a set of dynamical field variables. The metric tensor not only acts upon the matter fields, but is acted upon by them, via their stress-energy tensors.

Within a few years, Levi-Cività, Weyl and Cartan generalized the Christoffel symbols to the concept of affine connection. This concept served to make the relationship between the mathematical representations of various physical concepts much clearer. Just because it is not a tensor field, the connection field provides an adequate representation of the gravitational-cum-inertial field required by Einstein's interpretation of the equivalence principle. There is no (unique) decomposition of the connection field into an inertial connection plus a gravitational tensor field.

> ... what characterizes the existence of a gravitational field from the empirical standpoint is the non-vanishing of the [components of the affine connection- JS], not the non-vanishing of the [components of the Riemann tensor- JS]. If one does not think in such intuitive (anschaulich) ways, one cannot grasp why something like curvature should have anything at all to do with gravitation. In any case, no rational person would have hit upon anything otherwise. The key to the understanding of the equality of gravitational and inertial mass would have been missing. (Einstein to Max von Laue, September 1950.)

Discussions of the thermal ambience registered by accelerating particle detectors in Minkowski space-time, summarized, for example, in Sciama (1979), pp. 706-714, serve to emphasize the physical significance of the non-vanishing of the affine connection in certain frames of reference, even when the Riemann tensor vanishes. The current fashionability of gauge theories has also made elementary particle physicists well aware of the fundamental significance of connections in physics.

Attempts to quantize the gravitational field while interpreting the resulting formalism in a 'background' space-time (usually taken to be flat) are basically efforts to impose a decomposition of the connection field into an inertial connection, which is left classical, plus a gravitational field tensor, which is quantized. From Einstein's point of view, this is a violation of the injunction: 'What God has joined together let no man put asunder.'

Of course, there is no guarantee that Einstein's approach to gravitation is correct; but it is certainly worth exploring the possibility that it is; and that the attempt to reconcile gravitation and quantization involves something more fundamental than minute, observationally insignificant gravitational corrections to special-relativistically calculated effects.

GENERAL COVARIANCE

The main difficulty at this stage was to realize that, in a theory in which the metric tensor is not given *a priori*, the points of the space-time manifold (which, in the physical interpretation, are to be correlated with events) only derive their physical individuation from the metric field.

Although he considered the possibility of generally covariant field equations in 1913, Einstein rejected them because of an argument that seemed to show that, given the distribution of matter and energy everywhere outside some region of space-time, no generally covariant field equations could uniquely determine the gravitational field inside that 'hole.' For, if the stress-energy tensor outside the region determined *one* such solution G(x), then dragging that solution over the region by any point transformation that reduces to the identity outside and on the boundary of the 'hole' will result in *another* solution G'(x) (in the *same* coordinate system, as Einstein realized) which satisfies the same field equations, is identical with G(x) outside the 'hole,' but differs from it inside the 'hole.'

Only two years later, after he realized that its validity depends on a tacit assumption, was he able to move beyond this 'hole' argument to adopt generally covariant field equations. Norton (1984) tells the story in some detail. Einstein had been assuming that the physical identity of the points of the region, that is, the events corresponding to the argument *x* in G(x) and G'(x), was established prior to the specification of the metric. In any theory in which the metrical structure is given a priori, the physical identity of points of the space-time manifold is indeed established before any dynamical considerations. But, in general relativity the metrical structure forms part of the set of dynamical variables, which must be determined before the points of space-time have any physical properties.

In special relativity, for example, it is assumed that the individuation of the points of Minkowski space is established by a framework of rigid rods and clocks or some similar device before one begins to do electrodynamics. Hence, even though such fields are mathematically isomorphic, one can distinguish physically between Coulomb fields with the same charge in Minkowski space centered at different points in the same inertial frame, or at the origins of different inertial frames. In general relativity, on the other hand, it makes no sense to try to make a physical distinction between Schwarzschild fields with the same mass. All such Schwarzschild space-times must be identified as corresponding to a single gravitational field, since the points of their respective manifolds only inherit their physical individuation and properties from the metrical field. Speaking more generally, a space-time (manifold with a metric) corresponds to a gravitational field; but a gravitational field corresponds to an equivalence class of space-times. I know of only a few textbooks of general relativity that make this point clearly, notably Hawking and Ellis (1973), p. 56, and Sachs and Wu (1977), p. 27. Without understanding this point, it is impossible, for example, to make sense of the Cauchy problem in general relativity (see Hawking and Ellis, pp. 227-228).

This is the true significance of the concept of general covariance, as Peter Bergmann among others has long known.Unfortunately, this profound insight of Einstein was originally expressed in the language of coordinate systems, rather than in the coordinate-free language of modern formulations of differential geometry. This has been responsible for a great deal of misunderstanding of the concept.

Attempts to quantize the field equations of general relativity utilizing some variant of the techniques of special-relativistic quantum field theory often fail to take into account problems peculiar to generally covariant theories. For example, the physical interpretation of the usual quantum formalism depends heavily on the presence of a given space-time metrical structure. With its help, such concepts as the position of a measuring apparatus, the timing of a measurement, etc., may be introduced.

Without such a pre-existent metrical structure- indeed, when this metrical structure is included among the dynamical variables- the meaning of such concepts (or their replacement by others) becomes a major problem in interpreting the formalism.

It is often assumed that such problems of interpretation can be avoided by the introduction of a background metric, usually taken to be flat, and often also used as the starting point for an iterative approximation procedure. In addition to the problem of justifying its introduction, discussed above, there is the problem of interpreting this background metric itself. Even though the two are the same in mathematical form, Minkowski space as a particular solution to the field equations is conceptually distinct from Minkowski space as the given space-time structure of the special theory of relativity. All of the problems of physical individuation of the points of space-time discussed above apply to Minkowski space in the first sense (or to any other background solution) just as much as they do to any other solution to the field equations.

PROBLEMS WITH THE STANDARD MATHEMATICAL FORMULATION

The standard mathematical formulation of the general theory (see the texts mentioned above, for example) starts with a bare differentiable manifold, the points of which are interpreted as physical events, and then imposes various fields on this manifold. In particular, a second rank symmetric covariant tensor field is interpreted as the space-time metrical structure interrelating these events, and as the gravitational potentials. While this way of introducing the space-time structure is quite adequate in the case of non-generally covariant theories in which this structure is given *a priori*, it does not do full justice either to Einstein's relational concept of space-time, or to the actual practice of relativists in formulating and solving problems.

In Einstein (1952), he wrote:

> On the basis of the general theory of relativity ... space as opposed to `what fills space' ... has no separate existence ... If we imagine the gravitational field, i.e., the functions $g_{ik}$ to be removed, there does not remain a space of the type [of Minkowski space in special relativity- JS], but absolutely *nothing*, not even a `topological space.' For the functions g_{ik} describe not only the field, but at the same time also the

> topological and metrical structural properties of the manifold ... There is no such thing as an empty space, i.e. a space without a field. Space-time does not claim existence on its own, but only as a structural quality of the field.

In one of his last comments on the question, Einstein (1954), he noted:

> It required a severe struggle to arrive at the concept of independent and absolute space, indispensible for the development of theory, It has required no less strenuous exertions subsequently to overcome this concept- a process which is by no means as yet completed.

The standard mathematical treatment clearly does not fulfill Einstein's criterion: no metric, no anything. The manifold of space-time events is introduced prior to the introduction of any other structures.

Even more important, the standard treatment does not correctly describe the actual practice of relativists in solving the field equations. At least as early as the 1917 discussion between Einstein and Felix Klein about the topology of the spatial cross-sections of the Einstein static cosmological model, there has been some understanding by relativists that the topology of a solution to the field equations is not given a priori, but has to be investigated. Einstein had further experience with this question when, together with Nathan Rosen, he developed the `bridge' interpretation of the Schwarzschild metric.

Having lived through the exciting period during which the Kruskal interpretation of the Schwarzschild solution was developed and black hole physics rose to prominence, I need hardly remind this audience that we do not start out with a fixed manifold topology, but solve the field equations on a generic patch. Once such a solution has been found, the problem of obtaining its maximal global extension is a major one. It involves a lot of hard work, which generally includes the formulation of criteria for an acceptable extension in order to arrive at a unique answer. Questions such as maximal analytic extension, geodesic completeness (timelike and/or null), and the adjunction of ideal boundary points at infinity (null and/or spatial) have generated some of the most interesting research on general relativity in recent years.

Yet we persist in formalizing what we do as if the choice of manifold topology were made at the outset. It appears to me unwise to allow this gap between official ideology and daily practice. Not only does it mislead students and researchers in other fields attempting to understand the nature of our work; the gap may conceal important new insights into the nature of the theory. The correct mathematical formulation of a problem often suggests further and unexpected avenues of progress.

FIBER BUNDLES AND SHEAVES

Recently (Stachel 1986), I discussed in some detail how use of the fibre bundle concept can resolve the first of the problems mentioned above (no metric, no anything). I shall briefly summarize the advantages of this approach to generally covariant field theories. First of all, it deals with entire classes of solutions to the field equations, rather than with individual metrics, thus focussing attention on the fact that the space-time structure itself is not given, but must be determined as a cross-section of the bundle satisfying certain criteria (generally, field equations). Second, the use of jet extensions of the bundle gives a structure on which field equations may be formulated. Third, and most relevant to the problem at hand, points of space-time may be represented, not by points of the base manifold as is usually done, but by a mapping of points of a cross-section of the bundle into points of the base manifold. Events are then defined as sets of such mappings in the equivalence class of cross-sections corresponding to a unique gravitational field. Thus, if there is no gravitational field, there are no events.

However, the fibre bundle approach clearly does not solve the second problem discussed in the previous section. The topology of the base manifold is given *a priori*, so that a different fibre bundle must be introduced *a posteriori* for each topologically distinct class of solutions. The process of finding the global topology cannot be formalized within the fibre bundle approach. It appears that sheaf cohomology theory is the appropriate mathematical theory for dealing with the problem of going from local to global solutions, as noted by Wells (1980), p. 36. This is not the time nor the place, nor am I the person to discuss this suggestion in detail. Let me close by appealing to the mathematically more sophisticated members of our group to work on closing this outstanding gap between our creed and our deed.

Einstein, A. (1952), Relativity and the problem of space. In ibid., Relativity: the Special and the General Theory, pp. 135-157. New York: Crown.

Einstein, A. (1954). Foreword. In Jammer, M. Concepts of Space, pp. xiii-xvi. Cambridge, Mass.: Harvard University Press.

Hawking, S. & Ellis, G. (1973). The Large Scale Structure of Space-Time. Cambridge: Cambridge University Press.

Norton, J. (1984). How Einstein Found his Field Equations: 1912-1915. In Historical Studies in the Physical Sciences, vol. 14, ed. J. L. Heilbron, pp. 253-316. Berkeley: University of California Press.

Sachs, R. & Wu, H. (1977). General Relativity for Mathematicians. New York: Springer-Verlag.

Sciama, D. (1979). Black Holes and Fluctuations of Quantum Particles: An Einstein Synthesis. In Relativity, Quanta and Cosmology in the Development of the Scientific Thought of Albert Einstein, vol. 2, ed. F. de Finis, pp. 681-724. New York: Johnson Reprint Corporation.

Stachel, J. (1979). The Genesis of General Relativity. In Einstein Symposion Berlin, eds. H. Nelkowski et al, pp. 428-442. Berlin: Springer-Verlag.

Stachel, J. (1980). Einstein and the Rigidly Rotating Disk. In General Relativity and Gravitation One Hundred Years After the Birth of Albert Einstein, vol. 1, ed. A. Held, pp. 1-15. New York: Plenum Press.

Stachel, J. (1986). What a Physicist Can Learn from the Discovery of General Relativity. In Proceedings of the Fourth Marcel Grossmann Meeting on General Relativity, ed. R. Ruffini, pp.1857-1862.

Wells, R. O. (1980). Differential Analysis on Complex Manifolds. New York: Springer-Verlag.

ASTRONOMICAL AND SPACE EXPERIMENTS TO TEST RELATIVITY

J. H. Taylor
Joseph Henry Laboratories and Physics Department,
Princeton University, Princeton, NJ 08544, USA

INTRODUCTION

On astronomical distance scales, where the dominating forces are often gravitational, the effects of general relativity are more readily observed and measured than in the terrestrial laboratory. However, testing general relativity has always been difficult because the astronomical observer, constrained to the status of passive recorder of distant events and phenomena, often does not know (and may not be able to determine) some of the important parameters of his "apparatus." These parameters may include such diverse and unpleasantly messy things as the details of planetary surfaces, the unmeasured masses of asteroids, or wave propagation through the solar corona.

Despite this handicap, dedicated workers have made impressive progress over the past quarter century in devising and improving astronomical and space experiments to test relativity. I will assume that you are familiar with the results of many of these experiments. For sake of completeness I will briefly review them here, but as the results are readily available in the literature, and as I have little to add to their elucidation, I will not dwell on them at length. For an excellent and still almost-up-to-date review, see Will (1984).

Some very ambitious space experiments of great potential importance have been planned, or at least contemplated, in recent years, and development work for some of them is well underway. I refer especially to the plans to put high-precision gyroscopes into Earth orbit, in order to observe "magnetic" gravitational effects that should cause the gyroscopes to precess (see Anderson *et al.* 1982 for a review). Other fond hopes include a spacecraft in an eccentric orbit passing close to the sun, carrying a hydrogen maser clock and outfitted to measure gravitational redshift and the solar mass quadrupole moment, and a space-borne precision optical interferometer which could measure the second-order part of the bending of light at the limb of the sun. Such experiments have long development times and uncertain dependence on national funding goals and patterns. Some further details and progress reports will be presented in the afternoon symposia this week.

My principal topic today will follow a somewhat different line: the uses of pulsars as tools in experimental relativity. The discovery of pulsars in 1967, and their identification as rapidly spinning neutron stars, marked an important confirmation of a longstanding prediction involving relativistic gravitation theory. At about the same time, neutron stars were discovered in accreting X-ray binary systems as well. Neutron stars are intrinsically relativistic objects, and *any* measurements of their detailed nature and characteristics are potentially useful as probes of relativity theory. However, with current levels of understanding and precision of observational techniques, quantitative tests of relativity along these lines are impractical. In the physics of neutron stars, as in numerous problems in cosmology, one makes more progress by assuming that general relativity is correct, and proceeding to investigate the astrophysical consequences.

The reason that pulsars are important to relativity, and the basis for most of my remarks today, is the fact that these spinning neutron stars are natural flywheels which make extraordinarily good clocks. In much of modern experimental physics, and in astronomy as well, the most precise measurements are made of quantities that can be reduced to either time intervals or frequencies. High-precision comparisons of the time kept by pulsar clocks with terrestrial atomic clocks has revealed much about rotating neutron stars, and has also provided new tools for probing relativistic phenomena. Because many of these techniques and results are not widely known in the relativity community, I will describe them in some detail.

HISTORICAL OVERVIEW

Experimental tests of general relativity are usually described in terms of measurements of various "effects" that are small departures of the real physical world from the Newtonian limit. We talk about gravitational redshift, the perihelion shift of planetary orbits, and the bending and delay of photon trajectories as if these things were inherent characteristics of Einstein's theory. Of course, the opposite is really true: in the language of relativity the various "effects" do not exist; it is only in the context of non-relativistic mechanics that corrections to equations are necessary to make them conform with the facts. I will, however, follow usual practice and discuss relativity tests as measurable corrections to a non-relativistic (or partially non-relativistic) description of the world.

Gravitational Redshift

Attempts to observe the predicted gravitational redshift of light from the surface of the Sun, or from white dwarf stars, were made as early as 1927. Unfortunately, the results were generally inconclusive because of the kind of problem I mentioned earlier: these astrophysical laboratories are messy places, with some important conditions outside the observer's control and some important parameters not directly accessible to measurement. Accurate measurements of gravitational redshift were made first in the ground-based laboratory of Pound and co-workers (1960, 1965), who achieved an experimental accuracy of about 1% of the expected effect. More recently, Vessot *et al.* (1980) carried out an experiment using hydrogen maser clocks carried in sub-orbital space flight, reaching the 10^{-4} fractional level of precision. At all achieved levels of accuracy, results of gravitational redshift experiments have been in accord with the Einstein equivalence principle and hence with relativity theory.

Perihelion Advance of Planetary Orbits

The rate of perihelion advance of the orbit of Mercury has been known for seventy years to be in approximate agreement with relativity. Once again, however, the cleanliness of the astrophysical laboratory is open to some question. Dicke (1974, and references therein) first pointed out that the observed motion of Mercury is in satisfactory agreement with relativity *only* if the mass quadrupole moment of the Sun is small enough to contribute negligibly to the Newtonian precession. Efforts to measure the dimensionless solar quadrupole moment, J_2, either by observing the oblateness of the sun's visible disk or through analysis of its acoustical normal modes of oscillation, have not yet produced fully convincing and consistent proof that J_2 is small enough to be neglected — though it is tantalizingly close to this limit. The current state of affairs (see, e.g. Shapiro *et al.* 1976, Hellings 1983) is that the observed perihelion advance is within about 0.5% of the value expected in general relativity, *assuming* that $J_2 \lesssim 10^{-6}$.

Bending of Light

Surprisingly little improvement has been made in more than 60 years of solar eclipse observations designed to measure the bending of light rays near the limb of the sun. This fact is illustrated in Figure 1, in which the measured amount of bending, expressed as a fraction of the expected amount, is plotted as a function of date. The best optical measurements have achieved accuracies only a little better than 10% — again largely because of the necessarily messy aspects of the experiments, including difficult conditions at remote eclipse sites, long intervals between eclipse observations and control observations, and so on.

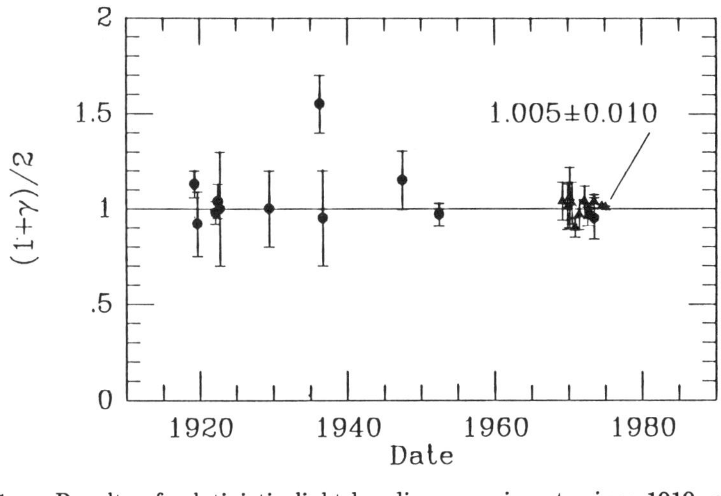

Fig. 1. – Results of relativistic light-bending experiments since 1919, expressed as fractions of the expected amount. The parametrized post-Newtonian parameter γ has the value 1.0 in general relativity. Triangles represent radio wavelength experiments; see Will (1984) for references to the original literature.

By the late 1960's, techniques of radio interferometry had matured sufficiently to provide a competitive alternative to the optical eclipse measurements. However, even the radio observations have their messy aspects, the principal difficulty being difficult-to-calibrate refractive effects in the solar corona. To deal with this problem, Fomalont & Sramek (1976) carried out a simultaneous dual-frequency experiment. Its results are in accord with relativity at the 1% level of accuracy.

Shapiro Time Delay

As first pointed out by Shapiro (1964), relativity predicts that photons propagating in a strong gravitational field should be retarded as well as deflected. As illustrated in Figure 2, experiments carried out between 1968 and 1980 confirmed the prediction — first by using radar signals passively reflected from planetary surfaces, and more recently with signals coherently transponded from electronic equipment aboard the Viking orbiters and landers near or on the surface of Mars. An important advantage in these experiments is the one mentioned earlier: the fundamental measured quantity is a time interval, which can be determined with very high precision. In this instance, the expected relativistic effect has a peak amplitude of more than 200 μs, and the experimental uncertainty in the Viking time-delay data was as small as 0.01 μs. Once again, though, at the required level of precision the solar system turns out to be a messy laboratory. To achieve the desired accuracy the solar corona had to be calibrated with the aid of multi-frequency transmissions, and various parameters of the solar-system model had to be estimated, simultaneously with the relativity parameter characterizing the propagation delay, from the same set of data. With such problems properly taken into account, Reasenberg *et al.* (1979) quote the impressive accuracy of 0.1% for the agreement of the excess time delay with general relativity.

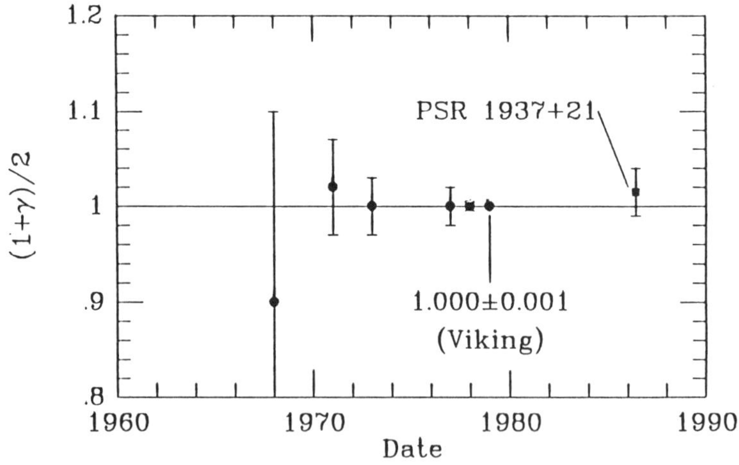

Fig. 2. – Measurements of the Shapiro time delay in the solar system, plotted as a function of date. Original references for all but the most recent point may be found in Will (1984).

Among the most comprehensive solar-system constraints on gravitation theories are those derived from multi-parameter fits for a general planetary ephemeris, using thousands of measurements of diverse kinds as input data and solving for hundreds of unknowns. The unknowns include masses, initial positions, and velocities of the moon, planets, and largest asteroids, as well as numerous "nuisance parameters" of no particular physical interest. In addition, the procedure estimates the relativistically interesting PPN parameters β and γ, the solar quadrupole moment J_2, and the apparent rate of change of the gravitational constant G. The results show that β and γ are unity, in agreement with relativity, to within tenths of a percent or better, and that G changes by less than a part in 10^{11} per year (Reasenberg 1983; Hellings 1983).

PULSAR TIMING AND RELATIVITY

The radio signal from a typical pulsar consists of a nearly periodic series of brief pulses of broad-band noise. Synchronous averages of the power in the signal converge to a stable waveform with a typical pulse-width equal to a few percent of the period. The pulse repetition period is identified with the rotation period of the neutron star, and the shape of the waveform is identified with magnetically-constrained directional beaming of the emitted signal.

Pulsar timing observations are usually made by synchronously integrating the waveform for several minutes or longer and recording the start time of the integration as measured by a local time standard. The local standard is ultimately calibrated against the best available standard of atomic time. The phase of the observed waveform, measured relative to a long-term average "standard profile" for the pulsar in question, yields an estimate of the time on the pulsar "clock" at a known atomic time.

From the point of view of an external inertial observer, time kept by atomic clocks on Earth proceeds at irregular rates because they undergo accelerations and the Earth moves through regions of differing gravitational potential. However, the solar-system ephemeris mentioned earlier can be used to transform proper times of observations to equivalent coordinate times at, say, the solar system barycenter, and then to the reference frame of the pulsar. These corrections are non-trivial, but can be done with accuracies $< 1\ \mu$s, at least over time scales of a few years, with the aid of ephemerides based entirely on *other* types of observations. The messiness of the solar system is, from the point of view of the pulsar observer, contained *a priori* on a reel of magnetic tape, and is largely buried in his experimental noise level. Rather than numbering in the hundreds, the unknown quantities in a pulsar timing experiment are typically fewer than a dozen.

Figure 3 illustrates the results of nearly four years of timing observations of one particular pulsar, PSR 1937+21. This is not a typical pulsar, but rather the fastest rotating (period = 1.56 ms) and most accurately timable pulsar presently known. The observations were made with the 305 m radio telescope at the Arecibo Observatory in Puerto Rico. Those taken between November 1982 and October 1984 have been described by Davis *et al.* (1985); the more recent ones are the unpublished work of Lloyd Rawley, Mike Davis, and me. The plotted points are post-fit residuals from a seven-parameter fit to the observed pulse arrival times. Fitted parameters include the initial phase, frequency, and frequency derivative of the pulsar's spin, plus its celestial coordinates and their (very small) rates of change.

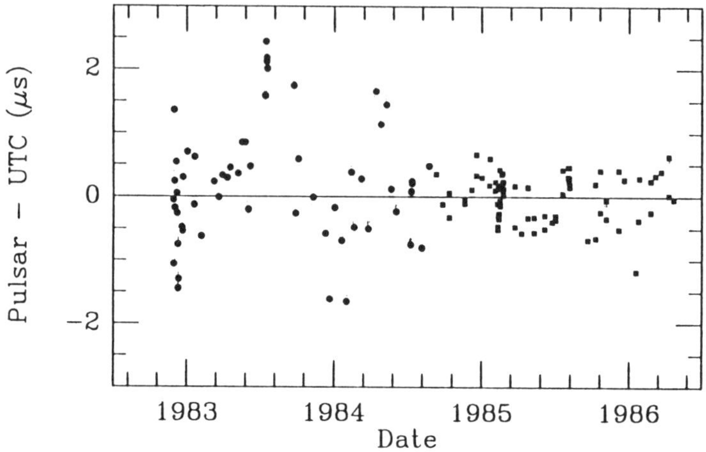

Fig. 3. – Timing residuals for the millisecond pulsar PSR 1937+21, obtained over nearly four years.

The data illustrated in Figure 3 show that our measurement uncertainty since late 1984 has been approximately $\delta t = 0.3$ μs, which corresponds to 0.07 degrees of longitude of the spinning star. Before October 1984 the uncertainties were several times larger, and less random, due to a problem now known to have been caused by the LORAN C navigation signals then used to calibrate the local atomic clock against standards kept by national timekeeping laboratories. Time transfer is now accomplished by means of Global Positioning Service satellites, and the offset of the Arecibo clock relative to UTC(NBS) is usually known to better than 0.1 μs. It is interesting to note that between the first and last measurements plotted in Figure 3, an interval of $T \approx 10^8$ s, *exactly* 68,811,216,436 pulse periods had elapsed. There are no pulse numbering ambiguities, even over gaps in the data as large as 10^9 pulses. The implied rotational stability of the pulsar is impressive: a crude estimator is $\delta t/T \lesssim 3 \times 10^{-15}$, which makes the pulsar stability comparable with that of the very best atomic clocks.

To illustrate the sensitivity of this data set to relativistic effects in the solar system, Figures 4 and 5 show what happens when two important terms are omitted in the transformation from terrestrial atomic time to barycentric coordinate time. Figure 4 shows the timing residuals obtained since October 1984, plotted as a function of lunar phase, with the monthly variation in solar gravitational potential at Earth intentionally ignored. Here is direct proof, based on a clock some 15,000 light years from the solar system, that clocks on Earth run more slowly when the moon is full — because at this time of the month we are deeper in the gravitational potential of the Sun!

In a similar way, Figure 5 illustrates the annual variation of the Shapiro time delay as signals from PSR 1937+21 propagate through the solar system toward Earth. We can, in fact, carry out Shapiro's experiment with these data; the result is the 1986 value for the parameter $(1+\gamma)/2$ that I have included in Figure 2. For this purpose the pulsar data are not really competitive with those from the Viking experiment, but the result provides an interesting and independent confirmation nevertheless.

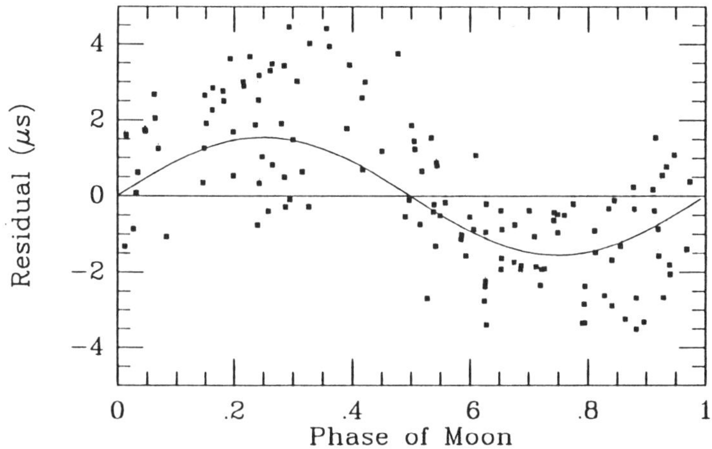

Fig. 4. – Timing residuals for PSR 1937+21, with the monthly term in the correction from proper atomic time to coordinate time intentionally omitted. The sinusoidal curve represents the omitted term.

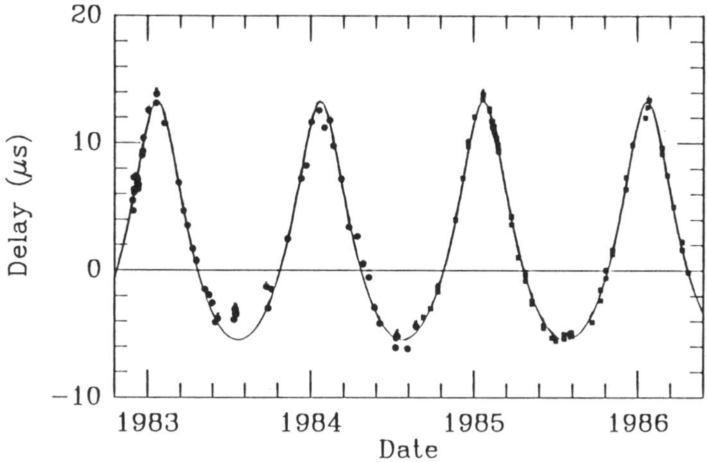

Fig. 5 . – Timing residuals for PSR 1937+21 with the Shapiro time delay in the solar system intentionally omitted. The smooth curve represents the omitted correction.

Cosmic Background Gravitational Radiation

Detweiler (1979) first pointed out that pulsar timing residuals can yield an upper limit to the energy density contained in a cosmic background of gravitational radiation. Results obtained by Romani & Taylor (1983) and Hellings & Downs (1983), based on slowly rotating pulsars observed for more than a decade, showed that at wavelengths of a few light years the energy density in gravitational waves is $\lesssim 5 \times 10^{-4}$ of the density required to close the universe. Hogan and Rees (1984), Witten (1984), and others have pointed out that upper limits much below this level will begin to place interesting constraints on possible conditions in the early universe.

The discovery of millisecond pulsars made this method of searching for gravitational radiation several orders of magnitude more sensitive overnight. Consequently, based on just two years of data on PSR 1937+21, Davis *et al.* (1985) were able to obtain an upper limit $\Omega_g < 5 \times 10^{-4}$ for gravitational wave frequencies in the range 1 to 3 y^{-1}. Today, with almost twice as much data and much of it of higher quality, the limits have become $\Omega_g < 5 \times 10^{-5}$ for $f \sim 1\ y^{-1}$ and $\Omega_g < 1.5 \times 10^{-6}$ for $f \sim 0.3\ y^{-1}$. It is worth noting that any claims to have *detected* a cosmic background of gravitational radiation, as opposed to specifying upper limits, will probably have to wait until a number of millisecond pulsars are available, so that inter-comparisons can be made for objects in different directions.

Relativity and Binary Pulsars

Of more than 400 catalogued pulsars, only seven are known to have orbiting companion stars. Their binary nature is revealed by periodic doppler shifts superimposed on their otherwise nearly periodic sequences of pulses. Binary systems containing pulsars, particularly if their orbital periods are short, can make excellent relativity laboratories. As it happens, though, only PSR 1913+16, the first binary pulsar system to be discovered, has so far proven to be so.

As is the case for timing observations of single pulsars, the important experimental data for binaries consist of measured phases of the pulsar waveform at known atomic times. For binary pulsars, in addition to astrometric and pulsar spin parameters, one must determine from the timing data quantities which specify the orbital motion to a precision consistent with the measurement uncertainties.

Timing observations of the PSR 1913+16 system have been made at more-or-less regular intervals since 1974 (Taylor *et al.* 1976, Taylor & Weisberg 1982, Weisberg & Taylor 1984). If the observed phases near a single epoch are corrected for solar-system effects, converted to doppler velocities, and plotted modulo the 7.75-hour orbital period, they yield radial velocity curves like those shown in the upper part of Figure 6. The maximum radial velocity of the pulsar exceeds 300 km s^{-1}, which suggests at once that the system is at least mildly relativistic. In fact, compared to the measurement uncertainties in the timing data, the largest relativistic effects turn out to be enormous. The evolution of the velocity curve's shape, which is a result of relativistic periastron advance of the orbit, is easily detectable over time spans as short as ten days. The measured rate of periastron advance is some 35,000 times

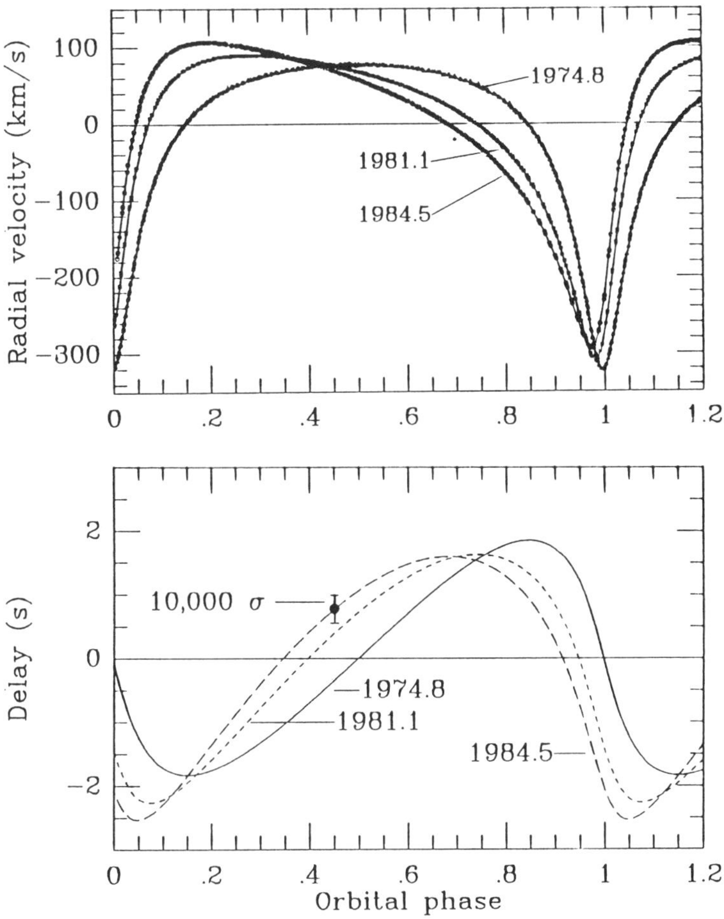

Fig. 6. – *Upper:* Radial velocity measurements of the binary pulsar PSR 1913+16 at three different epochs, plotted as a function of orbit phase. *Lower:* Time delay curves for the same epochs. Current pulse arrival-time measurements have uncertainties approximately 10^4 times smaller than the indicated error bar.

larger than that in the Mercury-Sun binary system, because of the larger masses and much shorter orbital period.

Because a pulsar's rotation is phase-coherent over very long time scales, it is advantageous to analyze the timing data in terms of radial displacements instead of radial velocities. As with single pulsars, one does this by keeping careful tally of the exact number of pulse periods that have elapsed between observations. Radial displacements in the PSR 1913+16 system are plotted as a function of orbital phase at the bottom of Figure 6. Again, the effects of precession of the orbit are easy to see. The light travel time across the orbit is about 4 s; for comparison purposes, and to emphasize the precision with which the largest relativistic effects in the system can be measured, a pseudo-measurement with 10,000-σ error bars is included in Figure 6.

Eight quantities are required in order to completely specify the gravitational dynamics of two point masses in isolated orbit: for example, the initial coordinates and vector velocities of one of them, and the magnitudes of the two masses. In the absence of measurable relativistic effects, doppler data for one member of an orbiting pair can be used to determine five of these parameters. Thus, if we ignore uninteresting rotations about the line of sight, two more parameters would be needed to fully specify the system. Relativistic effects in the orbit can provide the additional measurable parameters. Indeed, it has been shown that with good enough data, *more* than two additional independent parameters can be determined from binary pulsar timing data. Therefore tests of relativity, and of the degree of messiness of the system, should be possible (Blandford & Teukolsky 1975, Smarr & Blandford 1976, Epstein 1977, Haugan 1985, Damour & Deruelle 1986).

When PSR 1913+16 was first being observed in 1974, the measurement uncertainty for a five-minute timing observation was approximately 300 μs (Taylor *et al.* 1976). Improvements in equipment and techniques since then have reduced this uncertainty by a little more than an order of magnitude, to about 20 μs since 1981 (Taylor & Weisberg 1984). At this level of accuracy three post-Newtonian parameters can be measured to levels of a few percent or better, and a fourth can be estimated at the $\sim$ 10% level. Using the Blandford-Teukolsky-Epstein-Haugan formalism as the basis of our analysis, and all of the timing measurements acquired since 1974, Weisberg and I have shown that the system is, indeed, over-constrained by the data.

The five parameters that can be estimated from a Newtonian analysis are all known to five or more significant digits. As illustrated in Figure 7, the four relativistically measurable quantities are then more than sufficient to determine the remaining parameters, the masses of the two stars. All of the measured quantities are consistent with pulsar and companion star masses of 1.445 ± 0.007 and 1.384 ± 0.007 $M_\odot$, respectively. These masses, close to the Chandrasekhar limit for white dwarfs, look very much like neutron star masses. They have made happy the people who argued, on evolutionary grounds, that the PSR 1913+16 system must consist of two neutron stars.

Quantitative Test for Gravitational Radiation

If the companion of PSR 1913+16 is another neutron star — perhaps a "dead" pulsar, now spinning too slowly to generate the potential differences required to activate the radio emission process — then we have quantitative proof of the existence of gravitational radiation as predicted by relativity. The orbital dynamics are not messy at an important level of accuracy; the well-measured parameters provide a clean prediction (from the "quadrupole formula") for the rate $\dot{P}$ at which the orbital period should decay because of energy losses to gravitational waves; and the measured value of $\dot{P}$ is in excellent agreement with this prediction. Furthermore, although the accuracy of the measurement is not high, the observed value of the orbital inclination, $\sin i$, provides a consistency check confirming that all seems to be well.

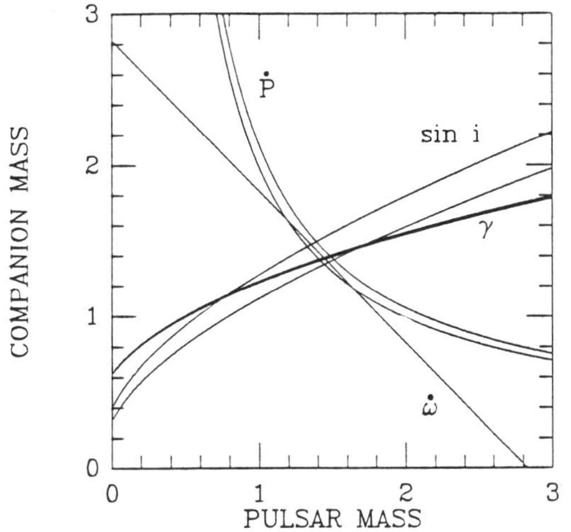

Fig. 7. – Constraints on the component masses in the PSR 1913+16 system imposed by the four relativistically measurable parameters $\dot{\omega}$, γ, $\dot{P}$, and $\sin i$.

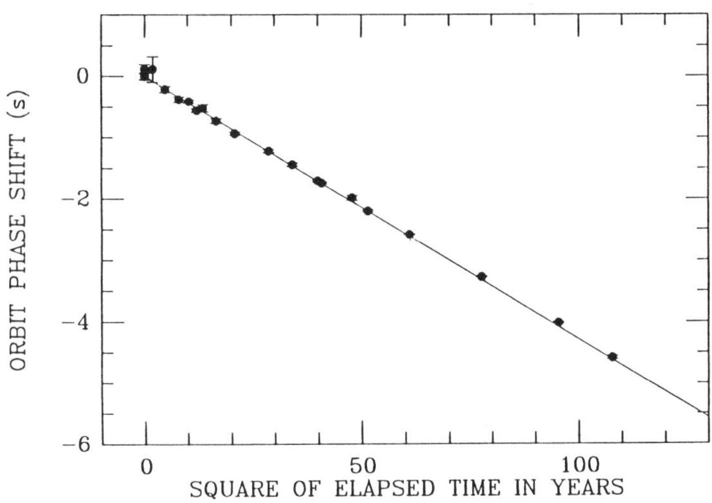

Fig. 8. – Measured orbital phase advance in the PSR 1913+16 system, plotted as a function of t^2 since 1974.8. The sloping line gives the expected rate of orbital decay from gravitational radiation.

For those who like to see and follow the progress of the orbital decay with time, Figure 8 shows that the pulsar is now passing periastron more than 5 s earlier than it would have been if the orbital period had remained fixed at its late 1974 value. It also shows that the orbital phase is, as predicted, advancing quadratically with time. The ratio of the observed rate of decay to that predicted by the quadrupole formula is presently 1.00 ± 0.04.

Is the binary pulsar system really clean enough that we can be sure that non-relativistic contributions to $\dot{\omega}$ are negligible, and our mass estimates reliable? And can we be sure that subtleties in the relativistic timing formula have not been overlooked? I believe there are increasingly good reasons to be confident that such worries are unfounded. One such reason is that improved data, and the lengthening span of the observations, have substantially reduced the parameter space available for messy solutions to the system — so much so that these solutions now seem quite contrived, and much less probable than the simplest solution. Another reason is the result of a careful re-analysis of the relativistic two-body problem that has recently been carried out by Damour & Deruelle (1985, 1986). They have produced a much simplified derivation of a timing formula for a relativistic binary pulsar system, with a level of accuracy more than sufficient for the present experimental data on PSR 1913+16.

Weisberg and I have used the Damour-Deruelle formalism to carry out a new and largely independent analysis of the pulsar's orbit. We have chosen a parameter set which allows a minimum, zero-redundancy solution for the seven quantities needed to specify the two-body dynamics, including gravitational radiation damping. One or two extra terms can then be introduced in order to test for "messiness" in the form of non-relativistic contributions to $\dot{\omega}$, or departures from the quadrupole-formula prediction in the rate of orbital decay, or both.

Results of this analysis are summarized in Table 1, the first part of which lists the values obtained (or assumed) in each of three solutions. Parameters include the total mass of the system M, the mass of the companion star m_2, a "Fractional Dirtiness" parameter $\Delta\dot{\omega}/\dot{\omega}_{GR}$, and the "Orbital Damping" parameter $\dot{P}/\dot{P}_{GR}$. In a clean system obeying general relativity, the last two terms will be 0.0 and 1.0, respectively.

We find that the minimal parameter set, solution 1, fits the data as well as any of the augmented sets. Solution 2, which assumes the quadrupole formula is correct and tests for evidence of messiness in the orbit, reveals none at the 2% level in $\dot{\omega}$. Solution 3, which assumes a clean system — as would be the case for a pair of orbiting neutron stars — shows that observed orbital damping agrees with the quadrupole formula at the 4% level of accuracy. In our view this test firmly and quantitatively establishes the existence of gravitational radiation at the level predicted by relativity.

A doubly redundant solution, in which *both* of the extra parameters are allowed to vary, would provide a test of gravitational radiation damping of the PSR 1913+16 system free of any assumptions about the compactness of the companion star. At presently achievable levels of accuracy such a solution is infeasible, but we are hopeful that a proposed major upgrading of the Arecibo radio telescope will make it possible in the near future. Together with opportunities afforded by additional millisecond pulsars, which are now being actively sought by a number of groups worldwide, this prospect promises a rosy future for astronomical contributions to relativity.

Table 1. Measured parameters of the PSR 1913+16 system.

Solution	Total Mass, M ($M_\odot$)	Companion Mass, m_2 ($M_\odot$)	Fractional Dirtiness $\Delta\dot{\omega}/\dot{\omega}_{GR}$	Orbital Damping $\dot{P}/\dot{P}_{GR}$
1.	2.82872 ±0.00012	1.384 ±0.007	0.000*	1.00*
2.	2.82 ±0.08	1.383 ±0.022	0.001 ±0.020	1.00*
3.	2.82872 ±0.00012	1.384 ±0.007	0.000*	1.00 ±0.04

* Parameter held fixed at this value.

REFERENCES

Anderson, J. T., Cabrera, B., Everitt, C. W. F., Leslie, B. C., & Lipa, J. A. (1982). *In* Proc. 2nd Marcel Grossman Meeting on General Relativity, ed. R. Ruffini, p. 939. (Amsterdam: North-Holland).

Blandford, R. D., & Teukolsky, S. A. (1976). *Astrophys. J.*, **207**, 580.

Damour, T., & Deruelle, N. (1985). *Ann. Inst. H. Poincaré (Physique Théorique)*, **43**, 107.

Damour, T., & Deruelle, N. (1986). *Ann. Inst. H. Poincaré (Physique Théorique)*, **44**, 263.

Davis, M. M., Taylor, J. H., Weisberg, J. M., & Backer, D. C. (1985). *Nature*, **315**, 547.

Detweiler, S. (1979). *Astrophys. J.*, **234**, 1100.

Dicke, R. H. (1974). *Science*, **184**, 419.

Fomalont, E. B., & Sramek, R. A. (1976). *Phys. Rev. Lett.*, **36**, 1475.

Epstein, R. (1977). *Astrophys. J.*, **219**, 92.

Haugan, M. P. (1985). *Astrophys. J.*, **296**, 1.

Hellings, R. W. (1983). *In* General Relativity and Gravitation, ed. B. Bertotti, F. deFelice, & A. Pascolini, p. 365. (Dordrecht: Reidel).

Hellings, R. W., & Downs, G. S. (1983). *Astrophys. J. (Letters)*, **265**, L39.

Hogan, C. J., & Rees, M. J. (1984). *Nature*, **311**, 109.

Pound, R. V., & Rebka, G. A. (1960). *Phys. Rev. Lett.*, **4**, 337.

Pound, R. V., & Snider, J. L. (1965). *Phys. Rev.*, **140**, B788.

Reasenberg, R. D., Shapiro, I. I., MacNeil, P. E., Goldstein, R. B., Breidenthal, J. C., Brenkle, J. P., Cain, D. L., Kaufman, T. M., Komarek, T. A., & Zygielbaum, A. I. (1979). *Astrophys. J. (Letters)*, **234**, L219.

Reasenberg, R. D. (1983). *Phil. Trans. Roy. Soc.*, **310**, 227.

Romani, R. W., & Taylor, J. H. (1983). *Astrophys. J. (Letters)*, **265**, L37.

Shapiro, I. I. (1964). *Phys. Rev. Lett.*, **13**, 789.

Shapiro, I. I., Counselman, C. C., & King, R. W. (1976), *Phys. Rev. Lett.*, **36**, 555.

Smarr, L. L. & Blandford, R. D. (1976). *Astrophys. J.*, **207**, 574.

Taylor, J. H., Hulse, R. A., Fowler, L. A., Gullahorn, G. E., & Rankin, J. M. (1976). *Astrophys. J. (Letters)*, **206**, L53.

Taylor, J. H., & Weisberg, J. M. (1982). *Astrophys. J.*, **253**, 908.

Vessot, R. F. C., Levine, M. W., Mattison, E. M., Blomberg, E. L., Hoffman, T. E., Nystrom, G. U., Farrell, B. F., Decher, R., Eby, P. B., Baugher, C. R., Watts, J. W., Teuber, D. L., & Wills, F. O. (1980). *Phys. Rev. Lett.*, **45**, 2081.

Weisberg, J. M., & Taylor, J. H. (1984), *Phys. Rev. Lett.*, **52**, 1348.

Will, C. M. (1984). *Physics Reports*, **113**, 345-422.

Witten, E. (1984). *Phys. Rev.*, **D30**, 272.

INFLATION IN THE UNIVERSE, CIRCA 1986

Michael S. Turner
Fermilab Astrophysics, Fermi National Accelerator Laboratory, Batavia, IL 60510 USA, and Departments of Physics and Astronomy and Astrophysics, Enrico Fermi Institute, The University of Chicago, Chicago, IL 60637 USA

Abstract. The hot big bang cosmology, or the standard cosmology as it is appropriately known, is a highly successful model, providing a reliable and tested accounting of the Universe from 0.01 sec after the bang until today, some 15 Gyr later. However, very special initial data seem to be required in order to account for the observed smoothness and flatness of our Hubble volume and for the existence of the small primeval density inhomogeneities required for the formation of structure in the Universe. Inflation offers a means of accounting for these special initial data which is based upon physics at sub-planck energy scales ($\ll m_{pl} \simeq 10^{19} GeV$) and is motivated by contemporary ideas in particle theory. Here I review the status of the 'Inflationary Paradigm,' *Circa* 1986. At present essentially all inflationary models involve a very weakly-coupled (quantified by the presence of a dimensionless parameter of order 10^{-12} or so) scalar field which is displaced from the minimum of its potential. Regions of the Universe where the scalar field is initially displaced from its minimum undergo inflation as the scalar field relaxes, resulting in a Universe today which resembles ours in regions much larger than our present Hubble volume ($\simeq 10^{28} cm$), but which on very large scales ($\gg 10^{28} cm$) may be highly irregular. The most conspicuous blemish on the paradigm is the lack of a compelling particle physics model to implement it. I also review some other unresolved issues, including the all-important confrontation between inflation and observational data.

SUCCESSES OF THE STANDARD COSMOLOGY

The standard cosmology is a remarkable achievement. Based upon the Friedmann-Robertson-Walker (FRW) homogeneous and isotropic cosmological model, it provides us with an accurate description of the evolution of the Universe from about $10^{-2} sec$ after the bang (when the temperature of the Universe was about $10 MeV$) until the present some $3 \times 10^{17} sec$ later (and temperature $2.75K$). Support for the standard cosmology is based upon a triad of observations. First, the isotropic Hubble flow and homogeneous distribution of galaxies; light from the most distant galaxies and QSO's (redshifts of order 3-4) left these objects only a few billion years after the bang and therefore test the model to within a few billion years of the bang. Second, there is the cosmic microwave background radiation (CMBR) whose spectrum is consistent with that of a black body at a temperature of $2.75 \pm 0.05K$ (Smoot et al. 1985; Peterson et al. 1985) and which is spatially uniform to about a part in 10^4

on angular scales from a few arcminutes to 180 degrees (Wilkinson 1986). [The only anisotropy unambiguously detected thus far is the dipole component whose magnitude is of order 10^{-3} and whose simplest interpretation is being due to our motion with respect to the cosmic rest frame.] The surface of last scattering for the CMBR is the Universe at $t \simeq 10^{13} sec$ and $T \simeq 1/3 eV$, and so it provides a probe of the standard model to within a few 100,000 yrs of the bang. Finally, there is the concordance of the cosmic abundances of D, ^{3}He, ^{4}He, and ^{7}Li with the predictions of big bang nucleosynthesis, providing that the present baryon-to-photon ratio $\eta \simeq (4-7) \times 10^{10}$ (equivalently, $0.014 \lesssim \Omega_B h^2 \lesssim 0.035$, where as usual $H_0 = 100 hkmsec^{-1}Mpc^{-1}$) (Yang et al. 1984; Boesgaard & Steigman 1985). According to the standard cosmology there was an epoch of primordial nucleosynthesis from about $t \simeq 0.01 sec - 300 sec$ ($T \simeq 10 MeV - 0.1 MeV$), and so the cosmic abundances of these light elements serve to test the model at times to within a fraction of a second after the bang. In addition, the standard cosmology provides a general framework for understanding how structure in the Universe evolved (see, e.g., Efstathiou & Silk 1983): once the Universe became matter-dominated ($t \simeq 10^{10} sec$ and $T \simeq 10 eV$), small ($\simeq 10^{-4}$) primeval density inhomogeneities grew via the Jeans (or gravitational) instability into the plethora of structure we observe today (galaxies, clusters of galaxies, superclusters, voids, etc.).

During its early history ($t \lesssim 10^{10} sec$) the energy density of the Universe was dominated by relativistic particle species in thermal equilibrium with a temperature $T \simeq (t/sec)^{-1/2} MeV$, and the cosmic scale factor $R(t) \propto t^{1/2}$. While the standard cosmology is only tested back to times of order $10^{-2} sec$, the standard model of particle physics, the $SU(3)_c \otimes SU(2)_L \otimes U(1)_Y$ gauge theory of the strong, weak, and electromagnetic interactions (believed to be valid at energies $\lesssim 1000 GeV$), and theoretical speculations about physics at very high ($\gtrsim 10^{14} GeV$) energies (e.g., grand unification, supersymmetry/supergravity, and superstring theories) allow us to extrapolate the model back to times as early as $10^{-43} sec$ and perhaps even earlier (see Fig. 1). [At times earlier than $10^{-43} sec$ (corresponding to temperatures $\gtrsim 10^{19} GeV$) quantum gravitational effects should become very important and extrapolation to times this early necessarily requires a quantum description of gravity.] The speculations have proven very interesting, from extra dimensions, to baryogenesis, to monopoles, to cosmic strings, to relic WIMPs, to phase transitions, and finally inflation, the subject of my talk. Of course, all of these interesting speculations could prove to be nothing more than that; however, nothing in our present knowledge of physics would tell us that such speculations are *a priori* wrong. Compare this to the situation some 20 years ago when it was thought that hadrons were fundamental: at times earlier than about $10^{-5} sec$ after the bang inter-particle distances should have been less than the size of a typical hadron, thus precluding sensible speculations about times earlier than this.

SHORTCOMINGS OF THE STANDARD MODEL

As successful as it is, the standard model has its shortcomings. They involve a number of very fundamental facts about the Universe we observe within

Figure 1 – The Complete History of the Universe, according to the Standard Cosmology and currently fashionable ideas in elementary particle physics.

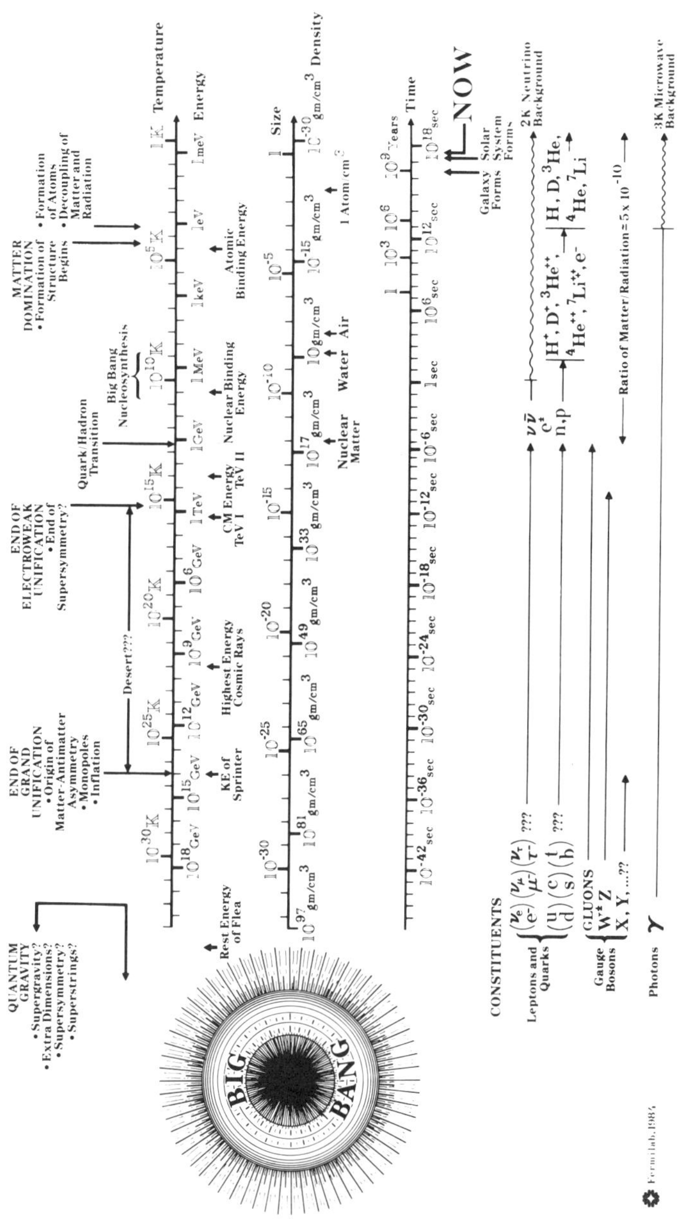

our Hubble volume, which it can accommodate, but by no means provides fundamental explanations for (in contrast, the standard cosmology provides a fundamental explanation for the abundance of the light elements). These *cosmological conundrums* are by this time very well known; they include: (1) The smoothness (isotropy and homogeneity) of our present Hubble volume (radius $\simeq H^{-1} \simeq 10^{28}cm$) on scales $\gg 10Mpc$, as evidenced by the uniformity of the CMBR and of the distribution of galaxies. The size of our Hubble volume is conveniently quantified by the entropy within it, which is dominated by the relic photon and neutrino seas and is of order 10^{88}. Because of the existence of particle horizons in the standard cosmology it is essentially impossible to account for a smooth volume this large as having evolved due to physical processes operating in the early Universe: when matter and radiation last interacted, the Hubble volume at the time contained an entropy of only about 10^{82}, so that particle interactions could not account for such a large smooth volume. (2) The origin and nature of the primeval fluctuations required to explain the rich array of structure in the Universe today; curvature fluctuations of order 10^{-4} on mass scales $\sim 10^8 M_\odot - 10^{15} M_\odot$ are required. In the standard cosmology, curvature fluctuations cannot arise spontaneously (again because of the existence of particle horizons; see, e.g., Bardeen 1980) and must be put in *ab initio.* [It is possible that the requisite fluctuations are isocurvature fluctuations and were created during the early history of the Universe, perhaps during a phase transition; a promising example which has attracted a great deal of attention lately is cosmic strings (see, e.g., Vilenkin 1985; Albrecht & Turok 1985; Turok 1985).] (3) The apparent flatness of our Hubble volume; the radius of curvature ($R_{curv} \equiv R(t)|k|^{-1/2} = H^{-1}/|\Omega - 1|^{1/2}$) in our vicinity must be at least comparable to the radius of our Hubble volume. Had the radius of curvature been of order the planck length ($\simeq 10^{-33}cm$) at the planck time ($\simeq 10^{-43}sec$), it would only be of order $0.1cm$ when the Universe reached a temperature of $3K$. [Put another way, in order that Ω still not be too different from unity, at the planck time it must have been equal to 1 to within a part in 10^{60}.] (This dilemma and the naturalness of the flat, Einstein–deSitter model has been emphasized by Dicke & Peebles 1979.) (4) The net baryon number within our Hubble volume, quantified as the baryon-to-entropy ratio, $n_B/s \simeq \eta/7 \simeq (7-10) \times 10^{-11}$. Of course one of the great successes of the InnerSpace/OuterSpace connection is baryogenesis, the modern theory of the origin of the baryon asymmetry, and it appears that now we at least have a framework for understanding the origin of this very fundamental quantity (for a review see Kolb & Turner 1983). (5) The dearth (thank goodness) of monopoles and other topological beasts which would have been produced in great excess during the earliest moments of the Universe ($t \lesssim 10^{-34}sec$) had the standard model been valid and if the interactions of nature are unified by a semi-simple gauge group, such as SU_5, SO_{10}, or E_6 (for a review see Preskill 1984). (6) The smallness of the present cosmological term. With the possible exception of supersymmetry, no symmetry forbids such a term in the Einstein-Hilbert action, and so on dimensional grounds one would expect such a term to be of order m_{pl}^2, corresponding to a vacuum energy of order m_{pl}^4. In any case, contributions to the vacuum energy of order M^4 arise due to spontaneous symmetry breaking (SSB) at the energy scale M. The measured expansion rate of the Universe restricts the present vacuum energy contribution to be $\lesssim 10^{-46} GeV^4$. Even the contribution from chiral symmetry breaking ($M \simeq few 100 MeV$)–a phenomenon that particle physicists think they know something about, violates this bound by some 42 orders-of-magnitude!

All of these cosmological facts can be accommodated by the standard model, but seemingly at the expense of highly special initial data (the possible exception being the monopole problem). In 1973, Collins and Hawking pointed out that the set of initial data which evolve to a Universe such as ours is of measure zero providing that the stress energy in the Universe has always satisfied the strong and dominant energy conditions. Over the years there have been a number of attempts to try to understand and/or explain this apparent dilemma of initial data. Inflation is the most recent attempt and I believe shows great promise. Let me begin by briefly mentioning the earlier attempts:

⋆ **Mixmaster Paradigm**–Starting with a solution with a singularity which exhibits the features of the most general singular solutions known (the so-called mixmaster model) Misner and his coworkers hoped that they could show that particle viscosity would smooth out the geometry. In part because horizons still effectively exist in the mixmaster solution 'the chaotic cosmology program' has proven unsuccessful (see, e.g., Misner 1968, 1972; Matzner & Misner 1972).

⋆ **Nature of the Initial Singularity**–Penrose (1979) explored the possibility of explaining the observed smoothness of the Universe by restricting the kinds of initial singularities which are permitted in Nature (those with vanishing Weyl curvature). In a sense his approach is to postulate a law of physics governing allowed initial data.

⋆ **Quantum Gravity Effects**–The first two solutions involve appealing to classical gravitational effects. A number of authors have suggested that quantum gravity effects might be responsible for smoothing out the space-time geometry (deWitt 1953; Parker 1966; Zel'dovich 1970; Starobinskii 1980; Anderson 1983; Hartle & Hu 1979; Fischetti et al. 1979). The basic idea being that anisotropy and/or inhomogeneity would drive gravitational particle creation, which due to back reaction effects would eliminate particle horizons and smooth out the geometry. Recently, Hawking and Hartle (1983) have advocated the Quantum Cosmology approach to actually compute the initial state. All of these approaches necessarily involve events at times $\lesssim 10^{-43} sec$ and energy densities $\gtrsim m_{pl}^4$.

⋆ **Anthropic Principle**–Some (Carr & Rees 1979; Barrow & Tipler 1986) have suggested (or in some cases even advocated) 'explaining' many of the puzzling features of the Universe around us (and in some cases, even the laws of physics!) by arguing that unless they were as they are intelligent life would not have been able to develop and observe them! Hopefully we will not have to resort to such an explanation.

The approach of inflation is somewhat different from previous approaches. Inflation (at least from my point-of-view) is based upon well-defined and reasonably well-understood microphysics (albeit, some of it very speculative). That microphysics is:

• Classical Gravity (general relativity), at least as an effective, low energy theory of gravitation

• 'Modern Particle Physics'–grand unification, supersymmetry /supergravity, field theory limit of superstring theories, etc. at energy scales $\lesssim m_{pl}$

As I will emphasize, in all viable models of inflation the inflationary period (at least the portion of interest to us) takes place well after the planck epoch, with the energy densities involved being far less than m_{pl}^4 (although semi-classical quantum gravity effects might have to be included as non-renormalizable terms in the effective Lagrangian). Of course, it could be that a resolution to the cosmological puzzles discussed above involves both 'modern particle physics' and quantum gravitational effects in their full glory (as in a fully ten dimensional quantum theory of strings).

I will not take the time or the space here to review the historical development of our present view of inflation; I refer the interested reader to the interesting paper on this subject by Lindley (1985). It suffices to say that Guth's very influential paper of 1981 was the 'shot heard 'round the world' which initiated the inflation revolution, and that Guth's doomed original model (Guth & Weinberg 1983; Hawking et al. 1982) was revived by Linde's (1982a) and Albrecht and Steinhardt's (1982) variant, 'new inflation'. I will focus all of my attention on the present status of the 'slow-rollover' model of Linde (1982a) and Albrecht & Steinhardt (1982).

BASIC MECHANICS OF NEW INFLATION

Stated in the most general terms, the current view of inflation is that it involves the dynamical evolution of a very weakly-coupled scalar field (hereafter referred to as ϕ) which is, for one reason or another, initially displaced from the minimum of its potential (see Fig. 2). While it is displaced from its mimimum, and is slowly-evolving toward that minimum, its potential energy density drives the rapid (exponential) expansion of the Universe, now known as inflation.

The usual assumptions which are made (often implicitly) in order to analyze inflation are:

• A FRW spacetime with scale factor $R(t)$ and expansion rate

$$H^2 \equiv (\dot{R}/R)^2 = 8\pi\rho/3m_{pl}^2 - k/R^2 \tag{1}$$

where the energy density is assumed to be dominated by the stress energy associated with the scalar field (in any case, other forms of stress energy rapidly redshift away during inflation and become irrelevant).

• The scalar field ϕ is spatially constant (at least on a scale $\gtrsim H^{-1}$) with initial value $\phi_i \neq \sigma$, where $V(\sigma) = V'(\sigma) = 0$, and stress energy tensor

$$T^{\mu}_{\nu} = diagonal(-\rho, p, p, p) \tag{2a}$$

Figure 2 – Stated in the most general terms, inflation involves the dynamical evolution of a scalar field which was initially displaced from the minimum of its potential, be that minimum at $\sigma = 0$ or $\sigma \neq 0$.

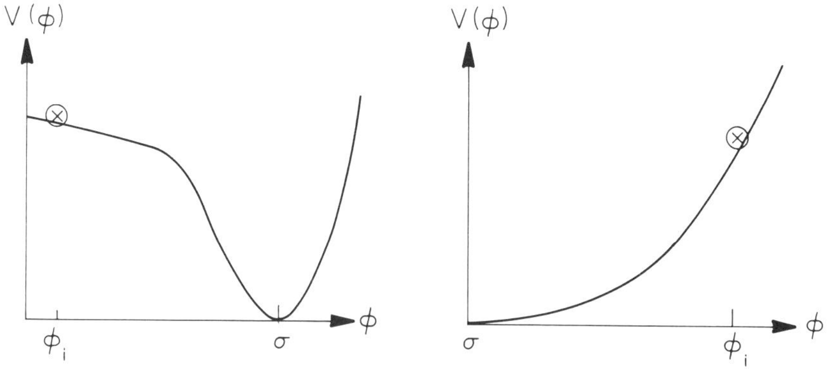

$$\rho = V + \dot{\phi}^2/2 \;\; (+\nabla\phi^2/2R^2) \tag{2b}$$

$$p = -V + \dot{\phi}^2/2 \;\; (-\nabla\phi^2/6R^2) \tag{2c}$$

(I have indicated the contribution of the spatial gradient terms for future reference.)

• The semi-classical equation of motion for ϕ provides an accurate description of its evolution; more precisely,

$$\phi(t) = \phi_{cl}(t) + \Delta\phi_{QM} \tag{3a}$$

$$\ddot{\phi}_{cl} + 3H\dot{\phi}_{cl} + \Gamma\dot{\phi}_{cl} + V'(\phi_{cl}) \;\; (-\nabla^2\phi_{cl}/R^2) = 0 \tag{3b}$$

where the quantum fluctuations (characterized by size $\Delta\phi_{QM} \simeq H/2\pi$) are assumed to be a small perturbation to the classical trajectory $\phi_{cl}(t)$, and Γ is the decay width of the ϕ particle. Throughout I use units where $\hbar = k_B = c = 1$; overdot indicates a derivative with respect to proper time and prime with respect to ϕ. From this point forward I will drop the subscript '*cl*'. I will return later to these assumptions to discuss how they have been or can be relaxed and/or justified.

The semi-classical evolution of ϕ naturally splits into three phases: slow-roll; coherent scalar field oscillations; and quantum fluctuations.

(1) *Slow-Roll* Assuming that the scalar potential is sufficiently flat (the requirement being that $9H^2 \gtrsim |V''|$) there will be a period when the motion of ϕ is friction-dominated, so that the $\ddot{\phi}$ term in Eqn(3b) can be neglected and $\dot{\phi}^2 \ll V$ can be neglected in Eqn(2b). The equation of motion for ϕ becomes

$$3H\dot{\phi} \simeq -V'$$

$$H^2 \simeq 8\pi V(\phi)/3m_{pl}^2$$

The growth of the scale factor during the slow-roll is approximately exponential since $H \simeq constant$, and

$$R_f/R_i = exp(\int H dt) \equiv exp(N) \tag{4a}$$

$$N \simeq -3\int H^2 d\phi/V' \simeq 8\pi m_{pl}^{-2}\int -V(\phi)d\phi/V'(\phi) \tag{4b}$$

The exponential growth or inflation occurs during the slow-roll phase.

(2) *Coherent Scalar Field Oscillations* As ϕ approaches σ the potential steepens sufficiently (or $V(\phi)$ becomes sufficiently small) so that the motion of ϕ is no longer friction-dominated (which occurs when $|V''| \gtrsim 9H^2$) and ϕ begins to oscillate about the minimum of the potential ($\phi = \sigma$) on a timescale ($\simeq |V''|^{-1/2} = m_\phi^{-1}$) short compared to the Hubble time. During this phase its oscillatory motion can be time-averaged, and the equation for its evolution becomes

$$\dot{\rho}_\phi + 3H\rho_\phi + \Gamma\rho_\phi = 0 \tag{5}$$

whose solution is $\rho_\phi \propto exp(-\Gamma t)/R^{-3}$–precisely that of an unstable, NR particle species. The coherent field oscillations behave like NR matter (as they should, since they correspond to the zero momentum mode of the field), and decay in a time Γ^{-1} ($= \tau_\phi$, the lifetime of the ϕ particle) due to particle creation by the oscillating ϕ field. From the particle point-of-view, the oscillations correspond to a very cold condensate of ϕ particles, which then decay. During this phase the scale factor $R(t) \propto t^{2/3}$. Assuming that the decay products thermalize quickly (or at least are relativistic) and neglecting any relativistic particles present before inflation as they've been exponentially diluted, the evolution of the energy density in radiation produced by the decay of the coherent field oscillations is governed by

$$\dot{\rho}_R + 4H\rho_R = \Gamma\rho_\phi \tag{6}$$

The evolution of ρ_ϕ, ρ_R, and the entropy per comoving volume ($\equiv S \propto R^3\rho_R^{3/4}$) are shown in Fig. 3. At early times, $t \ll \Gamma^{-1}$, $\rho_r \propto t^{-1}$ and $S \propto t^{5/4}$. The entropy per comoving volume levels off when most of the ϕ particles have decayed, at $t \simeq \Gamma^{-1}$, and thereafter the energy density in coherent field oscillations decreases exponentially. The reheating process is essentially complete at this time (save for the possible thermalization of the decay products of the ϕ particles) and the temperature is about

$$T_{RH} \simeq (\Gamma m_{pl})^{1/2} \tag{7}$$

Figure 3 summarizes the evolution of ϕ and the reheating process. For further discussion see Turner (1983).

(3) *Quantum Fluctuations* During inflation ϕ is on the flat part of its potential and can be treated as an effectively massless scalar field. The spectrum of de Sitter space quantum fluctuations is given by (Bunch & Davies 1978)

$$\Delta\phi_{QM}^2 \equiv |\delta\phi_k|^2 k^3/(2\pi)^3 = H^2/16\pi^3 \tag{8}$$

Figure 3 – The evolution of the energy density in the scalar field (ρ_ϕ), in radiation (ρ_R), and of the entropy per comoving volume, $S \propto R^3 \rho_R^{3/4}$.

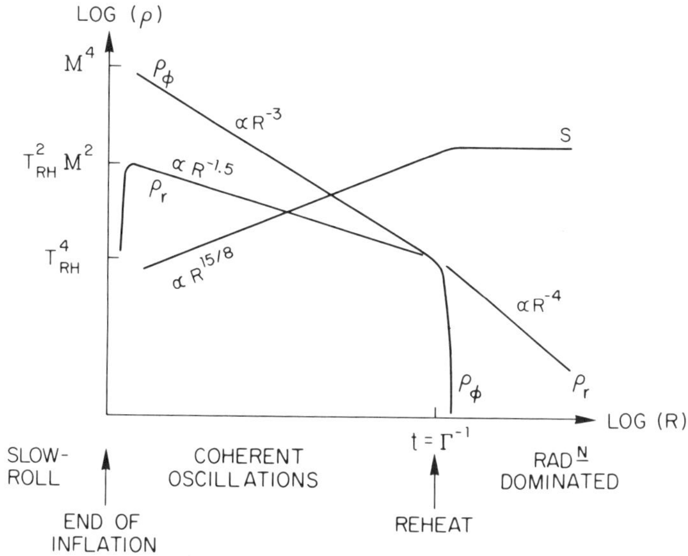

where $\delta\phi_k$ is the $k-th$ Fourier component of $\delta\phi$ and $k_{phys} = k/R(t)$ is the physical wavenumber of the mode with comoving wavenumber k. For scales $k_{phys} \gtrsim H^{-1}$ these perturbations are treated quantum mechanically; as k_{phys} becomes $\lesssim H^{-1}$, and a given scale crosses outside the 'physics horizon' (or Hubble radius) H^{-1} the fluctuation on that scale is taken to evolve classically thereafter (the quantum fluctuation is *assumed* to 'freeze in' as a classical perturbation in the metric). The evolution of the metric perturbations due to the classical fluctuations is straightforward to compute. They give rise to curvature (scalar mode) fluctuations of amplitude (Bardeen et al. 1983; Guth & Pi 1982; Starobinskii 1982; Hawking 1982)

$$(\delta\rho/\rho)_{HOR} \simeq k^{3/2}|\delta_k|/(2\pi)^{3/2} \simeq \begin{cases} (H^2/\pi^{3/2}\dot{\phi})|_{t_1} & (9a) \\ (H^2/10\pi^{3/2}\dot{\phi})|_{t_1} & (9b) \end{cases}$$

when they reenter the horizon. Perturbations which reenter when the Universe is radiation-dominated do so as pressure waves (in the baryon–photon fluid) of the amplitude indicated in Eqn(9a); those which reenter when it is matter-dominated do so as growing mode perturbations with the amplitude indicated in Eqn(9b).

Tensor mode metric (gravitational wave) perturbations also arise and are of dimensionless amplitude (Abbott & Wise 1984; Starobinskii 1979; Rubakov et al. 1982)

$$h_{GW}|_{HOR} \simeq (H/m_{pl})|_{t_1} \qquad (10)$$

Figure 4– The evolution of the physical wavelength of a given mode. Early during inflation $\lambda_{phys} \lesssim H^{-1}$ and the scale is inside the physics horizon and fluctuations are treated quantum mechanically. As the mode leaves the physics horizon ($\lambda_{phys} \geq H^{-1}$), the perturbation is assumed to 'freeze in' as a classical metric perturbation.

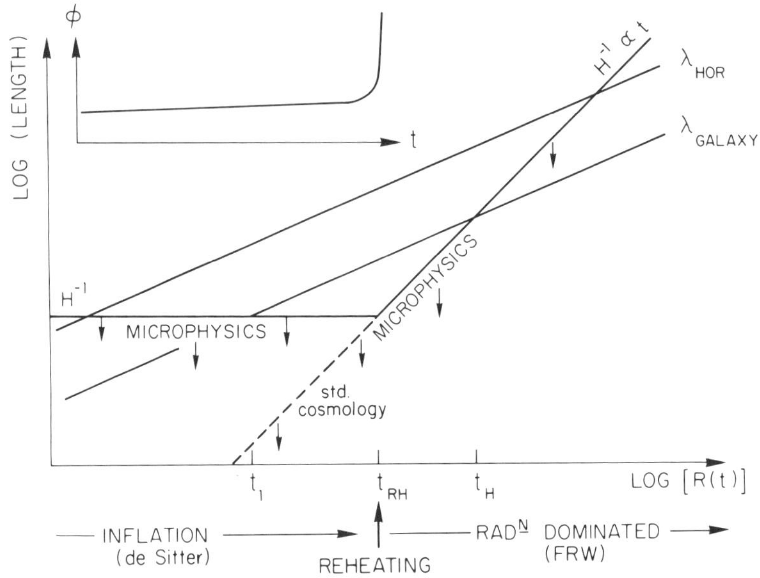

when they reenter the horizon after inflation. In both cases the quantities on the *rhs* are to be evaluated when the comoving scale of interest crossed outside the Hubble radius ($k/R \simeq H$), at $t = t_1$ (see Fig. 4). Normalizing $R_{today} = 1$, it follows that a given scale k ($\equiv 2\pi/\lambda$) crossed outside the horizon during inflation

$$N(\lambda) \simeq 45 + ln(\lambda/Mpc) + 2ln(M/10^{14}GeV)/3 + ln(T_{RH}/10^{10}GeV)/3 \qquad (11)$$

Hubble times (or e-folds) before the end of inflation (see Fig. 4). Thus the scales of astrophysical interest crossed outside the horizon 40 ± 10 or so Hubble times before the end of inflation. Although H and $\dot{\phi}$ can vary considerably during inflation, over such a small range of Hubble times both H and $H^2/\dot{\phi}$ are essentially constant (vary by less than a factor of 2 in all the models I have studied) and so a generic prediction of inflationary models is scale invariant curvature and tensor perturbations. [In models where there are other massless fields, quantum fluctuations of order $H/2\pi$ arise in these fields too during inflation and can give rise to isocurvature (often called isothermal) perturbations. A simple example being axion models where the quantum fluctuations in the axion field result in fluctuations in the local axion to entropy density ratio n_a/s (see Steinhardt & Turner 1983; Linde 1985; Seckel & Turner 1985).] The mechanics of inflation are described in much greater detail in the review by Turner (1986).

SUCCESSFULLY IMPLEMENTING INFLATION

Now that I have discussed the basic mechanics of inflation, how does one build a model which actually leads to a Universe which resembles ours in regions as large as our current Hubble volume? It's as easy as 1,2,3,4,5,6,7,8,...

(1) *Smoothness/Flatness*–After inflation there must be smooth/flat regions of the Universe which contain an entropy of at least 10^{88}. Assuming that ϕ was constant in a region of size of order the Hubble radius before inflation, it is straightforward to show that the number of e-folds of inflation required so that the patch contains an entropy of at least 10^{88} after reheating is

$$N \equiv \int H dt \simeq \int -3H^2 d\phi/V' \gtrsim 53 + 2ln(M/10^{14}GeV)/3 + ln(T_{RH}/10^{10}GeV)/3 \tag{12}$$

The number of e-folds required to solve the 'flatness problem' is equal to this plus $ln|1-\Omega_i^{-1}|$ where Ω_i is the value of Ω at the beginning of inflation. In general, if a given model inflates at all, it is not difficult to make it inflate enough to solve the smoothness/flatness problems.

(2) *Sufficiently Large Smooth Patch*–In order that the vacuum energy contribution to the energy density of the Universe dominates the gradient terms $((\nabla\phi)^2)$, ϕ must be smooth over a sufficiently large region. [If the gradient term dominates, the stress energy of the scalar field behaves like a fluid with $p = -\rho/3$ and the scale factor only grows as $R(t) \propto t$, not exponentially.] The condition that $V \gg (\nabla\phi)^2$ requires ϕ to be smooth over a region of size greater than $\phi_i/V^{1/2} \simeq (\phi_i/m_{pl})H^{-1}$.

(3) *Validity of Semi-Classical Description*–In order that the semi-classical equations of motion be self consistent $\Delta\phi_{QM} \sim H$ must be much less than ϕ_{cl}. I will return to this point later–in general it is not a difficulty.

(4) *Topological Beasts*–One must be careful not to produce any of the dangerous topological beasts such as domain walls, monopoles, etc. after inflation. This can be arranged by having the symmetry breaking stages which result in the formation of such objects occur before or early on during inflation.

(5) *Other Unwanted Garbage*–As is well-known the ratio of the energy density in nonrelativistic (NR) particles relative to that in relativistic (R) particles grows with time: $\rho_{NR}/\rho_R \propto R(t)$. Since our Universe did not become matter-dominated until rather recently, $T \simeq 10eV$ and $t \simeq 10^{10}sec$, the ratio of energy density in stable, NR particles (which do not come into thermal equilibrium, or annihilate) to that in R particles after inflation must be small:

$$\rho_{NR}/\rho_R \lesssim 10^{-19}(10^{10}GeV/T_{RH}) \tag{13}$$

–which is not always an easy thing to do. [Just ask any experimentalist about suppressing something by 19 orders-of-magnitude!] Examples of potentially dangerous forms of NR matter include: gravitinos, weakly-coupled scalar fields which may be put into oscillation after inflation (the so-called Polonyi problem; e.g., see Coughlan et al. 1983), etc. There are even stronger limits on unstable NR particles (particularly gravitinos) which follow by considering their effect on primordial nucleosynthesis (Ellis et al. 1984; Khlopov & Linde 1984; Scherrer & Turner 1987).

(6) *Metric Perturbations*–While the eventual development of structure in the Universe requires density perturbations of the order of $few \times 10^{-5}$ or so, the observed isotropy of the CMBR precludes scalar or tensor perturbations of size greater than about 10^{-4}. Achieving tensor mode perturbations of this size or smaller is not difficult; it only requires that inflation occur at an energy scale $M \simeq V^{1/4} \lesssim 10^{-2} m_{pl}$, putting an upper limit on the reheating temperature: $T_{RH} \lesssim 3 \times 10^{16} GeV$. The scalar perturbations are another matter; thus far, they have posed the most serious obstacle to constructing a successful model of inflation. To achieve curvature fluctuations of amplitude less than about 10^{-4} requires a dimensionless parameter in the scalar potential of the order of 10^{-12}. For example, for a potential of the form $V = \lambda\phi^4$, λ must be $\lesssim 10^{-13}$; for the potential $V = V_0 + \alpha\phi^2 - \beta\phi^3 + \lambda\phi^4$, $\alpha \lesssim 10^{-15} m_{pl}^2$, $\lambda \lesssim 10^{-11}$, $\beta \lesssim 3 \times 10^{-12} m_{pl}$; for the potential $V = m^2\phi^2$ (i.e., a non-interacting, massive scalar field), $m^2 \lesssim 10^{-9} m_{pl}^2$. Achieving density perturbations of an acceptable amplitude necessitates a very weakly-coupled scalar field; weakly-coupled to all fields in the theory or else radiative corrections would spoil the small coupling put in at tree level. Such a small coupling also implies that inflation takes place at an energy scale much less than the planck scale; typically, $V \simeq \lambda\phi^4 \simeq 10^{-12} m_{pl}^4$ or less.

(7) *Sufficient Reheating*–In order not to spoil the concordance of the predictions of primordial nucleosynthesis with the observed abundances, the Universe should be radiation-dominated when $t \simeq 0.01 sec$ and $T \simeq 10 MeV$ at the very latest, i.e., $T_{RH} \gtrsim 10 MeV$. This implies that Γ must be $\gtrsim 10^{-23} GeV$–which is not difficult to achieve, even for a very weakly-coupled scalar field. Baryogenesis, however, poses a more formidable challenge. It goes without saying that baryogenesis must follow inflation, as any baryon asymmetry produced before inflation is diluted away by the enormous entropy produced by inflation. If baryogenesis is to proceed in the usual way, T_{RH} must be greater than about a $m_H/10$, where m_H is the mass of the superheavy boson whose out-of-equilibrium decays produce the baryon asymmetry (see Kolb & Turner 1983). In most unified theories the longevity of the proton requires the masses of superheavy bosons whose interactions violate B-conservation to be greater than about $10^{10} GeV$, thereby requiring $T_{RH} \gtrsim 10^9 GeV$ at the very least. Because the scalar field must be very weakly-coupled (to produce density perturbations of an acceptable magnitude) Γ tends to be very small, and sufficient reheating is often very difficult to achieve. There is, however, an alternative method which does not require such a high reheating temperature: the direct production of the baryon asymmetry by the decays of the scalar field responsible for inflation ($\phi \to q's,\ l's$ with $\Delta B \neq 0$). In this case the baryon asymmetry produced is

$$n_B/s \simeq \epsilon T_{RH}/m_\phi \tag{14}$$

where ϵ is the magnitude of the requisite C, CP violation in the decay of the ϕ particle. Note that the asymmetry produced only depends upon the ratio of the reheating temperature to the mass of the ϕ particle, and so it is possible to have a relatively low reheating temperature and still a baryon asymmetry of the required magnitude.

(8) *Part of a Unified Model Which Predicts Sensible Particle Physics*–In order to avoid having the tail wag the dog so to speak, the scalar field should be part of a unified theory which predicts sensible particle physics.

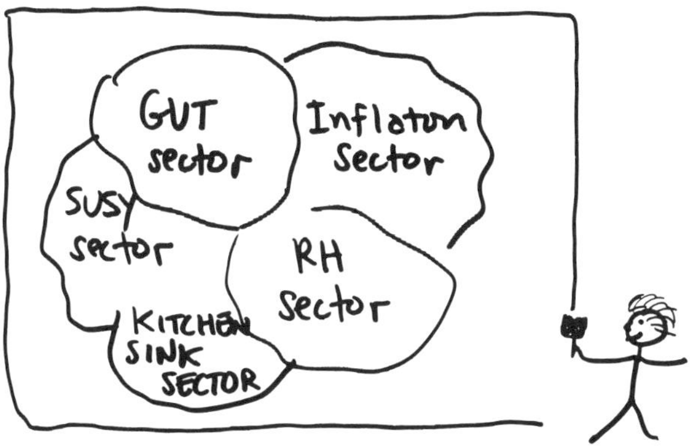

NB: JUST PUTTING A BOX AROUND IT DOES NOT MAKE IT A 'UNIFIED MODEL' !!

SPECIFIC MODELS

While the requirements on a successful model of inflation are straightforward they are not simple to satisfy simultaneously, and the path to a successful model is strewn with the remains of many an attractive model that failed for one reason or another. The toughest challenge has been the constraint imposed by the scalar density perturbations. At present there are no successful models which are so elegant as to be compelling (compelling here, meaning attractive to other than the authors of the model!), although there do exist a number of 'proof of existence' models. I will describe two particularly simple models here.

(1) *Shafi-Vilenkin-Pi Model*–This model (Shafi & Vilenkin 1984; Pi 1984) is based upon an SU_5 nonsupersymmetric GUT (although the gauge group could just as well be SO_{10} or E_6). The scalar field responsible for inflation is a complex Higgs gauge singlet, which in addition to being responsible for inflation, also breaks a Peccei-Quinn (PQ) symmetry and effects GUT symmetry breaking by inducing a negative mass squared for the adjoint Higgs representation which then breaks the GUT down to $SU(3) \otimes SU(2) \otimes U(1)$. The part of the Higgs potential which is relevant for inflation is a Coleman-Weinberg type potential

$$V(\phi) = B\sigma^4/2 + B\phi^4[ln(\phi^2/\sigma^2) - 1/2] \tag{15}$$

where $\sigma \simeq 10^{18} GeV$ is the vacuum expectation value of ϕ at low temperatures, which breaks PQ symmetry and induces GUT SSB; $B \simeq 10^{-15}$ arises due to radiative corrections from the coupling of ϕ to other scalar fields in the theory. Interestingly enough isocurvature axion fluctuations of similar magnitude to the curvature fluctuations also arise in this model (Seckel & Turner 1985).

(2) *Florida SUGR*–This model (Holman et al. 1984) is based on an effective low energy supergravity theory with a superpotential of the form, $I + S + G$, where the three pieces of the superpotential are responsible for inflation, supersymmetry breaking and GUT symmetry breaking respectively. The inflation piece of the superpotential takes a particularly simple form

$$I = (\Delta^2/M)(\phi - M)^2 \tag{16}$$

where Δ is the only adjustable parameter and corresponds to the intermediate scale and $M = m_{pl}/(8\pi)^{1/2} \simeq 2.43 \times 10^{18} GeV$. The resulting scalar potential for ϕ is given by

$$\begin{aligned} V_I(\phi) &= \Delta^4 exp(\phi^2/M^2)[\phi^6/M^6 - 4\phi^5/M^5 + 7\phi^4/M^4 - 4\phi^3/M^3 - \phi^2/M^2 + 1] \\ &\simeq \Delta^4[1 - 4\phi^3/M^3 + 6.5\phi^4/M^4 - 8\phi^5/M^5 + ... \end{aligned} \tag{17}$$

Achieving density perturbations of an acceptable magnitude requires that $(\Delta/M) \simeq 9 \times 10^{-5}$. Note that the coefficient of the ϕ^4 term in V_I is dimensionless and for this value of Δ is about 4×10^{-16}–the small dimensionless parameter which always arises. Of course, in this model it is directly related to the smallness of the intermediate scale relative to the planck scale, suggesting (or offering hope) that the very small parameter needed for successful inflation is related to fundamental physics. This model reheats to a temperature of about $10^6 GeV$ and baryogenesis is effected directly through the decays of the ϕ field.

OPEN (or semi-open) QUESTIONS

$\star$ **'Who is ϕ?'**–Inflationary models exist in which the scalar field ϕ: effects SSB of the GUT (Shafi & Vilenkin 1984; Pi 1984), effects SSB of SUSY (Ovrut & Steinhardt 1984a,b), induces Newton's constant (in a Landau-Ginzburg model of induced gravity) (Accetta et al. 1985; Spokoiny 1984), is $\sim ln(r_X/r_{XEQ})$ (where r_X is the radius of compactified extra dimensions) in theories with extra dimensions which become compactified (Shafi & Witterich 1983, 1985), is $\propto$ (scalar curvature $)^{1/2}$ (Starobinskii 1980; Mijic et al. 1986), is just some 'random' scalar field (Linde 1983), or is merely in the theory to effect inflation (Holman et al. 1984; Nanopoulos 1985). Given the number of different kinds of inflationary scenarios which exist, it seems as though inflation is generic to early Universe microphysics, occurring whenever a weakly-coupled scalar field finds itself displaced from the minimum of its potential. Clearly, a key question at this point is just how 'the inflation sector' of the theory fits into the Big Picture!

$\star$ **What Determines the Initial Value of ϕ?**–One thing is certain, and that is that ϕ must be very weakly-coupled, as quantified by its small dimensionless coupling constant. Because of this fact, it is almost certain that ϕ was not initially in thermal contact with the rest of the Universe and so ϕ_i is unlikely to be determined by thermal considerations (in the earliest models of new inflation, ϕ_i was determined by thermal considerations, however these models resulted in density perturbations of an unacceptably large amplitude). At present ϕ_i must be taken as initial data. Some have argued that ϕ_i might be determined in an anthropic-like way, as regions of the Universe where ϕ_i is sufficiently far displaced from equilibrium will undergo inflation and eventually occupy most of the physical volume of the Universe. Perhaps the wavefunction of the Universe approach will shed some light on the initial distribution of the scalar field ϕ. Or it could be that due to 'as-of-yet unknown dynamics' ϕ was indeed in thermal equilibrium at a very early epoch. It goes without saying that it is crucial that ϕ be initially displaced from its minimum.

⋆ **Validity of the Semi-Classical Equations of Motion for ϕ**– While it may seem perfectly plausible that ϕ evolves according to its semi-classical equations of motion, the validity of this assumption has troubled inflationists from the 'dawn of new inflation'. While a full quantum field theory treatment of inflation is very difficult and has not been effected, a number of specific issues have been addressed. Several authors have studied the role of inhomogeneities in ϕ, and have found that for the very weakly-coupled fields one is dealing with, mode coupling is not important and the individual modes are quickly smoothed by the exponential expansion of their physical wavelengths (Albrecht & Brandenberger 1985; Albrecht et al. 1985). I already mentioned the necessity of having ϕ smooth over a sufficiently large region so that the gradient terms in the stress energy do not dominate.

The effect of quantum fluctuations on the evolution of ϕ has been studied in some detail by Guth & Pi (1985), Fischler et al. (1985), Linde (1982b), and Vilenkin & Ford (1982). The basic conclusion that one draws from the work of these authors is that the use of the semi-classical equations of motion is valid so long as $\phi_{cl} \gg \Delta\phi_{QM} \simeq N^{1/2}H/2\pi$, which is almost always satisfied for the very flat potentials of interest to inflationists (at least for the last 50 or so e-folds which affect our present Hubble volume). [More precisely, the semi-classical change in ϕ in a Hubble time, $\Delta\phi_{Hubble} \simeq -V'/3H^2 \simeq -V'm_{pl}^2/(8\pi V)$, should be much greater than the increase in $< \phi^2 >_{QM}^{1/2}$, which is of order $H/2\pi$, due to the addition of another quantum mode; see Bardeen et al. 1983.] At present the validity of the semi-classical equations of motion seems to be reasonably well established.

⋆ **No Hair Conjectures**–While inflation has been touted from the very beginning as making the present state of the Universe insensitive to the initial space-time geometry, not much has been done to justify this claim until very recently. As I mentioned earlier, inflation is nearly always analyzed in the context of a flat, FRW cosmological model, making such a claim somewhat dubious. However, it has now been shown that all of the homogeneous models (with the exception of the highly-closed models) undergo inflation, isotropize and remain isotropic to the present epoch providing that the model would have inflated the requisite 60 or so e-folds in the absence of anisotropy (Turner & Widrow 1986; Jensen & Stein-Schabes 1986a).

The proof of this result involves three parts. First, Wald (1983) demonstrated that all homogeneous models with a positive cosmological term asymptotically approach deSitter (less the aforementioned highly-closed models which recollapse before the cosmological term becomes relevant). Wald's result follows because all forms of 'anisotropy energy density' decrease with increasing proper volume element, whereas the cosmological term remains constant, and so eventually triumphs. Of course, inflationary models do not, in the strictest sense, have a cosmological term, rather they have a positive vacuum energy as long as the scalar field is displaced from the minimum of its potential. Thus the dynamics of the scalar field comes into play: does ϕ stay displaced from the minimum of its potential long enough so that the vacuum energy comes to dominate? Due to the presence of anisotropy the expansion rate is greater than if there were only vacuum energy density, and so the friction felt by ϕ as it trys to roll (the $3H\dot{\phi}$ term) is greater and it takes ϕ longer to evolve to its minimum than without anisotropy. For this reason the Universe does become vacuum dominated before the vacuum energy disappears, and in fact the Universe inflates slightly longer in the presence of anisotropy (one or two e-folds) (Steigman & Turner 1983). Finally, is the anisotropy reduced sufficiently so that the Universe today is still nearly isotropic? As it turns out, the requisite 60 or so e-folds needed to solve the other conundrums reduces the growing modes of anisotropy sufficiently to render them small today.

Allowing for inhomogeneous initial spacetimes makes matters much more difficult. Jensen and Stein-Schabes (1986b) and Starobinskii (1983) have proven the analogue of Wald's theorem for spacetimes which are negatively-curved. Jensen and Stein-Schabes (1986b) have gone on to conjecture that spacetimes which have sufficiently large regions of negative curvature will undergo inflation, resulting in a Universe today which although not globally homogeneous, at least contains smooth volumes as large as our current Hubble volume.

Does this improve the situation that Collins and Hawking discussed in 1973? While the work of Jensen and Stein-Schabes (1986) seems to indicate that many inhomogeneous spacetimes undergo inflation and even leads one to speculate that the measure of the set of initial spacetimes which eventually inflate is non-zero, it is not possible to draw a definite conclusion without first defining a measure on the space of initial data. In fact, as Penrose (1979) pointed out there is at least one way of defining the measure such that this is not the case. Consider the set of all Cauchy data at the present epoch; intuitively it is clear that those spacetime slices which are highly irregular are the rule, and those which are smooth in regions much larger than our current Hubble volume are the exception. Defining the measure today, it seems very reasonable that the smooth spacetime slices are a set of measure zero. Now evolve the spacetimes back to some initial epoch (for example $t = 10^{-43} sec$). Using the seemingly very reasonable measure defined today and the mapping back to 'initial' spacetimes, one could argue that the set of initial data which inflate is still of measure zero. While I believe that this argument is technically correct, I also believe that it is silly. First, upon close examination of all of those initial spacetimes which led to spacetimes today without smooth regions as large as our present Hubble volume, one would presumably find that the scalar field in most would be very close to the minimum of its potential (in order that they not inflate)–not a very generic initial condition. Secondly, if one adopts the point-of-view of an evolving Universe which has an 'initial epoch' (and not everyone does), then there is a preferred epoch at which one would define a measure–the 'initial epoch,' and at that epoch I believe any reasonably defined measure would lead to the set of initial spacetimes which inflate being of non-zero measure.

Although it is not possible *yet* to claim rigorously that inflation has resolved the problem of the seemingly special initial data required to reproduce the Universe we see today (at least within our Hubble volume), I think that any fairminded person would admit that it has improved the situation dramatically. Extrapolating from the solid results that exist, it seems to me that starting with a general inhomogeneous spacetime, there will exist regions which undergo inflation and which today are much larger than our present Hubble volume, thereby accounting for the smooth region we find ourselves in. From a more global perspective, one might expect that on scales $\gg H^{-1}$ the Universe would be highly irregular.

⋆ **The Present Vanishingly Small Value of the Cosmological Constant**–Inflation has shed no light on this difficult and very fundamental puzzle (nor has anything else for that matter!). In fact, since inflation runs on vacuum energy so to speak, the fate of inflation hinges upon the resolution of this puzzle. For example, suppose there were a grand principle that dictated that the vacuum energy of the Universe is always zero, or that there were an axion-like mechanism which operated and ensured that any cosmological constant rapidly relaxed to zero; either would be a disaster to inflation, shorting out its source of power–vacuum energy. [Another possiblity which has received a great deal of attention recently is the possibility that deSitter space might be quantum mechanically unstable–of course, if its lifetime were at least 60 some e-folds that would not necessarily adversely affect inflation (Starobinskii 1980; Myhrvad 1983; Mottola 1985; Parker 1983; Ford 1985; Anderson 1985; Traschen & Hill 1986).]

INFLATION CONFRONTS OBSERVATION

No matter how appealing a theory may be, it must meet and pass the test of experimental verification. Experiment and/or observation is the final arbiter. One of the few blemishes on early Universe physics is the lack, thus far, of experimental/observational tests of the many beautiful and exciting predictions. That situation is beginning to change as the field starts to mature. Inflation is one of the early Universe theories which is becoming amenable to verification or falsification. Inflation makes the following very definite predictions (postdictions?):

• $\Omega = 1.0$ (more precisely, $R_{curv} = R(t)|k|^{-1/2} = H^{-1}/|\Omega - 1|^{1/2} \gg H^{-1}$)

• Harrison-Zel'dovich spectrum of constant curvature perturbations (and possibly isocurvature perturbations as well) and tensor mode gravitational wave mode perturbations

The prediction of $\Omega = 1.0$ together with the primordial nucleosynthesis constraint on the baryonic contribution, $0.014 \lesssim \Omega_B h^2 \lesssim 0.035 \lesssim 0.15$ (Yang et al. 1984), suggests that most of the matter in the Universe must be nonbaryonic. The simplest and most plausible possibility is that it exists in the form of relic WIMPs (*W*eakly-*I*nteracting *M*assive *P*articles, e.g., axions, photinos, neutrinos; for a review, see Turner 1987). Going a step further, these two original predictions then lead to testable consequences:

⋆ $H_0 t_0 = 2/3$ (providing that the bulk of the matter in the Universe today is in the form of NR particles)–The observational data on both H_0 and t_0 are far from being definitive: $H_0 \simeq 40-100 km sec^{-1} Mpc^{-1}$ and $t_0 \simeq 12-20 Gyr$, implying only that $H_0 t_0 \simeq 0.5 - 2.0$.

⋆ $\Omega = 1.0$–All of the dynamical observations suggest that the fraction of critical density contributed by matter which is clumped on scales $\lesssim 10 - 30 Mpc$ is only about: $\Omega_{\lesssim 30} \simeq 0.2 \pm 0.1$ (± 0.1 is not meant to be a formal error estimate, but indicates the spread in the observations) (see the recent review by Trimble 1987). If inflation is not to be falsified, that leaves but two options: (1) the observations are somehow misleading or wrong; or (2) there exists a component of energy density which is smoothly distributed on scales $\lesssim 10 - 30 Mpc$ (and therefore would not be reflected in the dynamical determinations). Candidates for the smooth component include: relic, light neutrinos, which by virtue of the large length scale ($\lambda_\nu \simeq 13 h^{-2} Mpc$) on which neutrino perturbations are damped by freestreaming, would likely still be smooth on these scales; relic relativistic particles produced by the recent decay of an unstable WIMP species (Turner et al. 1984; Dicus et al. 1977; Olive et al. 1985); a relic cosmological term (Turner et al. 1984; Peebles 1984); 'failed galaxies,' referring to a population of galaxies which have the same mix of dark matter to baryons, but are more smoothly distributed and are too faint to observe (at least thus far) (Kaiser 1984, 1986; Bardeen et al. 1986); a relic population of light strings–either fast moving non-intercommuting strings or a tangled network of non-Abelian strings (Vilenkin 1984). All of these smooth component scenarios have testable consequences (Charlton & Turner 1987)–their predictions for $H_0 t_0$ differ from 2/3; the growth of perturbations is different; the evolution of the cosmic scale factor $R(t)$ is different from the matter-dominated model and various kinematic tests (magnitude-redshift, angular size-redshift, lookback time-redshift, proper volume element-redshift, etc.) can in principle differentiate between them.

⋆ *Microwave Fluctuations*–Both the scalar and tensor metric perturbations, cf. Eqns(9.10), lead to fluctuations in the CMBR on large angular scales ($\gg 1°$). On such large scales causal processes (such as reionization) cannot have erased the primordial fluctuations, and so if ever present, they must still be there. The scalar perturbations (if they have anything to do with structure formation) must be of amplitude $\gtrsim few \times 10^{-6}$, which is within a factor of 10 or less of the current upper limits on these scales.

⋆ *Two Detailed Stories of Structure Formation*–The simplest possibility, namely that most of the mass density is in relic WIMPs ($\Omega_{WIMP} = 1.0 - \Omega_B \simeq 0.9$) leads to two very detailed scenarios of structure formation: hot dark matter (the case where the dark matter is neutrinos) and cold dark matter (essentially any other WIMP as dark matter). At present, the numerical simulations of these scenarios are sufficiently definite that it is possible to falsify them–and in fact, both of these simplest scenarios have difficulties (see the recent review by White 1986). In the hot dark matter case it is forming galaxies early enough. The large-scale structure which evolves in this case (voids, superclusters, froth) qualitatively agrees with what is observed; however, in order to get agreement with the galaxy-galaxy correlation function, galaxies must form very recently (redshifts $\lesssim 1$) in contradiction to all the galaxies (redshifts as large as 3.2) and QSO's (redshifts as large as 4.0) which are seen at redshifts $\gtrsim 1$.

With cold dark matter the simulations can nicely reproduce galaxy clustering, most of the observed properties of galaxies (masses and densities, rotation curves, etc) (Blumenthal et al. 1984; Davis et al. 1985). However the simulations do not seem to be able to produce sufficient large-scale structure. In particular, they fail to account for the amplitude of the cluster-cluster correlation function (by a factor of about 3), large amplitude, large-scale peculiar velocities, and voids. [In fairness I should mention that our knowledge of large-scale structure of the Universe is still very fragmentary, with the first moderate sized ($\sim 10^4$), 3-dimensional surveys having just recently been completed.] In order to account for $\Omega = 1.0$, galaxy formation must be biased (i.e., only density-averaged peaks greater than some threshold, typically $2 - 3\sigma$, are assumed to evolve into galaxies which we see today, the more typical 1σ peaks resulting in 'failed galaxies' for some reason or another; see Bardeen et al. 1986).

[The situation with respect to large scale structure is becoming more interesting every moment. Several groups have now reported large-amplitude ($600 - 1000 kmsec^{-1}$) peculiar velocities on large scales ($\sim 50h^{-1}Mpc$) (Burstein et al. 1986; Collins et al. 1986). Such large peculiar velocities are very difficult, if not impossible, to reconcile with either hot or cold dark matter (or even smooth component models) and the Zel'dovich spectrum (Vittorio & Turner 1987). If these data hold up they may pose an almost insurmountable obstacle to any scenario with the Zel'dovich spectrum of density perturbations. The frothy structure observed in the galaxy distribution by de Lapparent et al. (1986), galaxies distributed on the surfaces on large ($\sim 30h^{-1}Mpc$), empty bubbles, although somewhat more qualitative, also seems difficult to reconcile with cold dark matter.]

There are a number of observations/experiments which can and will be done in the next few years and which should really put the inflationary scenario to the test. They include improved sensitivity measurements of the CMBR anisotropy. The microwave background anisotropies predicted in the hot dark matter scenario are very close to the observational upper limits on angular scales of both 5 or so arcminutes and $\gtrsim$ few degrees (Vittorio & Silk 1984; Bond & Efstathiou 1984). With cold dark matter, the predictions are a factor of $3-10$ away from the observational limits (for the isocurvature spectrum, the quadrupole upper limit may actually rule out this possibility; see, Efstathiou & Bond 1986). An improvement in sensitivity to microwave anisotropies of the order of $3-10$ could either begin to confirm one of the scenarios or rule them both out, and is definitely within the realm of experimental reality (Wilkinson 1986).

The relic WIMP hypothesis for the dark matter can also be tested. While it was once almost universally believed that all WIMP dark matter candidates were, in spite of their large abundance, essentially impossible to detect because of the feebleness of their interactions, a number of clever ideas have recently been suggested (and are being experimentally implemented) for detecting axions (Sikivie 1983), photinos, sneutrinos, heavy neutrinos, etc (Goodman & Witten 1985). Results and/or limits will be forth coming soon. With the coming online of the Tevatron at Fermilab, the SLC at SLAC, and hopefully the SSC it is possible that one of the candidates may be directly produced in the lab. Experiments to detect neutrino masses in the eV mass range also continue.

A geometric measurement of the curvature of the Universe (which uses the dependence of the comoving volume element as a function of redshift) has recently been made by Loh and Spillar (1986). Their preliminary results indicate $\Omega = 0.9^{+0.7}_{-0.5}$ (95% confidence) (for a matter-dominated model). This technique appears to have great cosmological leverage and looks very promising–far more promising than the traditional approach of determining the density of the Universe through the deceleration parameter q_0.

Another area with great potential for improvement is 3d surveys of the distribution of galaxies. The largest redshift surveys at present contain only a few 1000 galaxies, yet have been very tantalizing, indicating evidence of voids and froth-like structure to the galaxy distribution (de Lapparent et al. 1986). The large, automated surveys which are likely to be done in the next decade could very well lead to a quantum leap in our understanding of the large scale features of the Universe and help to provide hints as to how they evolved.

The peculiar velocity field of the Universe is potentially a very valuable and direct probe of the density field of the Universe:

$$|\delta v_k| = |\dot{\delta}_k/k| \quad (= (\lambda H/2\pi)\delta_k \ for \ \Omega = 1) \tag{18}$$

$$(\delta v/c)_\lambda \simeq (\lambda/10^4 h^{-1}Mpc)(\delta\rho/\rho)_\lambda \tag{19}$$

where δ_k and δv_k are the $k-th$ Fourier components of $\delta\rho/\rho$ and $\delta v/c$ respectively. The very recent measurements which indicate large amplitude peculiar velocities on scales of $\sim 50h^{-1}Mpc$ are surprising in that they indicate substantial power on these scales, and are problematic to almost every scenario of structure formation. Should they be confirmed they will provide a very acute test of structure formation in inflationary models.

Of course, theorists are very accommodating and have already started suggesting alternatives to the simplest scenarios for structure formation. As I mentioned earlier, scenarios with a smooth component to the energy density have been put forward to solve the Ω problem. Cosmic strings present a radically different approach to structure formation with their non-gaussian spectrum of density fluctuations. [It is interesting to note that cosmic strings of the right 'weight' ($G\mu \simeq 10^{-6}$ or so, where μ is the string tension) seem to be somewhat incompatible with inflation, as they must necessarily be produced after inflation and require reheating to a temperature $\gtrsim \mu^{1/2} \simeq 10^{16} GeV$ which seems difficult.] Somewhat immodestly I mention a proposal Silk and I recently made: 'double inflation' (Silk & Turner 1987; Turner et al. 1987). While the Harrison-Zel'dovich spectrum is a beautiful prediction both because of its geometric simplicity and its definiteness, it may well be in conflict with observation because it does not seem to allow enough power on large scales to account for the recent observations of froth and large amplitude peculiar velocities. In the variant we have proposed there are two (or more) episodes of inflation, with the final episode lasting only about 40 e-folds or so, so that the amplitudes of perturbations on large scales are set by the first episode and those on small scales by the second episode. This enables one to have very large amplitude perturbations on small scales (of order 10^{-1}) and larger than usual amplitude perturbations on large scales (nearly saturating the large scale microwave limits), thereby providing enough power for the large scale structure which the recent redshift surveys and peculiar velocity measurements indicate. The large amplitude perturbations on small scales allow for very early galaxy formation (and reionization of the Universe, thereby erasing the CMBR fluctuations on small angular scales). If the second episode of inflation proceeds via the nucleation of bubbles, they might directly explain the froth-like structure recently reported by de Lapparent (1986).

EPILOGUE

Despite the absence of a compelling model which successfully implements the inflationary paradigm, inflation remains a very attractive means of accounting for a number of very fundamental cosmological facts by microphysics that we have some understanding of: namely, scalar field dynamics at sub-planck energies. The lack of a compelling model at present must be viewed in the light of the fact that at present we have no compelling, detailed model for the '*Theory of Everything*' and the fact that despite vigorous scrutiny there has yet to be a *No-Go Theorem* for inflation unearthed. It is my belief that the undoing of inflation (if it should come) will involve observations and not theory. At the very least The Inflationary Paradigm is still worthy of further consideration–and I hope that I have convinced you of that fact!

Due to space/time limitations my review of inflation has necessarily been incomplete, for which I apologize. I refer the interested reader to the more complete reviews by Linde (1984); by Abbott & Pi (1986); by Steinhardt (1984); by Brandenberger (1985); and by myself (Turner 1986). My prescription for successfully implementing inflation borrows heavily from the paper written by Steinhardt and myself (Steinhardt

& Turner 1984). This work was supported in part by the DoE (at Chicago) and by my Alfred P. Sloan Fellowship.

REFERENCES

Abbott, L. & Pi, S.-Y. (1986). *The Inflationary Universe*. Singapore: World Scientific.
Abbott, L. & Wise, M.B. (1984). *Nucl. Phys.*, **B244**, 541.
Accetta, F. et al. (1985). *Phys. Rev.*, **D31**, 3046.
Albrecht, A. & Brandenberger, R. (1985). *Phys. Rev.*, **D31**, 1225.
Albrecht, A. et al. (1985). *Phys. Rev.*, **D32**, 1280.
Albrecht, A. & Steinhardt, P.J. (1982). *Phys. Rev. Lett.*, **48**, 1220.
Albrecht, A. & Turok, N. (1985). *Phys. Rev. Lett.*, **54**, 1868.
Anderson, P. (1983). *Phys. Rev.*, **D28**, 271; **D29**, 615.
Anderson, P. (1985). *Phys. Rev.*, **D32**, 1302.
Bardeen, J.M. (1980). *Phys. Rev.*, **D22**, 1882.
Bardeen, J.M., et al. (1983). *Phys. Rev.*, **D28**, 679.
Bardeen, J.M., et al. (1986). *Astrophys. J.*, **304**, 15.
Barrow, J.D. & Tipler, F. (1986). *The Anthropic Cosmological Principle*. Oxford: Oxford University Press.
Blumenthal, G., et al. (1984). *Nature*, **311**, 517.
Boesgaard, A. & Steigman, G. (1985). *Ann. Rev. Astron. Astrophys.*, **23**, 319.
Bond, J.R. & Efstathiou, G. (1984). *Astrophys. J.*, **285**, L44.
Brandenberger, R. (1985). *Rev. Mod. Phys.*, **57**, 1.
Bunch, T. & Davies, P.C.W. (1978). *Proc. Roy. Soc. London*, **A360**, 117.
Burstein, D., et al. (1986). *Astrophys. J.*, in press.
Carr, B.J. & Rees, M.J. (1979). *Nature*, **278**, 605.
Charlton, J. & Turner, M.S. (1987). *Astrophys. J.*, **313**, 495.
Collins, C.A., et al. (1986). *Nature*, **320**, 506.
Collins, C.B. & Hawking, S.W. (1983). *Astrophys. J.*, **180**, 317.
Coughlan, G. et al. (1983). *Phys. Lett.*, **131B**, 54.
Davis, M., et al. (1985). *Astrophys. J.*, **292**, 371.
deLapparent, V., et al. (1986). *Astrophys. J.*, **302**, L1.
DeWitt, B. (1953). *Phys. Rev.*, **90**, 357.
Dicke, R.H. & Peebles, P.J.E. (1979). In *General Relativity: An Einstein Centenary Survey*, eds. S. Hawking & W. Israel. Cambridge: Cambridge University Press.
Dicus, D.A., et al. (1977). *Phys. Rev. Lett.*, **39**, 168.
Efstathiou, G. & Bond, J.R. (1986). *Mon. Not. r. Astron. Soc.*, **218**, 103.
Efstathiou, G. & Silk, J. (1983). *Fund. Cosmic Phys.*, **9**, 1.
Ellis, J.E., et al. (1984). *Phys. Lett.*, **145B**, 181.
Fischetti, M.V., et al. (1979). *Phys. Rev.*, **D20**, 1757.
Fischler, W. et al. (1985). *Nucl. Phys.*, **B259**, 730.
Ford, L. (1985). *Phys. Rev.*, **D31**, 710.
Goodman, M. & Witten, E. (1985). *Phys. Rev.*, **D31**, 3059.
Guth, A. (1981). *Phys. Rev.*, **D23**, 347.
Guth, A. & Pi, S.-Y. (1982). *Phys. Rev. Lett.*, **49**, 1110.
Guth, A. & Pi, S.-Y. (1985). *Phys. Rev.*, **D32**, 1899.
Guth, A. & Weinberg, E. (1983). *Nucl. Phys.*, **B212**, 321.
Hartle, J.M. & Hawking, S.W. (1983). *Phys. Rev.*, **D28**, 2960.

Hartle, J.M. & Hu, B.-L. (1979). *Phys. Rev.*, **D20**, 1772.
Hawking, S.W. (1982). *Phys. Lett.*, **115B**, 295.
Hawking, S.W., et al. (1982). *Phys. Rev.*, **D26**, 2681.
Holman, R., et al. (1984). *Phys. Lett.*, **137B**, 343.
Jensen, L. & Stein-Schabes, J. (1986a). *Phys. Rev.*, **D34**, 931.
Jensen, L. & Stein-Schabes, J. (1986b). *Phys. Rev.*, in press.
Kaiser, N. (1983). *Astrophys. J.*, **273**, L17.
Kaiser, N. (1986). In *Inner Space /Outer Space*, eds. E.W. Kolb, et al. Chicago: University of Chicago Press.
Khlopov, M.Yu. & Linde, A.D. (1984). *Phys. Lett.*, **138B**, 265.
Kolb, E.W. & Turner, M.S. (1983). *Ann. Rev. Nucl. Part. Sci.*, **33**, 645.
Linde, A.D. (1982a). *Phys. Lett.*, **108B**, 389.
Linde, A.D. (1982b). *Phys. Lett.*, **116B**, 335.
Linde, A.D. (1983). *Phys. Lett.*, **129B**, 177.
Linde, A.D. (1984). *Rep. Prog. Phys.*, **47**, 925.
Linde, A.D. (1985). *Phys. Lett.*, **158B**, 375.
Lindley, D. (1985). Fermilab preprint (unpublished).
Loh, E. & Spillar, E. (1986). *Astrophys. J.*, **307**, L1; *Phys. Rev. Lett.*, **57**, 2865.
Matzner, R. & Misner, C.W. (1972). *Astrophys. J.*, **171**, 415.
Mijic, M.B., et al. (1986). *Phys. Rev.*, **D34**, 2934.
Misner, C.W. (1968). *Astrophys. J.*, **151**, 431.
Misner, C.W. (1972). In *Magic Without Magic*, ed. J. Klauder. San Francisco: Freeman.
Mottola, E. (1985). *Phys. Rev.*, **D31**, 754; **D33**, 2136.
Myhrvad, N. (1983). *Phys. Lett.*, **132B**, 308.
Nanopoulos, D.V. (1985). *Comments on Astrophys.*, **X**, 219.
Olive, K., et al. (1985). *Astrophys. J.*, **292**, 1.
Ovrut, B. & Steinhardt, P.J. (1984a). *Phys. Rev. Lett.*, **53**, 732.
Ovrut, B. & Steinhardt, P.J. (1984b). *Phys. Lett.*, **147B**, 263.
Parker, L. (1976). *Nature*, **261**, 20.
Parker, L. (1983). *Phys. Rev. Lett.*, **50**, 1009.
Peebles, P.J.E. (1984). *Astrophys. J.*, **284**, 439.
Penrose, R. (1979). In *General Relativity: An Einstein Centenary Survey*, eds. S.W. Hawking & W. Israel. Cambridge: Cambridge University Press.
Peterson, J., et al. (1985). *Phys. Rev. Lett.*, **55**, 332.
Pi, S.-Y. (1984). *Phys. Rev. Lett.*, **52**, 1725.
Preskill, J. (1984). *Ann. Rev. Nucl. Part. Sci.*, **34**, 461.
Rubakov, V.A., et al. (1982). *Phys. Lett.*, **115B**, 189.
Scherrer, R.J. & Turner, M.S. (1987). *Astrophys. J.*, in press.
Seckel, D. & Turner, M.S. (1985). *Phys. Rev.*, **D32**, 3178.
Shafi, Q. & Vilenkin, A. (1984). *Phys. Rev. Lett.*, **52**, 691.
Shafi, Q. & Wetterich, C. (1983). *Phys. Lett.*, **129B**, 387.
Shafi, Q. & Wetterich, C. (1985). *Phys. Lett.*, **152B**, 51.
Sikivie, P. (1983). *Phys. Rev. Lett.*, **51**, 1415.
Silk, J. & Turner, M.S. (1987). *Phys. Rev.*, **D35**, in press.
Smoot, G.F., et al. (1985). *Astrophys. J.*, **291**, L23.
Spokoiny, B.L. (1984). *Phys. Lett.*, **147B**, 39.
Starobinskii, A.A. (1979). *JETP Lett.*, **30**, 682.
Starobinskii, A.A. (1980). *Phys. Lett.*, **91B**, 99.

Starobinskii, A.A. (1982). *Phys. Lett.*, **117B**, 175.
Starobinskii, A.A. (1983). *JETP Lett.*, **37**, 66.
Steigman, G. & Turner, M.S. (1983). *Phys. Lett.*, **128B**, 295.
Steinhardt, P.J. (1984). *Comments Nucl. Part. Phys.*, **12**, 273.
Steinhardt, P.J. & Turner, M.S. (1983). *Phys. Lett.*, **129B**, 51.
Steinhardt, P.J. & Turner, M.S. (1984). *Phys. Rev.*, **D29**, 2162.
Traschen, J. & Hill, C.T. (1986). *Phys. Rev.*, **D33**, 3519.
Trimble, V. (1987). *Ann. Rev. Astron. Astrophys.*, **25**, in press.
Turner, M.S. (1983). *Phys. Rev.*, **D28**, 1243.
Turner, M.S. (1986). In *Fundamental Physics and Cosmology*, eds. J. Audouze & J. Tran Thanh Van. Gif-sur-Yvette: Editions Frontieres; and in *The Architecture of Fundamental Interactions at Short Distances*, eds. P. Ramond & R. Stora. Amsterdam: North-Holland.
Turner, M.S. (1987). In *Dark Matter in the Universe*, eds. J. Kormendy & J. Knapp. Dordrecht: Reidel, p. 445.
Turner, M.S., et al. (1984). *Phys. Rev. Lett.*, **52**, 2090.
Turner, M.S., et al. (1987). *Astrophys. J.*, in press.
Turner, M.S. & Widrow, L. (1986). *Phys. Rev. Lett.*, **57**, 2237.
Turok, N. (1985). *Phys. Rev. Lett.*, **55**, 1801.
Vilenkin, A. (1984). *Phys. Rev. Lett.*, **53**, 1016.
Vilenkin, A. (1985). *Phys. Rep.*, **121**, 263.
Vilenkin, A. & Ford, L. (1982). *Phys. Rev.*, **D26**, 1231.
Vittorio, N. & Silk, J. (1984). *Astrophys. J.*, **285**, L39.
Vittorio, N. & Turner, M.S. (1987). *Astrophys. J.*, in press.
Wald, R.M. (1983). *Phys. Rev.*, **D28**, 2118.
White, S.D.M. (1986). In *Inner Space /Outer Space*, eds. E.W. Kolb, et al. Chicago: University of Chicago Press.
Wilkinson, D.T. (1986). In *Inner Space /Outer Space*, eds. E.W. Kolb, et al. Chicago: University of Chicago Press.
Yang, J., et al. (1984). *Astrophys. J.*, **281**, 493.
Zel'dovich, Ya.B. (1970). *JETP Lett.*, **12**, 307.

Some recent developments in general relativity*

Shing-Tung Yau
University of California, San Diego
and
University of Texas at Austin

I will focus my attention on some problems related to asymptotically flat space-time that occur in general relativity. Hence I will deal with a physical system which is isolated. The concept of asymptotically flat is not as well understood as what one would like. R. Penrose did make a nice proposal to give a definition of such space-time. However, besides stationary space-times, I have not found examples that satisfy his definition. This is a problem if one wants to study dynamics of gravitational waves.

There seems to be no major problem in understanding the question of asymptotical flatness for a space-like hypersurface. If M is such a hypersurface, the induced metric g_{ij} is required to be positive definite and in a neighborhood of each infinity, it should have the expansion

$$g_{ij} = \delta_{ij} + O(r^{-\alpha})$$

where r is the geodesic distance from a fixed point and $\alpha > 0$. One expects to assume $|dg_{ij}| = O(r^{-\alpha-1})$ and the scalar curvature to be integrable also. If $\alpha > \frac{1}{2}$, one can define total mass [1,5,6] and quite a few general properties hold. It should be noted that, a priori, there can be two different (asymptotically flat) coordinate systems defined at infinity. Under fairly general conditions, Bartnik [2] proved that in the neighborhood of each infinity, asymptotically flat coordinate systems are equivalent and the total mass is uniquely attached to each regime at infinity.

Optimally one should be able to define asymptotical flatness without introducing any coordinate system at infinity. It should be possible to assume only that

* This research was partially supported by NSF No. DMS84-09447 and ONR No. N-00014-85-K-0367.

the curvature tensor decays in a suitable way and that the topology at infinity is correct.

We should bear in mind that the asymptotic structure of the initial spacelike hypersurface should be defined in such a way that its qualitative behaviour is preserved under the evolution equation. Of course, the basic physical quantities including total energy, total linear momentum and total angular momentum should be finite also. For spatial infinity, the well-known short time existence theorem is perhaps good enough for most purposes.

In order to understand the dynamical part of space-time, we are forced to study the question of the global existence of the Einstein equation. A sound understanding of the asymptotic nature of the space-time should depend on such an existence theorem. While it is widely believed that the Penrose diagram gives a good description of asymptotically flat space-times, it is clear that the high differentiability of the boundary of the Penrose diagram is presumably unjustified. The difficulties in understanding the Penrose diagram are that all the known asymptotically flat space-times are stationary and provide no information about the dynamical nature of a general space-time.

The best approach is to study the global solution for the Einstein equation. Unfortunately, this is a very difficult problem. Part of the behaviour of such solutions was proposed by Penrose and was called the cosmic censorship. In order to isolate the effect of gravity on the structure of singularities, it is more appropriate to study vacuum solutions. When the initial data on the initial hypersurface is close to being trivial or when the initial data is close to that induced by the Minkowski data, the global existence problem for the vacuum equation can be considered as a problem of demonstrating the stability of the Minkowski space-time. Notice that the stability here is different from the one understood in quantum gravity and hence is not implied by the positive mass conjecture. Under reasonable assumptions on the initial data, Christodoulou, Klainerman and the author are studying this problem and have made good progress on it. It will shed light on the understanding of the Penrose diagram also.

For a general space-time, the global existence theorem is very difficult. Perhaps a good definition of weak solutions for the Einstein equation will be very helpful. Such a definition was proposed by Christodoulou for spherical symmetric space-time. He demonstrated the global existence theorem for such weak solutions and demonstrated the regularity of the weak solutions when the initial data is small. Unfortunately a uniqueness result for weak solutions is missing (unless one assumes that one of these weak solutions is smooth). One expects that cosmic censorship is valid for the class of spherical symmetric space-time. Christodoulou also demonstrated a remarkable theorem on the asymptotic nature of spherical space-time. He showed that when the Bondi mass of the light cone tends to zero as the light cone tends to future infinity, the space-time tends to the flat Minkowski space-time. Otherwise the Bondi mass has residue at infinity and the space-time will tend to the Schwarzschild solution. This last result tells us how spherical space-time collapses to a black hole. Perhaps a quantitative understanding of such result would be interesting.

It is highly desirable to generalize the theory of Christodoulou to axially symmetric space-time. Clearly the answer would be far more complicated because spherical symmetry has a strong constraint on the type of singularities that can occur. However, the asymptotic behaviour of a general space-time may not be too complicated. It is perhaps true that when Bondi mass tends to zero, the space-time tends to flat space-time and when the Bondi mass has residue at future infinity, the space-time tends to some type of superposition of the Kerr solutions. A proof of the latter claim is clearly very desirable.

The cosmic censorship is basically a problem of understanding the dynamic nature of space-time. In this regard, the works of Bondi-Sachs, Penrose, Newman and others clearly will play an important role. In studying the evolution of the nonlinear hyperbolic system, it is always important to understand physical quantities that are conserved during the evolution. In fact, the energy method was used extensively in the quasilinear hyperbolic equation. An important feature of the energy method is that the energy is nonnegative and that it is zero only if the system is "trivial." In general relativity, the energy is more complicated and there are two masses: the ADM mass and the Bondi mass. It turns out that both satisfy the requirement mentioned above.

The proof of the positivity of the ADM mass is a long story; many contributions were made on special space-time by Brill, Deser, Marsden, Choquet-Bruhat, etc. Finally in 1978, Schoen and the author found a complete demonstration of the mass conjecture. About two years later, E. Witten, building on some previous observations on supergravity by Deser and Teitelbaum, found a beautiful elegant proof of the conjecture also. The proof by Schoen and myself is actually rather simple except that it is more geometric in nature. There are several advantages to our proof that Witten's argument does not achieve. (For higher dimension spin manifolds, Witten's argument seems to be better in many aspects.) I believe that some of these points are relevant to the global existence problem.

In studying the global solution which evolved from nonsingular initial data, the three dimensional spatial hypersurface will encounter singularities of a different sort. These singularities occur in the interior of the slice and so the total mass is still well defined. It seems that the argument using minimal surfaces has a better chance of dealing with singularities. For example, if the singularity can be surrounded by a close surface whose mean curvature (with respect to the inward normal) is not less than the trace of the second fundamental form on this surface, the mass conjecture holds. One can also allow singularity to appear on this surface as long as all curves tending to the singularity have infinite length.

A by-product of the proof of the mass conjecture by using minimal surfaces is our demonstration of the existence of the black hole due to condensation of mass. Given a compact spacelike three dimensional hypersurface M (with boundary) in a space-time so that the mean curvature of ∂M is not less than either the trace of the second fundamental form of M taken on ∂M or its negative, we prove that a trapped surface must exist in M if the local mass density is not less than $3\pi^2/2\rho^2$ where ρ is the largest radius of a solid donut that can be embedded into M. (Notice that the boundary condition on ∂M is always valid if M is a big region on a three dimensional asymptotically flat spatial slice.) Ó Murchadha made some calculations and suggested a smaller constant than the one above. Coupling the famous singularity theorem of Penrose and Hawking, the above theorem confirms the statement that when matter condenses in a fixed region, that region collapses to a black hole. It also implies roughly that when a neutron star collapses, it collapses to a black hole. In our statement, we do not include the contribution of gravitational interaction. This is not possible at this moment because we do not have a good theory of local mass yet.

The proof of the mass conjecture also demonstrates another phenomenon. A black hole not only sucks in matter, it also sucks in the topology of the space-time in the following sense. Given any three dimensional spatial asymptotically flat hypersurface M of a space-time which satisfies the local energy condition, there exist mutually disjoint trapped surfaces $\Sigma_1, \Sigma_2, \ldots$ and Σ_n on M so that $M \setminus \left(\cup_{i=1}^{n} \Sigma_i \right)$ is homeomorphic to the complement of several disjoint balls in a compact three dimensional manifold which is homotopic to the connected sum of several copies of $S^2 \times S^1$ and the lens spaces.

It has been speculated for a long time that maximal slice foliation in a space-time will reflect the nature of singularities. This was in fact used in our work with Christodoulou and Klainerman. The existence of maximal slice foliation for a large class of general asymptotically flat space-time was not known until recently when R. Bartnik, basing his work on the fundamental work of Gerhardt, proved its existence by delicate estimates. For space-time satisfying a local energy condition, the induced metric on the maximal slice has nonnegative scalar curvature. This fact implies that the maximal slice must have the special topology mentioned above. Hence one can construct examples of space-time evolved from some non-singular initial data on a three dimensional manifold with complicated topology which admits no maximal slice.

Finally, let us discuss the idea of local mass. Such a concept will be extremely useful in understanding the global existence theorem. There were several proposals including one given by Hawking and another one by Penrose. The one by Hawking is not necessarily positive. The definition by Penrose is rather attractive except that it is difficult to compute. Christodoulou and the author found that the Hawking mass of a two dimensional surface is well behaved if such a surface Σ is round in the following sense. There exists a three dimensional spatial manifold

M so that Σ bounds a volume in M and among all surfaces Σ' near Σ which bounds the same volume as Σ, Σ has the smallest area. We demonstrated that the Hawking mass is positive if $\int_\Sigma R_M \geq 0$ where R_M is the scalar curvature of M. Note that this last assumption is automatic if M is a maximal slice. Huisken and the author proved the existence of an abundance of these round surfaces in the following way. Given a three dimensional asymptotically flat slice M so that the metric is spherical at infinity up to the second order, there is a unique foliation of M by round spheres at infinity. This seems to be an interesting fact by itself. The uniqueness comes from the positivity of the total mass. It seems to be relevant to the angular momentum of these manifolds. Ashtekar has made many contributions on the subject of total angular momentum.

We hope that eventually we will have a better understanding of the Einstein equation.

References

1. Annowitt, R., Deser, S., and Misner, C., in Gravitation: An Introduction to Current Research, L. Witten (ed.), Wiley, New York, 1962.
2. Bartnik, R., Total mass of an asymptotically flat manifold, to appear.
3. Brill, D., and Deser, S., *Ann. Phys.* (N.Y.) **50**, 548-570 (1968).
4. Ó Murchadha, N., Total energy-momentum in general relativity, to appear.
5. Schoen, R. and Yau, S.-T., *Comm. Math. Phys.* **65**, 45-76 (1979).
6. Schoen, R. and Yau, S.-T., *Comm. Math. Phys.* **79**, 231-260 (1981).
7. Witten, E., *Comm. Math. Phys.* **80**, 381-402 (1981).

ALGEBRAIC COMPUTING

Inge Frick
Institute of Theoretical Physics, University of Stockholm,
Vanadisv. 9, S-113 46 Stockholm, Sweden

In this report from the session on Algebraic Computing (AC) in General Relativity (GR), I will briefly summarize the papers and give some general comments on the status of AC in GR.

The session on AC attracted 13 papers and 4 computer demonstrations. All of these, except 2 late papers, were presented at the session.

SUMMARY OF PAPERS AND COMPUTER DEMONSTRATIONS

Papers on AC in GR can be divided into two groups, but often with a considerable overlap between the two. The first group consist of papers that discuss a program and what GR calculations it can do. They were all placed in the session on AC. The second group consist of papers that discuss a problem and what programs can be used to solve it. They were only placed in the session on AC if the use of AC was central to the paper, otherwise they were placed in the session where their problem belonged e.g. Exact Solutions.

General Programs

A. Krasinski discussed the development of ORTOCARTAN since 1983. ORTOCARTAN is a system that can compute the curvature tensor etc. in an orthonormal frame. General descriptions of ORTOCARTAN can be found in (Krasinski & Perkowski 1981, Krasinski 1983 a,b). ORTOCARTAN is available as a tape containing versions for IBM(like) and CDC computers. There is also a UNIVAC version. Improvements since 1983 include: pattern matching with wild cards so that e.g. if A is a wild card then (log A) will match any logarithm; more powerful algorithm for exponentiation of rational numbers; better way to use auxiliary programs that e.g. invert matrices and compute the Lie derivative.

C.C. Dyer in two talks described muTENSOR and the GR facilities in MAPLE.

MuTENSOR (Harper & Dyer 1985) is available in three different versions that run in REDUCE, MAPLE and muMATH which means that it is available on a large number of machines. MuMATH runs on micro computers while REDUCE and MAPLE run on larger machines. MuTENSOR manipulates indexed objects where the indices might be coordinate, tetrad, spinor etc. indices. It has built in facilities for calculating the curvature tensor etc. but one can also directly, in a format similar to text books, tell the system to perform addition, outer product, contraction, covariant and ordinary differentiation, raising or lowering of indices and symmetry operations.

There are three facilities for GR calculations in MAPLE (Char et al. 1983, 1985). There is muTENSOR (see above) and there are two programs for calculations with the Newman Penrose (NP) formalism. The first program computes the spin coefficients and curvature components from a given null tetrad (Allen et al. In preparation,n Fee & McLenaghan 1977). The second program is described by Czapor below.

S.R. Czapor talked and gave a computer demonstration of a MAPLE package for symbolic calulations in the NP formalism (Czapor & McLenaghan In preparation). This package treats the NP equations as symbolic objects, not as explicit tensor components. It can perform (symbolic) differention using the NP differentiation operators and it can do complex conjugation, recognizing real and imaginary objects. An example of its use can be found in (Czapor & McLenaghan 1982).

J.E. Åman gave a demonstration of SHEEP and CLASSI. SHEEP (Frick 1977) is a computer algebra system with special facilities for manipulation of tensor components and other indexed objects. For an example of its use see (d'Inverno & Frick 1982). CLASSI (Åman 1982) is an implementation in SHEEP of the algorithms for invariant classification of geometries (see classification project below). The demonstration covered classification and use of the library of GR programs that compute e.g. the curvature tensor, working with various frames, spinors, NP formalism etc.

L. Hörnfeldt demonstrated STENSOR. STENSOR (Hörnfeldt 1980, 1986), that runs in SHEEP or MACSYMA, allows tensors, spinors, (non-comuting) operators etc. to be defined with input quite similar to mathematical text and then to be manipulated as symbolic objects i.e. not only as components. Side-relations (Hörnfeldt 1982) like A+B=C can be handled in a sophisticated way. STENSOR can also generate code for SHEEP (see below).

R.A. d'Inverno gave a talk describing GFRAME and VFRAME, two tensor packages that run in SHEEP. They both implement algorithms for calculating the curvature tensor etc. in a general (neither coordinate nor constant) frame. GFRAME is hand coded in LISP, the language underlying SHEEP. To test STENSOR, VFRAME was written in the more natural tensor definition format of STENSOR, and the final LISP code was automatically generated by the STENSOR tensor compiler. The conclusion one can draw from this test is that it takes a much shorter time to develop a package like GFRAME with the help of STENSOR than doing it directly in LISP. Furthermore, the generated code is in some cases faster.

The Classification Project

The classification project (Åman et al. 1985) is so far the best example in GR of a research project that would have been impossible without Algebraic Computing. This project is concerned with the equivalence problem, which is the problem of deciding whether two metrics are locally isometric, i.e. whether a set of ten second order differential equations have a solution. There is an algorithm due to Cartan (Karlhede 1980) that reduces the problem to whether a set of non-differential equations have a solution. This is done by finding a sufficient set of invariants consisting of components of the Riemann tensor and its covariant derivatives. For a concrete description of the procedure see e.g. (Cohen et al. 1984).

M. MacCallum presented three papers related to the equivalence problem: the roots of a quartic, a minimal set of derivatives and the present status of the project.

An important step in the algorithm mentioned above, is to determine the Petrov type of the Weyl tensor and the Plebanski or Segre type of the Ricci tensor. This means determining the multiplicities of the roots of a quartic. The classical method to do this tests the discriminant, a 6th degree polynomial in the quartic's coefficients. This is inefficient when the coefficients are long formulae. In (Åman et al. 1984 and In preparation) a new method is presented that carefully looks at the special cases in such an order that higher degree expressions are computed only when it is necessary. The new method has been shown to be slightly less efficient on easy cases but much better on difficult ones. The Plebanski or Segree classification (Joly & MacCallum In preparation) has not been implemented on a computer before.

The equivalence algorithm determines up to what order covariant derivatives of the Riemann tensor are needed to characterize a metric. Because of the Ricci and Bianchi identities, only some derivatives of each order are needed. In (MacCallum & Åman 1986), a minimal set of algebraicaley independent derivatives is given and the number of necessary components is discussed for different cases.

The invariants produced by the equivalence algorithm give an invariant characterization of each metric. They will be used to construct a database of known metrics (MacCallum 1983). There exists a preliminary version of the program that produces the database. It uses the results mentioned above and can now automatically transform to a standard form for algbraically special metrics.

M.E. Araujo presented a SHEEP/CLASSI implementation (Araujo & Skea In preparation) of a scheme for the invariant characterization of the Bianchi universe models. The scheme uses an invariant basis which is transformed using a time-dependent automorphism of the Bianchi group. This allows one to build a unified picture of the dynamics and considerably simplifies the field equations. The invariant characterization can be included in the classification database.

Homothetic Killing tensors

G. Prince described and demonstrated a set of REDUCE programs that assist in finding homothetic killing tensors. In (Prince & Crampin 1984 a,b) Projective Differential Geometry techniques were used to find geodesic conservation laws based on the tangent bundle group actions. A homothetic transformation is a generalization of isometry that involves a constant scale change. Homothetic Killing tensors are the related generalization of the concept of Killing tensors. Associated to each Homothetic Killing tensor is a conserved quantity. It is of great interest to find these objects in known solutions of the Einstein equations.

Kaluza-Klein and Super Gravity

The following two papers were received too late for an oral presentation.

Essen and Hörnfeldt have studied the embedding of an n dimensional ellipsoid in an n+1 dimensional space. STENSOR (Hörnfeldt 1986) has been used to study this problem for n up to 6. Formulae for arbitrary n could then be deduced. Spherical coordinates were used and STENSOR was used to generate the second fundamental form and then the curvature tensor. The STENSOR trigonometric simplifier (Hörnfeldt 1982) was used to simplify the enormous trigonometric expressions generated.

K. Shaker-Jomaa has written a program in STENSOR that can evaluate the Lagrangian density of an arbitrary Kaluza-Klein ansatz. As an application, a new ansatz for coset spaces of type G/H, was studied. The program is included in the STENSOR distribution.

ALGEBRAIC COMPUTING, DEVELOPMENT SINCE 1983

At GR10, Cohen et al. (Cohen et al. 1984) surveyed the status of AC in GR. The survey by MacCallum (MacCallum 1984) is from the end of the same year (1983). I will here note some points that somewhat update these surveys to cover the status of AC in GR as of summer 1986.

SMP is now available but doesn't seem to have GR facilities. As can be seen from the present report, MAPLE should be added to the list of general computer algebra systems that have GR capabilities and there is still another REDUCE package for GR.

The price for a good computer for e.g. REDUCE is now USD 5.000 and will soon be down to USD 3.000.

I will end by pointing out an important development that has just started, namely the use of computer graphics in AC. As AC is normally done interactively it is very important that it is easy to talk to the machine. In the future the following can be expected: input and output to and from the computer will look like mathematical text; with a pointing device e.g. a "mouse" you will be able to point at subexpressions to do substitutions and other rearrangements; you will also be able to use your "mouse" to select rearrangeing rules from menus suited to your problem. Good graphics machines (and software) are now starting to come down in price - quite good machines are available for less than USD 10.000. The machines for USD 3.000 mentioned above actually have graphical capabilities, but they do not have all software yet.

PAPERS AND DEMONSTRATIONS AT THE SESSION

Krasinski, A. (Paper) The Program ORTOCARTAN - Developments since 1983.

Dyer, C.C., Allen, S. & Harper, J.F. (Paper) The muTENSOR Computer System for General Relativity.

Dyer, C.C. & McLenaghan, R.G. (Paper) General Relativity Calculations in MAPLE.

Czapor, S.R. & McLenaghan, R.G. (Paper and demonstration) A MAPLE Package for Performing Calculations in the Newman-Penrose Formalism.

Åman, J.E. & Frick, I. (Demonstration) SHEEP and CLASSI.

Hörnfeldt, L. (Demonstration) STENSOR.

Joly, G.C. (Paper) Interacting with GFRAME.

Holmes, G. & d'Inverno, R.A. (Paper) Interacting with VFRAME.

MacCallum, M.A.H., d'Inverno, R.A., Joly, G.C. & Åman, J.E. (Paper) Quartics and Computer-aided Classification of the Riemann Tensor in General Relativity.

MacCallum, M.A.H. & Åman, J.E. (Paper) Algebraically Independent n-th Derivatives of the Riemannian Curvature Spinor.

MacCallum, M.A.H., Skea, J.E.F., d'Inverno, R.A. & Åman, J.E. (Paper) Progress on the Equivalence Problem.

Araujo, M.E. & Skea, J.E.F. (Paper) The Automorphism Groups for Bianchi Universe Models and Computer-aided Invariant Classification of Metrics.

Prince, G. (Paper and demonstration) The search for Homothetic Killing Tensors.

Essen, H. & Hörnfeldt, L. (Paper) Curvature of n-Dimensional Ellipsoids embedded in R^{n+1}.

Shaker-Jomaa, K. (Paper) Computing Kaluza-Klein Ansatz.

REFERENCES

Allen, S., Fee, G.J. & McLenaghan, R.G. (In preparation). General Relativity and the Newman-Penrose Formalism in Maple. University of Waterloo, Waterloo, Ontario, Canada.

Åman, J.E. (1982) Manual for CLASSI - Classification Programs for Geometries in General Relativity, preprint, Institute of Theoretical Physics, University of Stockholm, Stockholm, Sweden.

Åman, J.E., d'Inverno, R.A., Joly, G.C. & MacCallum, M.A.H. (1984). Quartic Equations and Algorithms for Riemann Tensor Classification. In Proc. EUROSAM 84, ed. J. Fitch, Lecture Notes in Computer Science, Vol. 174, pp. 47-58. Berlin: Springer-Verlag.

Åman, J.E., d'Inverno, R.A., Joly, G.C. & MacCallum, M.A.H. (1985). Progress on the Equivalence Problem. In Proc. EUROCAL'85, ed. B.F. Caviness, Lecture Notes in Computer Science, Vol. 204, pp. 89-98. Berlin: Springer-Verlag.

Åman, J.E., d'Inverno, R.A., Joly, G.C. & MacCallum, M.A.H. (In preparation). Quartic Equations and Algorithms for Riemann Tensor Classification.

Araujo, M.E. & Skea, J.E.F. (1986) The Automorphism Groups for Bianchi Universe Models and Computer-aided Invariant Classification of Metrics. Preprint, Queen Mary College, London, United Kingdom.

Char, B.W., Geddes, K.O., Gentleman, W.M. & Gonnet, G.H. (1983). The Design of MAPLE : A Compact, Portable and Powerful Computer Algebra System. In Proc. EUROCAL'83, ed. J.A. van Hulzen, Lecture Notes in Computer Science, Vol. 162, pp. 101-15. Berlin: Springer-Verlag

Char, B.W., Geddes, K.O., Gonnet, G.H. & Watt, S.M. (1985). MAPLE User's Guide. Waterloo, Ontario, Canada: WATCOM Publications.

Czapor, S.R. & McLenaghan, R.G. (1982). Orthogonal transitivity, invertibility and null geodesic separability in type D vacuum solutions of Einstein's field equations with cosmological constant. J. Math. Phys. 23, no.11 2159-67.

Czapor, S.R. & McLenaghan, R.G. (In preparation) NP: A Maple Package For Performing Calculations in the Newman-Penrose Formalism. Department of Applied Mathematics. University of Waterloo, Waterloo, Ontario, Canada.

d'Inverno, R.A. & Frick, I. (1982). Interacting with SHEEP. Gen. Rel. Grav. 14, no.10, 935-64.

Fee, G.J. & McLenaghan, R.G. (1977). An efficient algorithm for the computer calculation of the components of the connection and the curvature for a given null tetrad. GR8 Conference Abstracts, University of Waterloo, Waterloo, Ontario, Canada.

Frick, I. (1977). SHEEP User's Guide. Institute of Theoretical Physics, University of Stockholm, Report 77-14. Stockholm, Sweden.

Harper, J.F. & Dyer, C.C. (1985). The muTENSOR User Manual. Department of Astronomy and David Dunlap Observatory, University of Toronto, Canada.

Hörnfeldt, L. (1980). Indicial Tensor and Spinor Calculus with Computer. GR9 Conference Abstracts, Friedrich Schiller University, Jena, GDR.

Hörnfeldt, L. (1982). A Sum-Substitutor used as a Trigonometric Simplifier. In Proc. EUROCAM'82, ed. J. Calmet, Lecture Notes in Computer Science, Vol. 144, pp. 188-9. Berlin: Springer-Verlag

Hörnfeldt, L. (1986). STENSOR reference manual. Preprint. Institute of Theoretical Physics, University of Stockholm, Stockholm, Sweden.

Joly, G.C. & MacCallum, M.A.H. (In preparation) Computer-aided classification of the Ricci tensor in General Relativity. Queen Mary College, London, United Kingdom.

Krasinski, A. & Perkowski, M. (1981). ORTOCARTAN - A New Computer Program for Analytic Calculations in Genaral Relativity. Gen. Rel. Grav. 13, no.1, 67.

Krasinski, A. (1983 a). ORTOCARTAN - A Program for Algebraic Computations in General Relativity. SIGSAM Bulletin, 17, no.3,4, 12-8.

Krasinski, A. (1983 b). The Program ORTOCARTAN for Algebraic Calculations - New Developments. In GR10 Conference Abstracts, ed. B. Bertotti, F. de Felice & A. Pascolini, pp. 433-5. Rome: CNR.

MacCallum, M.A.H. (1983). Proposed format for recording the invariant characterisation of exact solutions. In GR10 Conference Abstracts, ed. B. Bertotti, F. de Felice & A. Pascolini, pp. 301-3. Rome: CNR.

MacCallum, M.A.H. (1984). Algebraic computing in general relativity. In Classical General Relativity, ed. W.B. Bonnor, J.N. Islam & M.A.H. MacCallum, pp. 145-71. Cambridge: Cambridge University Press.

MacCallum, M.A.H. & Åman, J.E. (1986). Algebraically independent nth derivatives of the Riemannian curvature spinor in general spacetime. Class. Quantum Grav., 3, no.6, 1133-41.

Prince, G.E. & Crampin, M. (1984a) Projective Differential Geometry and Geodesic Conservation Laws in General Relativity. I: Projective Actions. Gen. Rel. Grav., 16, no.10, 921-42.

Prince, G.E. & Crampin, M. (1984b) Projective Differential Geometry and Geodesic Conservation Laws in General Relativity. II: Conservation Laws. Gen. Rel. Grav., 16, no.11, 1063-75.

REPORT ON SYMPOSIUM "ALTERNATIVE THEORIES OF GRAVITATION"

H.F.M. Goenner
Institute for Theoretical Physics
University of Göttingen
D-3400 Göttingen, West Germany

The subject of "alternative theories of gravitation" is a branch of scientific speculation neither sanctioned nor accepted to the same degree as other such theoretical reasoning offered during this conference. This might be explained by the lack of empirical checks, by present technology, on the theories suggested. And, by the diversity in methods and quality of the different approaches taken.

In view of the 71 contributed abstracts falling into 20 categories obviously only a few speakers could actually present their work. I began by reviewing some motivations and constructive possibilities. Then, three talks by *J.D. McCrea, M. Francaviglia* and *R.B. Mann* followed reflecting subjects of high interest and characterizing research going on presently.

In the classical context, new theories of gravitation are built to express in a preciser form than hitherto used, the fundamental principles underlying General Relativity (GR). Also, cosmological models are invoked, in particular those freeing space-time of singularities. The quest for a proper definition of the local energy-momentum density of the gravitational field goes on. Exact solutions of alternative theories are looked for. Finally, differing theories are classified and compared, an endeavour deserving further attention (cf. the third talk below).

With the growing interest in unified field theories the difficulty of separating out classical and quantum arguments increases. Take the issue of gravitation as a gauge field, attempts in the geometrization of quantum aspects of matter (spin, vacuum fluctuations) or references to correction terms arising from renormalization within quantum field theory. The failure up to now of establishing quantum gravity proper also induces authors to invent alternative theories of gravitation.

We now wander through the landscape of alternative theories of gravitation along four roads leaving at distinct locations but meeting then and crossing each other. Each indicates a fundamental change with respect to GR. Along the first way the dynamics of GR is altered.

Fig. 1 Change of Dynamics

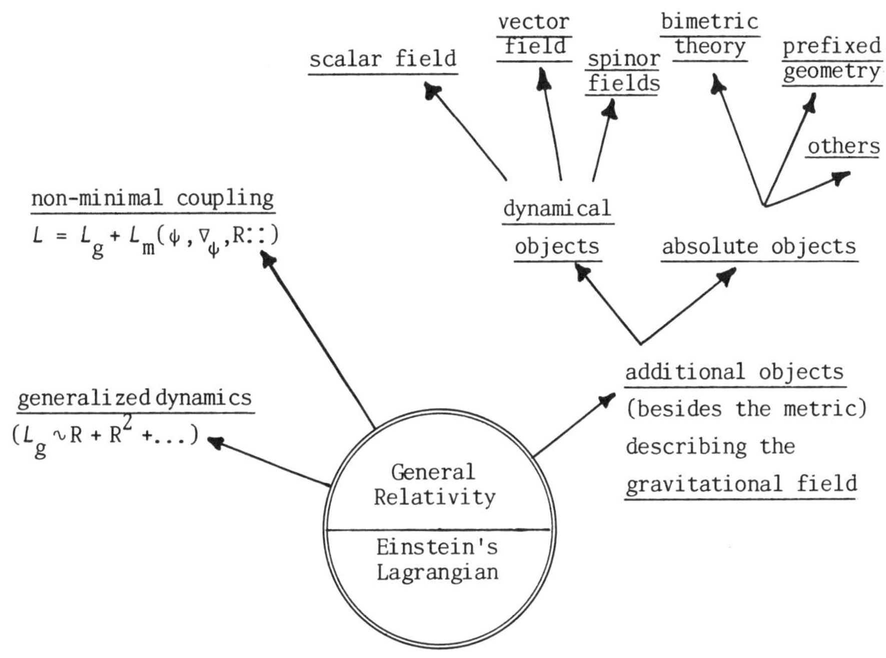

In Fig. 1, the higher derivative gravity theories mentioned in Isham's and Duff's lectures are contained. We note that the dimensional reduction of Kaluza-Klein theories with Lagrangians non-linear in the curvature is leading to non-minimal coupling within space time. We could also include, in Fig. 1, the f-g theory, a 2-tensor theory describing gravitation plus strong interactions (Huang & Raut), or the Einstein-Mayer theory introducing spinor fields and its generalizations (Rosen & Tauber).

The second road leads us to the spectrum of geometries. Fig. 2 cannot but sketch part of the wealth of available structures.

Fig. 2 Change of Geometry

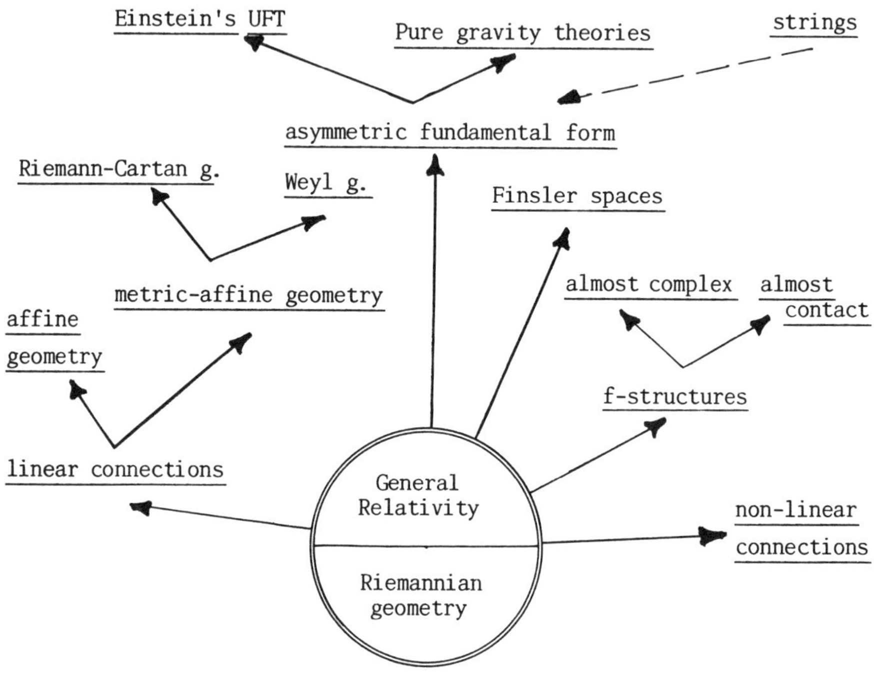

Gauge theories of gravitation, notably the Poincaré gauge field theory (PGT) discussed in the next talk appear within Riemann-Cartan or metric-affine geometry. Almost contact structures being defined only for odd dimensions, they eventually may become important in Kaluza-Klein theories.

This remark brings us to the third path: the change of dimension, Riemannian (or another) geometry in a higherdimensional space as the arena for a field theory unifying gravitation with other fundamental interactions. The main idea is to incorporate the internal gauge groups into the diffeomorphism group of a higherdimensional space. It is here that we leave both pure gravity theory and the classical domain. From string theory in 26 or even 506 dimensions, in the weak energy limit, geometrical structures arise resembling Einstein's unified field theory (UFT).

Fig. 3 Change of dimension

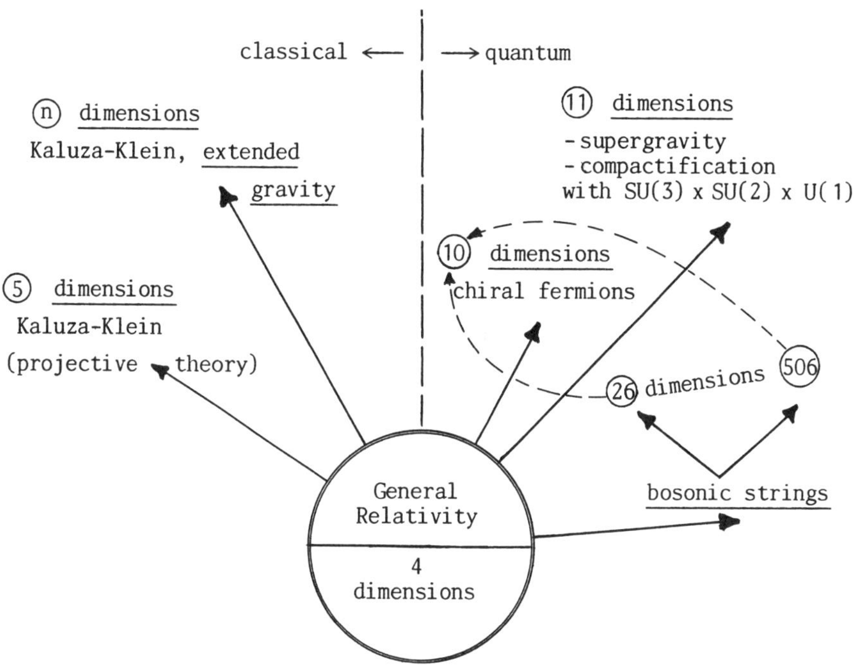

Along the last road the field of numbers in the tensor algebra is generalized from the reals to other division or split-division algebras as we shall see in the last of the following talks. The entries in Fig. 4 concerning twistors and H-spaces again leave the framework of alternative theories of gravitation.

Fig. 4 Change of algebra

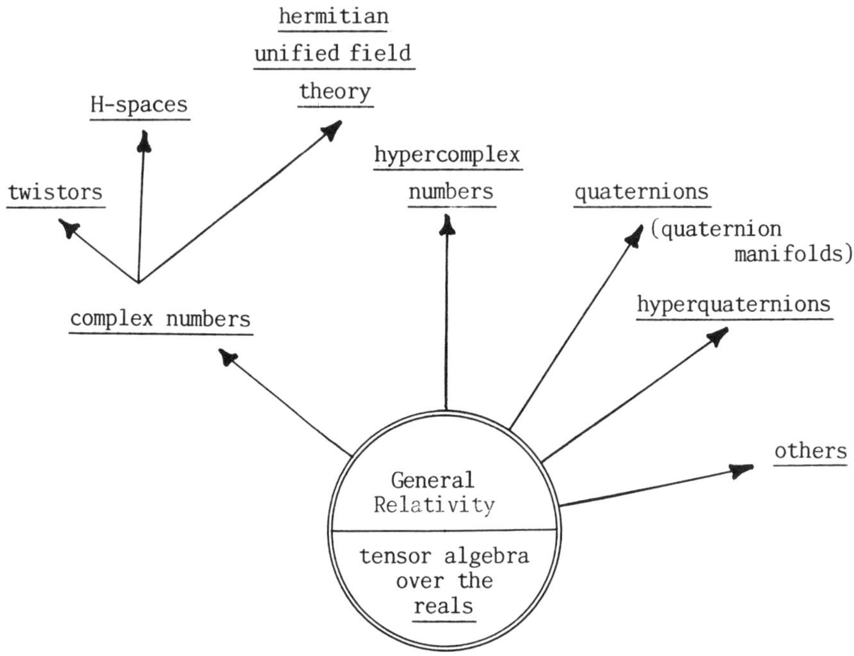

Theories of gravitation *combining* these four approaches have also been discussed in the literature. Asymmetrical Kaluza-Klein theory (Kalinowski 1983, Moffat 1984), Kaluza-Klein with Einstein Lagrangian plus dimensionally continued Euler forms (Wetterich 1982, Müller-Hoissen 1985, 1986), UFT over the hypercomplex numbers (Crumeyrolle 1967), to name just a few!

J.D. McCrea then gave a lucid presentation both of the foundations of PGT (Hehl 1980, Baekler et al. 1986) and of the numerous contributed abstracts. Here, space time is taken to be a manifold with Riemann-Cartan geometry. The dynamical variables are an orthonormal basis of 1-forms ϑ^a and the connection 1-forms $\omega^a{}_b$ with non-zero torsion. In the PGT $\omega^a{}_b$ represent the Lorentz gauge potentials whereas ϑ^a is connected to the translational gauge potentials. The latter point is still under debate (Sardanashvily, Hayashi & Shirafuji). The gravitational Lagrangian

L_f chosen by analogy to the Yang-Mills-Lagrangian and for obtaining the simplest quasilinear 2nd order field equations is of the form:
L_f = const + curvature +(curvature)^2 + (torsion)^2 or, more precisely (Hehl & Mc Crea 1986):

$$L_f = \frac{1}{\mu l^4} \varepsilon - \frac{1}{2} \theta^a \wedge H_a - \frac{1}{2} \Omega^{ab} \wedge H_{ab} + \frac{A_o}{4 l^2} \Omega^{ab} \wedge \varepsilon_{ab} \qquad (1)$$

with the volume form ε, $\varepsilon_{ab} := *(\vartheta_a \wedge \vartheta_b)$ and

$$H_a := -\frac{1}{l^2} *(\sum_{i=1}^{3} A_i \, {}^i\theta_a) \qquad (2)$$

$$H_{ab} := -\frac{A_o}{2 l^2} \eta_{ab} - \frac{1}{\varkappa} *(\sum_{j=1}^{6} B_j \, {}^j\Omega_{ab}) \qquad (3)$$

with ${}^i\theta_a$ and ${}^j\Omega_{ab}$ being the irreducible components of the torsion form θ^a and curvature form $\Omega^a{}_b$. Obviously, besides the coupling constants l and $\varkappa$, in L_f 10 constant parameters μ, A_i, B_j occur destroying the uniqueness of the usual Yang-Mills Lagrangian.

Part of the work in the field aims at establishing guidelines for the choice of these constants. Arguments come from the consideration of limits (Fedorov & Minkevich, Garecki, Frolov), e.g. for spinless macroscopic matter the testable consequences of GR should be recovered; from the absence of ghosts and tachyons in the theory linearized around flat space time (Kuhfuß & Nitsch 1986), the Cauchy initial value problem (work in progress by a Göttingen-Köln cooperative effort) and the Hamiltonian formulation (Nikolic, Frackiewicz).

A number of contributed abstracts dealt with cosmological models and related questions in Einstein-Cartan and other theories (Banerji & Sinha, Nurgaliev & Ponomariev, Pronin, Matravers). Mc Crea circumspectly avoided a stand concerning the applicability of PGT to the scale of cosmology. I believe that many workers use the high symmetry of the Robertson-Walker metric just in order to be able to find exact solutions without much enthusiasm about a cosmological interpretation (Goenner & Müller-Hoissen 1984). Also, the possible reinterpretation of PGT as a Riemannian tensor-tensor theory of gravitation (metric and torsion) supports its applicability to problems of cosmological significance.

A further line of study is the search for exact solutions in order to arrive at a physical interpretation of PGT. It is both helped and hindered by the arbitrariness of the constants in L_f. With a proper choice and an additional simplifying assumption like, for example, the *modified double duality* Ansatz (Baekler et al. 1981, Benn et al. 1981, Hehl & Lenzen):

$$\Omega_{ab} - \overset{*}{}\Omega^{*}_{ab} = \frac{\varkappa}{l^2}\, \vartheta_a \wedge \vartheta_b \tag{4}$$

exact solutions can be obtained, e.g. spherically-symmetric and Taub-NUT-like vacuum and electrovac solutions (Baekler et al. 1983, McCrea 1984). The relevance of such solutions is overshadowed by our lack of understanding of the physical meaning of constraints such as given by (4). It is fair to say, though, that PGT or, more generally, theories of gravitation with torsion is the subject most actively studied at present.

Next, *M. Francaviglia* reported on what he termed the "dynamical universality" of GR among metric theories of gravitation (Ferraris, Francaviglia & Magnano). The results are part of a larger program aimed at showing that, in a suitable sense, alternative metric theories are dynamically equivalent to GR (Ferraris 1983, 1986, Ferraris & Kijowski 1982, Magnano, Ferraris & Francaviglia 1986).

What is shown, in fact, is that GR plays the role of a "Hamiltonian image" of a large class of metric-affine or even affine theories of gravitation. This is obtained by defining a (new) metric density from the Lagrangian L

$$\bar{g}^{ab} := \frac{\partial L}{\partial R_{(ab)}} \cdot \left| \det \frac{\partial L}{\partial R_{(ab)}} \right|^{-1/2} \tag{5}$$

and then transforming, through a Legendre transformation, L into a new Lagrangian of the form of the Einstein-Hilbert Lagrangian plus additional terms interpreted as matter Lagrangian. The Einstein Lagrangian itself is a fixed point of this transformation.

As an example within Riemannian geometry, take Lagrangians depending on the Ricci-tensor R_{ab} only, i.e. $L(j^2 g) = f(R_{ab}, g_{cd})\sqrt{-g}$ where $j^K T$ denotes the k-th order jet prolongation of T. If the definition (5) of $\bar{g}^{ab}$ can be inverted to obtain $R_{ab} = R_{ab}(\bar{g}, g)$ a Hamiltonian K may be postulated by

$$H(g,\bar{g}) := \bar{g}^{ab} R_{ab}(\bar{g},g)\sqrt{\bar{g}} - L^{*}(g,\bar{g}) \tag{6}$$

where

$$L^{*}(g,\bar{g}) := \sqrt{-g}\, f\,[\, R_{ab}(\bar{g},g)\,,\, g_{cd}\,] \tag{7}$$

The new Lagrangian is taken to be

$$\hat{L}(\bar{g},j^2 g) := \bar{g}^{ab} R_{ab}(j^2 g)\sqrt{\bar{g}} - H(g,\bar{g}) \tag{8}$$

In the particular case given by

$$f = R_{ab} R_{cd}\, g^{ac} g^{bd} \tag{9}$$

we have

$$\sqrt{\bar{g}}\; \bar{g}^{ab} = R_{cd}\, g^{ac}\, g^{db} \quad . \tag{10}$$

and its inverse

$$R_{cd}(\bar{g},g) = \sqrt{\frac{\bar{g}}{g}}\; g_{ca}\, g_{db}\, \bar{g}^{ab} \tag{11}$$

Thus

$$\hat{L}(\bar{g},j^2 g) = \sqrt{\bar{g}}\,[\bar{g}^{ab} R_{ab}(j^2 g) - \tfrac{1}{2}\sqrt{\bar{g}/_g}\; \bar{g}^{ab}\bar{g}^{cd}\, g_{ac}\, g_{bd}\,] \tag{12}$$

In a final step, in the general formalism, $R_{ab}(j^2 g)$ is replaced by $\bar{R}_{ab}(j^2 \bar{g})$, by a known formula, such that

$$\hat{L}'(j^2\bar{g},j^2 g) = \sqrt{\bar{g}}\; \bar{g}^{ab}\, \bar{R}_{ab}(j^2\bar{g}) + L_{mat}(g,\bar{g}) \quad . \tag{13}$$

The old metric g now appears as a spin-2 field on the matter side. By this process, the Brans-Dicke theory is recast into Einstein's theory plus a scalar field, a result found earlier by Schücking (Jordan 1955) while

Weyl's theory corresponds to Einstein's theory with a Proca field.

From a mathematical point of view, these results are very interesting because a map from solutions of GR to solutions of certain metric-affine (affine) theories is established. From the point of view of physics not too much seems to be gained for the following three reasons:

i) The new metric $\bar{g}$, in general, does not have Lorentz signature, i.e. cannot be interpreted as a causal metric.

ii) The Hamiltonian formalism does not work if the definition of $\bar{g}$ cannot be inverted (i.e. if the velocities cannot be expressed by the canonical momentum). There exist large classes of theories of gravitation for which this situation obtains, a particular case being the theory of gravitation with non-minimal coupling I suggested (Goenner 1976, 1984a).

iii) In general, the physical meaning of the matter fields generated by the process of Hamiltonian imaging is unclear. Will they satisfy an energy condition as they should?

Thus, while, *mathematically*, the process of Hamiltonian mapping establishes the dynamical equivalence of classes of theories with GR, it also leads from one theory with a particular *physical* interpretation to *another* with an entirely *different physical* interpretation.

The last lecture of the symposium given by *R.B. Mann* dealt with the fourth way of generalizing GR, i.e. with the method of algebraic extension. The tangent bundle of the *real* space time manifold is modified to accommodate a particular algebraic structure J(X). The tensor algebra now is taken not over the real functions but over functions in a division or split-division algebra, e.g. the complex numbers or quaternions. Mann requires this extension to satisfy three postulates:

i) If, in the general formalism, the algebra is taken to be the real numbers, we must get back to GR.

ii) The algebraic structure J(X) must be compatible with the geometry of space time.

iii) Physical observables must be real valued.

The second postulate interconnects the algebraic structure J, fundamental form h (known as the hermitian structure tensor, in the complex case) and the connection ∇. To wit, for the complex and hypercomplex algebras, respectively,

$$g(J(X)\ ,\ J(Y)) = \pm g(x,y) \tag{14}$$

$$\nabla_y\ J(X) - J(\nabla_y X) = 0 \tag{15}$$

$$\nabla_z\ h(X,Y) = 0 \tag{16}$$

h is split into a real, symmetric part g, the metric, and a skew-symmetric part taking values in the particular algebra considered. From these postulates *five* distinct algebraic structures (real, complex, hypercomplex, quaternion and hyperquaternion) are deduced (Mann 1984).

The (real) Lagrangian is taken to be of the form

$$L = \frac{1}{4\ G}\ \sqrt{-h}\ h^{\mu\nu}\ (A_1\ R^{\alpha}{}_{\mu\alpha\nu} + A_2\ R^{\alpha}{}_{\alpha\mu\nu}) \tag{17}$$

with numerical parameters A_1, A_2. In addition to the Ricci-scalar, due to $h^{[\mu\nu]} \neq 0$, the homothetic curvature now is to be included also. If the ensuing theories are linearized and quantized, a spin-2 and a spin-0 particle emerge (graviton, skewon). The only cases free of ghosts are given by the real and hypercomplex algebras (Kelly & Mann 1985). A hypercomplex structure was introduced, in the context of unitary field theories of the type of Einstein's and Schrödinger's, by Crumeyrolle (Crumeyrolle 1967). Later, it was suggested in the framework of a pure gravitational theory (Kunstatter et al. 1983), and its theoretic and observational consequences were studied in detail (Moffat 1986).

These observational consequences, in particular, the possibility of explaining a retrograde motion of the periastron, were of prime importance for Mann's interest in algebraic extension. It should be kept in mind, however, that the observed periastron shift of the eclipsing binary system DI Herculis has been explained within Moffat's *real* non-symmetric gravitational theory already (Moffat 1986, Moffat & Woolgar 1985). If one of the algebraically extended theories seems to be acceptable at all, it is the one based on the hypercomplex algebra. From (17) at least three different such theories result according to the dynamical variables chosen. If the Lagrangian is varied with respect to both the hypercomplex hermitian structure tensor *and* the connection, two distinct cases arise depending on $A_2 \neq -1/4$ and $A_2 = -1/4$. The first case coincides with Moffat's theory. If only the structure tensor is varied, $A_2 = 0$ must hold in order that ghosts be ruled out.

Due to the richness in geometrical objects of these non-symmetric gravitational theories, their physical interpretation is neither straightforward nor entirely unambiguous (Goenner 1984 b).

References

Contributed abstracts to the symposium are indicated, within the article, by author's name *without* the year of publication.

Baekler, P. et al. (1981). Vacuum solutions with double duality properties of a quadratic Poincaré gauge field theory. In Second M. Grossmann Meeting, ed. R. Ruffini pp 413-453. Amsterdam: North Holland Publ. Co.

Baekler, P. et al. (1983). Vacuum solutions with double duality properties of the Poincaré gauge field theory II. In Proc. 3rd M. Grossmann Meeting, ed. Hu Ning pp. 107-128. Amsterdam: North Holland Publ. Co.

Baekler, P. et al. (1986). Nonmetricity and torsion: Facts and fancies in gauge approaches to gravity. To appear.

Benn, I.M. et al. (1981). Double dual solutions of generalized theories of gravitation. GRG J. 13, 581-589.

Crumeyrolle, A. (1967). Variétés différentiables à structure complexe hyperbolique. Application à la théorie unitaire relativiste des champs. Riv.Mat.Univ. Parma (2) 8, 27-53.

Ferraris, M. (1983). Affine unified theories of gravitation and electromagnetism. In Proc. Journées relativistes 1983, Torino, eds. S. Benenti, M. Ferraris & M. Francaviglia, pp. 125-153. Bologna: Pitagora Editrice.

Ferraris, M. (1986). Gravitational theories with quadratic Lagrangians. In Atti 6° Convegno Naz.Rel.Gen. e Fisica della Gravitazione, eds. R. Fabbri & M. Modugno, Bologna, pp. 127-136.

Ferraris, M. and Kijowski, J. (1982). On the equivalence of the relativistic theories of gravitation. GRG J. 14, 165-180.

Goenner, H. (1976). A metric theory of gravitation without minimal coupling of matter and gravitational field. Z.Naturf. 31a, 1451-1456.

Goenner, H. (1984a). Theories of gravitation with nonminimal coupling of matter and the gravitational field. Found. Phys. 14, 865-881.

Goenner, H. (1984b). Unified field theories: from Eddington and Einstein up to now. In: Proc. A. Eddington Centenary Symposium, eds. V. De Sabbata & T.M. Karade. Singapore: World Scientific.

Goenner, H. and Müller-Hoissen, F. (1984). Spatially homogeneous and isotropic spaces in theories of gravitation with torsion. Class. Quantum Grav. 1, 651-672.

Hehl, F.W. (1980). Four lectures in Poincaré gauge field theory. In Cosmology and Gravitation, eds. P.G. Bergmann & V. De Sabbata pp. 5-61. New York: Plenum Press.

Hehl, F.W. & McCrea, J.D. (1986). Bianchi identities and the automatic conservation of energy-momentum and angular momentum in general-relativistic field theories. Found.Phys. 16, 267-293.

Jordan, P. (1955). Schwerkraft und Weltall, pp. 207-213 Braunschweig: F. Vieweg u. Sohn.

Kalinowski, M.W. (1983). The nonsymmetric Kaluza-Klein theory. J. Math. Phys. 24, 1835-1845.

Kelly, P.F. & Mann, R.B. (1985) Ghost properties of algebraically extended theories of gravitation. Class. Quantum Grav. to appear.

Kuhfuß, R. & Nitsch, J. (1986). Propagating modes in gauge field theories of gravity. GRG J., to appear.

Kunstatter, G. et al. (1983). Geometrical interpretation of a generalized theory of gravitation. J. Math. Phys. 24, 886-889.

Magnano, G. et al. (1986). Non-linear gravitational Lagrangians. Preprint, Istituto di Fisica Matematica "J.L. Lagrange", Universita di Torino, GRG J., to appear.

McCrea, J.D. (1984). The use of "Reduce" in finding exact solutions of the quadratic Poincaré gauge field equations. In: Classical General Relativity, eds. W.B. Bonnor, J.N. Islam & M.A.H. MacCallum, Cambridge: University Press, pp. 173-182.

Mann, R.B. (1984). Five theories of gravitation. Class Quantum Grav. 1, 561-572.

Moffat, J.W. (1984). Generalized theory of gravitation: Found. Phys. 14, 1217-1251.

Moffat, J.W. (1986). Experimental consequences of the nonsymmetric gravitational theory including the apsidal motion of binaries. Canad.J.Phys. 64, 178-184.

Moffat, J.W. & Woolgar, E. (1985). The apsidal motion of a binary star in the nonsymmetric gravitation theory. Can.J.Phys. 63, 428-435.

Müller-Hoissen, F. (1985). Spontaneous compactification with quadratic and cubic curvature terms. Phys.Letts. 163B, 106-110.

Müller-Hoissen, F. (1986). Dimensionally continued Euler forms: Kaluza-Klein cosmology and dimensional reduction. Class.Quantum Grav. 3, 665-677.

Wetterich, C. (1982). Spontaneous compactification in higher dimensional gravity. Phys.Letts. 113B, 377-381.

ASTRONOMICAL AND SPACE EXPERIMENTS

R.D. Reasenberg
Smithsonian Astrophysical Observatory, Cambridge, MA, USA

Historically, the solar system has been the principal arena for testing theories of gravitation. There, in turn, the laws of Newton and of Einstein have successfully confronted reality. [For an extensive discussion of tests of general relativity, see Will (1981); for a more recent but more limited review, see Reasenberg (1986).] We seem now to be at the end of a golden age of solar-system tests, those that have addressed general relativity and related theories. The classical effects predicted by general relativity -- redshift, perihelion advance, light deflection, and Shapiro time delay -- have been measured to be consistent with general relativity to within uncertainties ranging from 10^{-2} to 10^{-4}. However, with the exception mentioned below, the annual rate of growth of the data sets and the prospect for new experiments are not good. That exception is a deflection experiment planned by B. E. Corey and I. I. Shapiro (private communication, 1986) which is expected to yield an estimate of the deflection coefficient uncertain by 10^{-3}. This will be an order of magnitude improvement over the present most reliable determination, that of Fomalont & Sramek (1977), which yielded a 0.9% uncertainty. The new experiment will use Very Long Baseline (radio) Interferometry (VLBI) observations at three frequencies to reduce the unmodeled contribution from the solar plasma. Current technology apparently does not permit further improvement on the near term.

One astronomical test that currently shows useful improvement from year to year is the timing of the arrival of the pulses from PSR 1913+16 (Taylor & Weisberg, 1984). This test has already provided significant information including strong indirect evidence for the existence of gravitational radiation. The continuing improvement of this test rests on two factors: (a) the data set is relatively young (cf., planetary radar) and therefore an additional year's observations is significant, and (b) there has been rapid improvement in the accuracy of the pulse timings. This subject is discussed in greater detail by Taylor in these proceedings.

Table 1 lists the papers presented at the two sessions of the Symposium on Astronomical and Space Experiments. Here we concentrate on two subsets of these papers. The first includes new ways to look at or analyze existing data sets. Of these inexpensive means of gaining new knowledge, three papers are discussed below, those of Nordtvedt, Haugan

Table 1.

Session A

R. D. Reasenberg, R. W. Babcock, J. F. Chandler, and I. I. Shapiro
The Light Deflection Experiment from an Astrometric Satellite

M. Cerdonio, G. A. Prodi, C. Rovelli, and S. Vitale
Feasibility of Laboratory Detection of Rotation-Like Gravitational Fields

W. M. Fairbank, D. Bardas, J. V. Breakwell, W. S. Cheung, D. B. DeBra, R. DiEsposti, C. W. F. Everitt, S. Feteih, R. Hacker, M. Jacobs, M. Keiser, B. King, M. Larsen, J. A. Lipa, T. Lydic, J. M. Lockhart, M. Macgirvin, B. Muhlfelder, J. Paff, B. W. Parkinson, J. P. Turneaure, R. A. Van Patten, C. Warren, M. S. Wooding
Recent Progress on the Gravity Probe B Orbiting Gyroscope Test of General Relativity

R. W. Hellings
Relativistic Effects in VLBI

D. G. Torr and H. Peng
Effect of the Gravitomagnetic Field on the Synchronization of Clocks

---

Session B

B. Lesche and M. L. Bedran

A Test of the Equivalence Principle with Polarized Light

D. F. Bartlett and D. van Buren
Asymmetry of the Moon and the Equivalence of Active and Passive Gravitational Mass

M. P. Haugan and Z. Zhou
The Binary Pulsar PSR 1913+16: Possible Companions and Newtonian Perturbations

M. A. Vincent and P. L. Bender
Mercury Relativity Satellite

H. G. Preston and J. Weber
The Period of Orbit as a Test of the General Theory of Relativity

Nordtvedt*
The Sun's Spin Axis -- Test of Post-Newtonian Gravity to a Part in 10^6

---

*Nordtvedt's abstract was submitted too late to be included in the regular abstract volumes.

and Zhou, and Bartlett and van Buren. The second subset includes proposals for new experiments that use existing technologies or reasonable projections of existing technologies. These proposals, which are by Fairbank et al., Cerdonio et al., Reasenberg, et al., and Vincent and Bender, would be far more expensive to bring to fruition; they are discussed later.

PREFERRED FRAMES

The Parametrized Post-Newtonian (PPN) metric (see Will, 1981) contains ten free parameters not found in general relativity. Of these, there are three (α_1, α_2, α_3) that are nonzero only if there are preferred-frame effects. From the analysis of gravimeter data, Warburton and Goodkind (1976) have set a bound consistent with general relativity: $|\alpha_2| < 4 \times 10^{-4}$. Nordtvedt discussed a new and more stringent bound on α_2. He assumes that the present angular separation of the spin angular momentum vector of the Sun and collective orbital angular momentum vector of the planets, 6 deg, represents an upper bound to a change from some more closely aligned configuration at the initial epoch of the system about 5×10^9 years ago. If this bound on the change in the Sun's orientation is taken as a bound on the solar precession due to a preferred-frame torque on the Sun, then $|\alpha_2| < 4 \times 10^{-7}$. Since this is small compared to the Newtonian potential at the solar limb, 2×10^{-6}, the test addresses a post-post-Newtonian ($1/c^4$) effect. Further, it should be possible to derive additional constraints from the precession of fast pulsars which have been observed for several years. The lack of pulsar precession already provides a sufficient constraint on α_2 to preclude the Sun having precessed completely around one or more times.

PULSAR MODELS

The binary pulsar, PSR 1913+16, has provided a relatively new and important test of general relativity. Taylor (in these proceedings) discusses the results of the analysis of the arrival times of the pulses. As the data set grows and the measurement accuracy improves, the analysis will require an increasingly accurate model. Such a model will need either to exclude with reason or to include an increasingly rich set of phenomena.

The present analysis of arrival times, which is based on a relativistic two-point-mass model of the binary pulsar system, shows no difficulty. However, a vigorous search for alternative models is an important part of the analysis. Haugan discussed his analyses of the system's behavior with a non-point-like companion. In particular, he considered an extended companion subject to both tidal and rotational deformations. His analysis is distinguished by the inclusion of the resulting rich set of periodic terms in addition to the previously considered secular terms. The periodic terms, which are at harmonics of the system's orbital period, become increasingly important as the model considered becomes more complex.

The present data set and the above enhanced model yield two useful results. First, the companion cannot be a helium star. Second, if the companion is a white dwarf star, it must not be spinning fast. Further computer simulations (Haugan, private communication, 1986) have shown that with the constraint on the masses, the distortion of the spinning companion is bounded by $\alpha_6 < 2.2$ where α_6 is the distortion parameter introduced by Smarr & Blandford (1976). These results strengthen the conclusion that the analysis of the arrival times of the pulses from PSR 1913+16 has shown the existence of gravitational radiation.

ACTIVE-PASSIVE MASS RATIO

Neither general relativity nor Newtonian gravity distinguishes between active and passive gravitational mass. It is assumed that $M_a/M_p = 1$ for all kinds of matter where M_a is the source of a gravity field and M_p is acted on by the field. This assumption was put to a classical test by Kreuzer (1968) who used a Cavendish balance and a Teflon cylinder (76% flourine by weight) neutrally buoyant in an organic liquid that was 74% bromine by weight. The Teflon was moved alternatively toward and away from the Cavendish test mass but no change in the resulting force was detected. Kreuzer concluded that $S(F, Br) < 5\times10^{-5}$, where $S(i, j) = M_{ai}/M_{pi} - M_{aj}/M_{pj}$.

Bartlett discussed a reanalysis of existing astronomical data that serve to place a more stringent limit on S. The analysis is based on Bondi's observation that a single body having two parts for which $S \neq 0$ will undergo self acceleration. The Moon is known to have an asymmetry which results in a 2 km displacement between the center of mass and the center of figure along a line about 14° from the mean Earth direction. The surface facing Earth is iron rich; the other, aluminum rich. If the surface chemistry is indicative of the composition in the near-surface region and if at greater depth the body is spherically symmetric, then if $S(Fe, Al) \neq 0$, the Moon should undergo a secular change in orbital period (in addition to the contribution from its tidal interaction). After making allowance for nonstatistical errors, Bartlett and Van Buren conclude that $|S(Fe, Al)| < 4\times10^{-12}$. Following Will (1981, pp. 214, 215), they translate their results to the PPN language to obtain $|\zeta_3| < 10^{-8}$ under the assumption that the violation is associated with the electrostatic energy of the nucleus.

STANFORD GYRO

The present tests of general relativity are all associated with the diagonal elements of the metric tensor. The off-diagonal terms give rise to the so-called gravitomagnetic effects, associated with the rotation of the source. A quarter century ago, G.E. Pugh (1959) and, independently, L.I. Schiff (1960) discussed an experiment in which a gyroscope would precess with respect to an inertial frame defined by distant stars as a result of the field of the Earth. At the same time, W.M. Fairbank and R.H. Cannon started to work on what was to be known as the Stanford Relativity Gyro Experiment.

In the present form of the experiment, also known as the NASA Gravity Probe B (GP-B), the instrument comprises four gyroscopes of unprecedented quality and an inertial reference mass, all mounted in a quartz block which is optically contacted to a quartz telescope. The instrument is held at about 1.8K inside a drag-free spacecraft in polar Earth orbit at an altitude of 650 km. Two relativistic effects will be measured: the geodetic precession of 6.6 arcsec per year is due to the motion of the instrument around the Earth; and the Lense-Thirring precession of 0.042 arcsec per year in the orthogonal direction is due to the rotation of the Earth. The experiment poses severe technological challenges. Workers attempting to meet these challenges have produced three noteworthy spin-off experiments: (1) the technology of a "drag-free" spacecraft was demonstrated with the DISCOS satellite; (2) Cabrera (1982) is continuing his search for magnetic monopoles using a superconductive detector; and (3) Worden (1983, 1987) is developing a test of the equivalence principle that uses superconductive levitation of concentric cylindrical test masses of different materials.

Fairbank discussed the present state of GP-B. The two effects are expected to be measured with uncertainties of 10^{-4} and 10^{-2}, respectively. In support of this work, the GP-B team has developed new technologies that include: (1) a method for reading out the direction of spin of the gyroscope with milliarc-sec precision by means of the London moment in a spinning superconductor, (2) techniques for fabricating and measuring spherical gyro rotors of extreme roundness and homogeneity, (3) a method for controlling superfluid helium under zero-g conditions in space (later applied in IRAS), (4) the development of ultra-low magnetic field technology, (5) the development of a star-tracking telescope with milliarc-sec precision, (6) the development of a multi-level low-g electrical suspension system for the gyroscope, (7) the development of technologies for precise pointing and drag-free control of the spacecraft, and (8) the development of an accurate method for subtraction and processing of the gyroscope and telescope signals.

At present, the Stanford Team, in cooperation with Lockheed Missiles and Space Corporation, is developing a complete integrated system which will be used for an extended set of tests and, later, will serve as the reference for the construction of the flight instrument. The flight instrument will be carried into Earth orbit by the Shuttle in the early 1990s to verify the instrument performance. It will then be incorporated into an independent spacecraft which will be flown a few years later for a nominal mission of two years. During the second year, a series of instrument evaluation tests will be performed to verify the accuracy of the relativity experiment and bound the magnitude of systematic errors.

ALTERNATIVE FRAME-DRAGGING TEST

For as long as the Stanford group has been working on their space-based instrument, consideration has been given to a ground-based version (Fairbank, private communication). Among the proposals are recent suggestions to use a Foucault pendulum at the South Pole (Braginsky, Polnarev, and Thorne, 1984) or a laser gyro (Ezekiel and Balsamo, 1977; Scully, 1979). In either case, the connection to the inertial frame defined by the distant stars presents a serious problem. If this external connection is by means of an optical telescope intimately connected to the instrument, then the severe problem of atmospheric refraction needs to be addressed. If the connection is by means of radio interferometry, then geophysical questions need to be addressed. Buffington (1985) has proposed to use a ronchi-ruling astrometric optical telescope to observe the stars.

Cerdonio reported on progress at Trento in developing a ground-based frame-dragging experiment. In particular, the group has been developing two new kinds of cryogenic detectors for the local "dragged" frame. Progress on the first detector, which is based on the magnetic field induced by rotation, is discussed by Cerdonio, Goller, and Vitale (1985). The second type of detector being studied experimentally is a superfluid He^4 analog of the rf SQUID. Its theoretical basis is discussed by Cerdonio and Vitale (1984).

POINTS

The light deflection test was one of the early triumphs of general relativity. The accuracy of this test improved when radio interferometry was substituted for photographic astrometry. The most reliable result, that of Fomalont and Sramek (1977), has an uncertainty of 0.9% in the estimate of an ad hoc coefficient of the deflection effect. Robertson and Carter (1984) have analyzed a large VLBI data set accumulated for geodetic purposes. They report a deflection coefficient of 1.004±0.25% where the error is statistical and little discussion of possible systematic effects is included in the paper.

Reasenberg discussed POINTS, a small astrometric interferometer for operation in Earth orbit with a nominal measurement uncertainty of 5 microarc-sec. Although originally intended to perform a second-order deflection test, POINTS would have a profusion of other uses including a deep search for extra-solar planetary systems, a direct determination of the parallax of RR Lyrae and Cepheid variables and thus an improvement of the distance scale, a sharpening of the mass-color-luminosity relation through the determination of stellar masses and parallaxes, and a study of the mass and mass distribution of the Galaxy.

The key design problem of a microarcsecond astrometric interferometer is the control of systematic error. In addition to the use of stable materials, this question is addressed by means of real-time metrology and bias estimation-correction during the post-measurement data analysis. The former is based on a novel approach called Full-Aperture

Metrology (FAM) by which the starlight interferometer forms a part of a laser interferometer. The laser signal illuminates the entire surface of each optical element that is used to process the starlight. Central to FAM is a holographic optical element, a zone plate on a focusing mirror, which scatters a small fraction of the light illuminating its surface from the principal focus to an auxiliary focus.

It has been shown that an instrument bias can be detected in the post analysis of the data essentially by means of 360 deg closure. This approach works particularly well with POINTS because the instrument measures the angle between target stars separated by about 90 deg.

MERCURY ORBITER

Early radar ranging to the inner planets yielded a substantial improvement in their ephemerides and radius estimates and ushered in an era of rapidly improving solar-system tests of relativity. Radio ranging to spacecraft bypassed the difficult problem of planetary topography, but the first such targets were interplanetary probes that suffered from unmodelable accelerations. Later, the Mars orbiters and landers provided more useful targets for measuring the solar system by radio ranging. In a parallel effort, the spacecraft ranging systems were improved. The ranging to the Vikings had a nominal uncertainty of 2 m and some capability to correct the data for the effect of solar plasma.

Bender presented a next logical step in the advance of radio ranging systems and targets. If the proposed Mercury orbiter were able to yield Earth-Mercury normal points uncertain by 10 cm, a two-year mission would result in a hundredfold improvement in our knowledge of the PPN parameters β and γ; it would permit J_2, the solar quadrupole moment coefficient, to be determined, independent of solar models, with an uncertainty under 10^{-7}; and $\dot{G}$, the secular rate of change of the apparent "constant" of gravity, would be determined with an uncertainty of 10^{-13} G yr^{-1}. A longer mission would yield increased accuracy, particularly for the estimate of $\dot{G}$ (Ashby et al., in preparation, 1987), although it would be difficult to reduce the error below about 10^{-14} G yr^{-1}, which corresponds to the estimated rate of solar mass loss: $\dot{M}/M \approx 10^{-14}$.

Vincent and Bender have investigated the mission characteristics of the Mercury orbiter by means of a series of simulations. They assume a 4-hour, near-polar orbit; their observations are made in 8-hour passes, three times per week, with a Doppler accuracy of 10^{-14} in 10 minutes and a 3 cm ranging accuracy once per pass. The simulated data from 40 passes are used to determine 40 sets of spacecraft elements and a set of Mercury gravity model coefficients. Additional, higher frequency gravity coefficients were "considered" in the solution. The resulting range normal points (one per pass) had an uncertainty ranging from 3.6 to 5.1 cm and a mean of 4.2 cm. By including an allowance for other sources of error not explicitly included in the simulation, they conclude normal points of 10 cm accuracy are achievable

DISCUSSION

The present rapid advance of technology is opening new and significant avenues to experimental general relativity. New approaches could yield substantial improvements in the classical tests as well as studies of predicted phenomena heretofore not addressed experimentally. Inexpensive powerful computers, from micros to supers, make possible experiment control, data analysis, and simulations of astrophysical systems that were not practical just a few years ago.

Access to space has been, and should continue to be, a significant factor in the advance of the tests of general relativity since it makes possible: (1) better targets to observe, such as the Vikings and the Mercury Orbiter; (2) a site for improved observing systems, such as POINTS; and (3) experiments that require a quiet "weightless" environment, such as GP-B. A critical question continues to be cost; space technology has not become cheap. However, with major resources being invested by several countries in the development of space capabilities, the opportunity exists to "ride along." This will not be an easy ride. We need to learn from our colleagues in particle physics who have mastered the art of persuading governments to pay for expensive fundamental physics with little hope of near-term practical results.

If we are to take advantage of the present favorable technological climate, we must, as a community, openly evaluate the various potential experiments, considering both scientific merit and technical feasibility. In this area, there is a need for more work in phenomenology and on the question of the significance of proposed experiments for our understanding of the nature of gravitation. The combination of phenomenology and experiment could bring us soon into one of the great eras in gravity research.

ACKNOWLEDGEMENTS

I thank N. L. Murphy for her careful preparation of the text. This work was supported by the Smithsonian Institution, both directly and through the Research Opportunities Fund.

REFERENCES

Braginsky, V.B., Polnarev, A.G. & Thorne, K.S. (1984). Foucault pendulum at the South Pole: proposal for an experiment to detect the Earth's general relativistic gravitomagnetic field. Phys. Rev. Lett., 53, 863.

Buffington, A. (1985). An astrometric program to define coordinate frames for general relativity experiments. BAAS, 17, 898.

Cabrera, B. (1982). First results from a superconductive detector for moving magnetic monopoles. Phys. Rev. Lett., 48, 1378.

Cerdonio, M. Goller, A. & Vitale, S. (1985). In SQUID 85 Superconducting Quantum Interference Devices and Their Applications, eds. H.D. Halbohm & H. Lubigg. Berlin: W. deGruyter.

Cerdonio, M. & Vitale, S. (1984). Superfluid He^4 analog of the rf superconducting quantum interference device and the detection of inertial and gravitational fields. Phys. Rev., B29, 484.

Ezekiel, S. & Balsamo, S.R. (1977). Passive ring resonator laser gyroscope. Appl. Phys. Lett., 30, 478.

Fomalont, E.B. & Sramek, R.A. (1977). The deflection of radio waves by the Sun. Comments on Astrophys., 7, 19.

Pugh, G.E. (1959). WSEG Research Memo 11, U.S. Dept. of Defense.

Reasenberg, R.D. (1986). Solar-system tests in transition. In Proceedings of the Ninth Course of the International School of Cosmology and Gravitation: Topological Properties and Global Structure of Space-Time, eds. P.G. Bergmann & V. deSabbata, pp. 177-198. New York: Plenum Press.

Schiff, L.I. (1960). Possible new experimental test of general relativity theory. Phys. Rev. Lett., 4, 215.

Scully, M.O. (1979). Suggestion and analysis for a new optical test of general relativity, In Proceedings of the 4th International Conference on Laser Spectroscopy, eds. H. Walther & K.W. Rothe, pp. 21-30. New York: Springer-Verlag.

Smarr, L.L. & Blandford, R.D. (1976). The binary pulsar, physical processes, possible companions, and evolutionary histories. Ap. J., 207, 574.

Taylor, J.H. & Weisberg, J.M. (1984). Observations of post-Newtonian timing effects in the binary pulsar PSR 1913+16. Phys. Rev. Lett., 52, 1348.

Warburton, R.J. & Goodkind, J.M. (1976). Search for evidence of a preferred reference frame. Astrophys. J., 208, 881.

Will, C.M. (1981). Theory and Experiment in Gravitational Physics. Cambridge: Cambridge University Press.

Worden, P.W. (1987). Almost exactly zero: the equivalence principle. In Near Zero, ed. C.W.F. Everitt. Dover, in press.

Worden, P.W. (1983). Cryogenic equivalence principle experiment, discussion and present status. In Proceedings of the Third Marcel Grossmann Meeting on General Relativity, ed. Hu Ning. Science Press and North Holland.

ASTROPHYSICAL COSMOLOGY

J.M. Bardeen
University of Washington, Seattle, WA 98195

OVERVIEW

Introduction

The last several years have seen a tremendous ferment of activity in astrophysical cosmology. Much of the theoretical impetus has come from particle physics theories of the early universe and candidates for dark matter, but what promise to be even more significant are improved direct observations of high z galaxies and intergalactic matter, deeper and more comprehensive redshift surveys, and the increasing power of computer simulations of the dynamical evolution of large scale structure. Upper limits on the anisotropy of the microwave background radiation are gradually getting tighter and constraining more severely theoretical scenarios for the evolution of the universe.

The inflation paradigm

Inflation continues to be very attractive as a paradigm in cosmology (Turner 1987), in spite of the lack of any one particularly compelling detailed model for inflation based on well-founded particle physics. Fortunately, the astrophysical consequences of inflation are to a large degree independent of details. The present universe should be spatially flat, and coupled with the normal theoretical preference for a zero cosmological constant this means the cosmological density parameter Ω should equal one (Guth 1981). Deviations from homogeneity are generated by quantum fluctuations in the scalar field driving the inflation. These evolve into adiabatic density perturbations with an initially scale-independent spectrum (Starobinsky 1982; Guth & Pi 1982; Hawking 1982; Bardeen et al 1983). The final density perturbation spectrum, which is responsible for galaxy formation and the present large scale structure of the universe, is easily calculated once the matter content of the universe is specified (Peebles 1982; Bond & Szalay 1983). The linear density perturbation theory can be used as an initial condition for numerical simulations of the nonlinear gravitational clustering.

The presumption that $\Omega = 1$ means that most of the matter in the universe must be nonluminous, nonbaryonic (according to conventional models of nucleosynthesis), and must be distributed more uniformly than the light. Various dynamical estimates of mass-to-light ratios on scales from galaxy clusters to superclusters (see Peebles 1986a for a review), when extrapolated to the universe as a whole, give values for Ω in the range 0.1-0.3. However, recent direct determinations of Ω from the distribution of galaxy counts with redshift (Loh & Spillar 1986) and from the dipole anisotropy of galaxies in the IRAS infrared survey (Yahil et al 1986) do at least tentatively suggest that the true global Ω is close to one.

There are many candidates for dark matter, ranging from massive neutrinos to more exotic kinds of hypothetical particles such as axions, photinos, weakly interacting Gev mass fermions, lumps of quark matter, primordial black holes, etc. Of course, at least some of the dark matter may well be baryonic. The cosmologically dominant dark matter may or may not be the same as the dark matter associated with galactic halos. While there are many interesting questions concerning possible direct detection of some types of dark matter, the effects on cosmology largely depend on the primordial velocity dispersion, hence the classification into "hot", "warm", or "cold" dark matter (Bond & Szalay 1983). Hot dark matter (massive neutrinos) has a large enough velocity dispersion that free streaming damps out primordial perturbations on scales less than clusters of galaxies, leading to a "top down" scenario for galaxy formation. Cold dark matter has a small enough velocity dispersion to allow primordial density perturbations to survive and grow on subgalactic scales. Galaxy formation proceeds "bottom-up", with hierarchical clustering (Blumenthal et al 1984).

Hot dark matter received a lot of attention in the early 1980's (Centrella & Melott 1983; Klypin & Shandarin 1983; White et al 1983), but has fallen out of favor in recent years. The numerical simulations indicate that it is difficult to reconcile the present observed strength of clustering of galaxies with a reasonably early epoch of galaxy formation, and the experiment of Lubimov et al (1980) claiming a measurement of the electron neutrino mass has not been confirmed. However, for counter arguments see Melott (1986).

Cold dark matter (CDM) has the problem that dynamical segregation of the dark matter from baryons is not expected on scales much larger than galaxies. To reconcile cold dark matter with $\Omega = 1$ requires that galaxies form more efficiently in overdense regions (protoclusters) than in underdense regions (protovoids). A particular scheme for accomplishing this biasing is to associate galaxies with peaks in the smoothed linear density perturbation field whose height relative to the root mean square fluctuation level exceeds a threshold determined by the requirement that the comoving number density of these peaks be equal to the number density of bright galaxies in the present universe. Numerical simulations (Davis et al 1985) and quasi-analytic studies (Bardeen et al 1986) have had rather good success in fitting the clustering properties of "galaxies" on scales of a few megaparsecs with the actual galaxy-galaxy correlations measured, for instance, in the CfA redshift survey (Davis & Peebles 1983). There are even reasons to believe this kind of biasing may arise naturally, in that high peaks collapse to bound objects with high virial velocities which are able to keep their gas despite the heating from supernova explosions (Dekel & Silk 1986; Frenk et al 1985).

White et al (1987) argue that the biased CDM simulations show much the same frothy structure of voids, sheets, and filaments seen in redshift surveys of the distribution of galaxies on scales of tens of megaparsecs (Davis et al 1982; de Lapparent et al 1986), though it is a bit disturbing that the new deeper survey shows no tendency to fill in the voids with fainter galaxies. Other indications that more power is present in density perturbations on these large scales than predicted in the standard CDM scenario are harder to argue away. The correlation amplitude for Abell clusters seems to be a factor of 18 larger than that of galaxies (Bahcall & Soneira 1983; Klypin & Kopylov 1983) and to remain positive out to separations of 100 h^{-1} Mpc. Statistical biasing, as first suggested by Kaiser (1984), only predicts a factor of 5 enhancement, and the cluster-cluster correlations should be negative beyond 50 h^{-1} Mpc. Recent determinations of a streaming velocity relative to the Hubble flow of 700 km/sec or more for a sphere of radius about 60 h^{-1} Mpc (Burstein et al 1986; Collins et al 1986) conflict with a maximum of about 200 km/sec expected from the standard biased CDM models (Vittorio et al 1986). There are suggestions of even larger peculiar velocities for the Abell clusters (Bahcall et al 1986).

The conflict with observations on large scales may eventually go away as sources of systematic error in the selection of data samples or in the methods of distance determination are better understood. However, at face value it would seem we are forced to consider more complicated versions of the inflation-inspired scenarios or go to some entirely different scheme for the generation of large scale structure. Some possible variations include composite models, in which the large scale structure still arises from inflation-generated density perturbations with a flat spectrum, but there is more than one kind of dark matter. Examples include hot-cold or hot-warm models (Achilli et al 1985; Ikeuchi & Norman 1987) which can have more power on large scales as in the hot models (Melott 1986), but avoid some of the difficulties of the pure hot models with the time of galaxy formation. A non-zero cosmological constant shifts the break in the evolved density perturbation spectrum associated with the transition to a matter-dominated universe to larger scales, and if baryons are a substantial fraction of the non-relativistic matter a feature in the density perturbation spectrum associated with the baryon-radiation Jeans mass at recombination may add even more power on large scales (see Dekel 1984). Another possibility is to complicate the model of inflation in order to break the scale invariance of the primordial density perturbation spectrum (Kofman & Linde 1987; Silk & Turner 1987), though to get a feature on the desired scale requires very special values of parameters and initial conditions. Unfortunately, once the models are opened up in these ways the beauty of the simple inflation-inspired CDM model, with its small number of free parameters, is lost.

Alternative paradigms

Inflation is nice because it simultaneously is a theory for the large scale homogeneity of the universe and the existence of structure on smaller scales. However, there are alternative ways of trying to understand the structure in the universe in which there is little or no connection between the present distribution of galaxies and clusters of galaxies and primordial density perturbations. One of these is the theory of galaxy formation based on explosions, starting from small seeds (Cowie & Ostriker 1981), and on the largest scales generating the voids in the redshift surveys. Unless the explosions occur at quite early times (high z) they have difficulty producing the larger voids seen. Still, explosions may play some role, at least in heating the intergalactic medium (Ikeuchi and Ostiker 1986).

An approach to large scale structure and galaxy formation based on cosmic strings has attracted increasing interest over the last few years. Cosmic strings are topologically stable relics of a phase transition in the very early universe (Kibble 1976). Loops can form by reconnection of self-intersecting pieces of string and act as seeds for accretion of matter, forming galaxies and clusters of galaxies (see a review by Vilenkin 1985). There is no need for primordial density perturbations generated by inflation. While inflation might still be desirable to explain large scale homogeneity and isotropy, the existence of strings and inflation tend to be incompatible (Vishniac et al 1986). Particularly interesting from the point of view of large scale structure are claims the strings naturally give rise to the correct number densities of galaxies and clusters and galaxy-galaxy and cluster-cluster correlation functions of the right slope and amplitude with only one free parameter, the mass per unit length μ (Turok & Brandenberger 1986). This is a rapidly developing area of research and detailed calculations and simulations of the quality of those done for conventional models have yet to be made. There are suggestions that large scale sheets of galaxies might arise from cosmic string wakes (Stebbins et al 1986), but the large scale streaming velocities seem hard to explain, as in conventional models. Other potential difficulties with the string scenario are discussed by Peebles (1986b).

CONTRIBUTED PAPERS

Brief reports follow on some of the more interesting papers contributed to the Astrophysical Cosmology Symposium.

Multiple weak gravitational lensing

Gravitational lensing can be an important probe of the lumpiness of the distribution of matter in the universe. L.M. Oattes presented a paper coauthored by C.C. Dyer on the effects of multiple weak lensing on the apparent brightness of distant sources, with density inhomogeneities modeled by a "Swiss Cheese" universe, a zero-pressure Friedmann model in which randomly placed spherical regions have all their mass concentrated at the center. The lensing is calculated by integrating the optical scalar equations along the line of sight from source to observer, in order to properly take into account the nonlinear interaction between the shear and convergence of the beams. The distribution of apparent luminosities for standard candle sources is found to be highly non-Gaussian. Most of a random sample of sources have their apparent luminosity decreased by the lensing, but there is a high amplification tail. The effects demonstrated need to be taken into account when considering the statistics of high redshift quasars.

Cosmic microwave anisotropy

Robert Schaefer discussed his work with L.F. Abbott (Abbott & Schaefer 1986) on a general gauge-invariant analysis of large scale microwave background anisotropy. On the assumption that the perturbations generating the anisotropy obey Gaussian statistics, they calculate mean square amplitudes and dispersions for the lowest several angular multipoles of the fractional perturbation in the microwave background temperature. The photons are treated as a perfect fluid up until recombination, and the Sachs-Wolfe formula, rewritten in gauge-invariant variables, is used to extrapolate to $\delta T/T$ at the present. Results are obtained for scalar, vector, and tensor perturbations in both open and closed background spacetimes.

Particularly interesting is the comparison between the dipole and quadrupole moments from scalar (density) perturbations. The dipole moment depends on the local motion of the observer, but as long as short wavelength perturbations are uncorrelated with long wavelength perturbations, a lower bound to the mean square value of the dipole moment is obtained by considering only wavelengths longer than a cutoff wavelength, which Abbot and Schaefer take to be 60 h^{-1} Mpc. Requiring the long wavelength contribution to the dipole to be less than the observed value puts an upper limit on the expected quadrupole anisotropy from scalar perturbations, as a function of the cosmological density parameter Ω. The upper limit is about 1×10^{-5} for $\Omega = 1$ and a flat Harrison-Zeĺdovich initial density perturbation spectrum and is higher for $\Omega < 1$ and $\Omega > 1$.

Multiply-connected universes

An amusing look at an unconventional type of cosmological model was presented by Li Zhi Fang. He points out that if the real universe is assumed to be multiply-connected with points identified to give a toroidal topology, as is done for mathematical convenience in numerical simulations of large scale structure, density perturbations can have higher amplitudes without violating microwave background constraints. Also, correlations are suppressed on scales larger than the toroid and quasars would have multiple images.

PANEL DISCUSSION

An informal panel discussion was held to highlight some of the important current issues in astrophysical cosmology. Participants included Michael Turner, Franco Occhionero, and Bernard Carr. Turner emphasized the testability of inflationary models, in particular the predictions for Ω and the density perturbation spectrum. Occhionero discussed work done jointly with R. Scaramella on hybrid warm-cold dark matter models. They calculate the number density of peaks in the density perturbation field as a function of mass along the lines of Schaeffer and Silk (1985) and relate this to the luminosity function of galaxies. A model with 10% cold dark matter and 90% warm dark matter (particle mass about 800 eV) is consistent with quasars forming at redshifts in the range 2-3 and with the Schechter luminosity function. Carr remarked on the coincidence that the baryons and the non-baryonic dark matter should have comparable mass densities and what constraints this places on fundamental physics.

SUMMARY

Astrophysical cosmology should continue to be exciting and stimulating, with rapid developments in both observation and theory, for the foreseeable future. Completion of a redshift survey of extragalactic IRAS sources in 1987 may allow us to begin to converge on an observational value for the global Ω. As optically selected galaxy redshift surveys probe more deeply and cover wider areas of the sky the statistics of the large scale distribution of galaxies will become more sharply defined. If the large scale streaming velocities are real, we should already be close to seeing anisotropy in the microwave background on angular scales of a few degrees. Perhaps we will have a clear positive detection by the time of GR12. The Hubble Space Telescope should tell us if there is intergalactic gas in the voids as expected in the biased CDM scenarios. As more high z galaxies are studied and more is learned about the gas clouds seen in absorption against quasars we should be able to understand much more about the nature and timing of galaxy formation, and perhaps begin to have good data on the evolution of galaxy clustering and galaxy correlations, which is a critical test of many of the theoretical models. No doubt there will be many more new questions than answers.

ACKNOWLEDGEMENT

The author acknowledges support from a Department of Energy contract No. DE-AC06-81ER-40048 and would like to thank J.R. Bond for many stimulating conversations.

REFERENCES

Abbott, L.F. & Schaefer, R.K. (1986). Astrophys. J., 308, 510.
Achilli, S., Occhionero, F., & Scaramella R. (1985). Astrophys. J.,299, 577.
Bahcall, N.A. & Soneira, R.M. (1983). Astrophys. J., 270, 20.
Bahcall, N.A., Soneira, R.M., & Burgett, W.S. (1986). Astrophys. J., 311, 15.
Bardeen, J.M., Steinhardt, P., & Turner, M.J. (1983). Phys. Rev. D., 28, 679.
Bardeen, J.M., Bond, J.R., Kaiser, N., & Szalay, A.S. (1985). Astrophys. J., 304, 15.
Blumenthal, G.R., Faber, S.M., Primack, J.R., Rees, M. (1984). Nature, 311, 517.
Bond, J.R. & Szalay, A.S. (1983). Astrophys. J., 274, 443.
Burstein, D., Davies, R.L., Dressler, A., Faber, S.M., Lynden-Bell, D., Terlevich, R., & Wegner, G. (1986). In Galaxy Distances and Deviations from Universal Expansion, ed. B.F. Madore & R.B. Tully. Dordrecht: Reidel.
Centrella, J. & Melott, A.L. (1983). Nature, 305, 196.
Collins, C.A., Joseph, L.D., & Robertson, N.A. (1986).Nature, 320, 506.
Davis, M., Huchra, J., Latham, D.W., & Tonry, J. (1982). Astrophys. J., 253, 423.
Davis, M. & Peebles, P.J.E. (1983). Astrophys. J., 267, 465.
Davis, M., Efstathiou, G., Frenk, C.S., & White, S.D.M. (1985). Astrophys. J., 292, 371.
Dekel, A. (1984). Astrophys. J., 284, 445.
Dekel, A. & Silk, J. (1986). Astrophys. J., 303, 39.
de Lapparent, V., Geller, M.J., & Huchra, J.P. (1986). Astrophys. J., 302, 1.
Frenk, C.S., White, S.D.M., Efstathiou, G., & Davis, M. (1985). Astrophys. J., 317, 595.
Guth, A. (1981). Phys. Rev. D., 23, 347.
Guth, A. & Pi, S.-Y. (1982). Phys. Rev. Lett., 49, 1110.
Hawking, S. (1982). Phys. Lett., 115B, 295.
Ikeuchi, S. & Ostriker, J.P. (1986). Astrophys. J., 301, 522.
Ikeuchi, S. & Norman, C.A. (1987). Astrophys. J., 312, 485.
Kaiser, N. (1984). Astrophys. J., 284, L9.
Kibble, T. (1976). J. Phys., A9, 1387.

Klypin, A.A. & Kopylov, A.A. (1983). Sov. Astr. Lett., 9, 41.

Klypin, A.A. & Shandarin, S.F. (1983). Mon. Not. Roy. Astr. Soc., 204, 891.

Kofman, L. & Linde, A. (1987). Nuc. Phys., B282, 555.

Loh, E.D. & Spillar, E.J. (1986). Astrophys. J., 307, L1.

Lubimov, V.A., Novikov, E.G., Nozik, V.Z., Tretyakov, E.F., & Kosik, V.S. (1980). Phys. Lett., 94B, 266.

Melott, A. (1986). In Galaxy Distances and Deviations From Universal Expansion, ed. B.F. Madore & R.B. Tully. Dordrecht: Reidel.

Ostriker, J.P. & Cowie, L.L. (1981). Astrophys. J., 243, L127.

Peebles, P.J.E.(1982). Astrophys. J., 259, 442; 263, L1.

Peebles, P.J.E.(1986a). Nature, 321, 27.

Peebles, P.J.E.(1986b). Preprint.

Schaeffer, R. & Silk, J. (1985). Astrophys. J., 292, 319.

Silk, J. & Turner, M.S. (1987). Phys. Rev. D., 35, 419.

Starobinsky, A.A. (1982). Phys. Lett., 117B, 175.

Stebbins, A., Brandenberger, R., Veeraraghavan, S., Silk, J., & Turok, N. (1987). Astrophys. J., to be published.

Turner, M.J. (1987). In these proceedings.

Turok, N. & Brandenberger, R.H. (1986). Phys. Rev. D., 33, 2175.

Vilenkin, A. (1985). Phys. Rep., 121, 263.

Vishniac, E.T., Olive, K.A., & Seckel., D. (1986). CERN preprint no. 4554/86.

Vittorio, N., Juskiewicz, R., & Davis, M. (1986). Nature, 323, 132.

White, S.D.M., Frenk, C.S., & Davis, M. (1983). Astrophys. J., 274, L1.

White, S.D.M., Frenk, C.S., Davis, M., & Efstathiou, G. (1987). Astrophys. J., to be published.

Yahil, A., Walker, D., & Rowan-Robinson, M. (1986). Astrophys. J., 301, L1.

REPORT ON THE SYMPOSIUM ON ASYMPTOTIA, SINGULARITIES AND GLOBAL STRUCTURE

C.J.S. Clarke
Department of Mathematics, The University,
Southampton, SO9 5NH, England

Abstract. The following is a personal account of the submissions and presentations which struck me most forcibly. Names followed by dates [e.g. Jones (1976)] are background references listed at the end; names without dates refer to the conference submission, for which I have listed a reference if the authors have supplied one, or otherwise the authors' affiliation.

1 FOUNDATIONAL ISSUES

There has for a long time been continuing speculation as to whether there is a need to consider alternative models alongside, or instead of, the conventional Lorentz manifold picture of space-time, motivated often by problems arising from quantisation attempts; the Twistor programme arose in part from such speculation. (Penrose, 1971). In this symposium the only radical contribution in this category was from Lisker, who proposed representing causal relations by algebras of functions. But the mathematical conclusions from this work are unclear to me. More conventionally, Bleyer and Liebscher investigated the consequences of a quantum-induced generalisation of the normal Lorenzian light-cone structure, with potentially observable spectroscopic effects. But apart from these there were no new alternatives to the Lorentzian model.

2 GLOBAL STRUCTURE AND CONJUGATE POINT THEORY

One way of handling global questions is by constructing mathematical spaces consisting, for instance, of all space-times in a given class (an example from another symposium being provided by the work of Swift, d'Inverno and Vickers where the idea is applied in the context of exact solutions). Canarutto considered spaces of connections in order to provide a general framework for stability analysis, and used these to demonstrate the stability, in a certain sense, of bundle-incompleteness.

Apart from work on singularities, the only contribution on the global topology of space-time as such was Galloway's work on the circumstances under which space-time "splits" into the metric direct product $(R \times V, -dt^2 + h)$ (where (V,h) is a Riemannian space). It turns out that, under rather general subsidiary conditions, either space-time is incomplete or it splits.

Work by Hukum Singh gave criteria for determining when a cut-point of a space-like submanifold (cf. Beem and Ehrlich, 1981) was the farthest cut-point.

3 SINGULARITIES

3.1 Particular examples

One motivation for studying, and for arguing away, singularities stems from the colliding wave solutions of Khan and Penrose (1971) and their later generalisations. If the singularities found there were a general feature of wave-interactions, then the universe would be full of singularities. On the other hand one might hope that waves without the high symmetry of the known solutions either produced no singularities or ones that could be incorporated into the space-time in some distributional sense. Thus interest attached to the detailed analyses of the plane-wave singularities by Helliwell and Konkowski, and of the analogous singularities in colliding matter shells by Dray and 't Hooft. The latter work clarified, in a sequence of mind-bending drawings, the causal relations between the different parts of the singularity; while the former showed that the part of the plane-wave singularity that could be reached by geodesics that encountered bounded curvature was in the boundary of the part where the curvature was unbounded, perhaps suggesting that one should think of it as a single directional singularity analogous to the Curzon type.

But this idea was shaken in a surprising analysis of the Curzon singularity by Scott, who constructed a smooth (C^∞) extension through it. Previously it was thought that some curves terminated at the singularity encountering bounded curvature; but in the extension all these geodesics can be extended into a region on the other side, showing that they do not in fact run to the same boundary point as those geodesics that feel unbounded curvature. The metric in the extension cannot be determined by analyticity; indeed, it can be taken to be flat.

In a cosmological context, there has been a great deal of work on homogeneous (Bianchi) singularities; but one suspects that inhomogeneous singularities may be qualitatively different; so it was very useful to have from Sussman a description of the singularities in the class of spherically symmetric, shear-free, (in general inhomogeneous) perfect fluid solutions - a very wide class. One problem is that the equation of state is space-dependent (in comoving coordinates) and one does not yet know over what regions, and in what solutions, the equation of state is realistic. The "big bang" singularity can be timelike in these models, which would thus provide a counter example to the strong cosmic censorship hypothesis if the matter could be made realistic.

The Einstein equations can be modified by the inclusion of terms representing the effect of quantum processes, and a spherically symmetric solution to these was presented to another symposium by Tomozawa. He discovered that the singularity was avoided by the effective repulsive nature of quantum corrections at small length-scales.

3.2 Cosmic censorship

Progress here has been undramatic since GR10. It is known that some singularities are naked (e.g. Christodoulou, 1984) and some are not; and there are theorems that classify the possibilities by showing that certain sorts of singularity (those with a lot of curvature) cannot be naked. The contributions added detail to this picture, but without bringing one much nearer to the basic question of whether or not the naked sort forms in an evolution from generic regular initial data. On the classification side, the question as to whether a *strong curvature singularity* could be naked was answered in the affirmative (both for weak and strong cosmic censorship) by Hollier's analysis of a case of the Vaidya metric, but qualified by Krolak's proof that such singularities cannot be (weakly) naked if they violate causal simplicity.

3.3 Singularity theorems

There were a number of results of the form "space-time is incomplete if ..." (in addition to that of Galloway in 2 above). Allison gave criteria for the completeness of warped products (space-times $(a,b) \times V$ with metrics $-f(x)dt^2 + h$, where (V,h) is a Riemannian space and f is a function on V): for example, a sufficient condition for timelike completeness when $(a,b) = (-\infty,\infty)$ is that V is noncompact and $f(x) \to \infty$ as $x \to \infty$ (the point at infinity in the one-point compactification). Krolak and Rudnicki produced a rather general theorem on the way singularities could arise from causality violation.

Many of the singularity theorems start with the occurrence of closed trapped surfaces. This motivated an analysis of such surfaces by Kupeli, who has obtained relationships between their topology and the curvature of the 4-dimensional space-time in which they are embeded.

There was a potentially interesting example from Gertsenstein and Melkumova analysing the meaning of the singularity theorems for charged dust; but the authors were not able to attend to present the details. Gavrilov stated a result that plane wave metrics (defined geometrically in terms of particular coordinates) were always geodesically complete provided the coordinates were the whole of $\mathbb{R}^4$, and gave canonical forms for such metrics.

4 ASYMPTOTICALLY FLAT SPACE-TIMES

4.1 The existence problem

The most central problem here is to determine whether or not there exist general radiating asymptotically-simple and empty space-times with material sources. If this is not so, then this class of space-times is clearly not appropriate for modelling isolated sources such as stars, and we must look for a more general class of space-times. Moreschi proposed such a class by weakening the definition of asymptotic flatness, specifying a weaker approach of the Riemann tensor to zero than in the usual formulation.

A most interesting and novel approach to the existence problem was provided by the work of Cagnac and Noutchegueme, who showed the existence of solutions with arbitrary sufficiently small prescribed values of $T^{ij} - (1/2)g^{ij}T^{\alpha}_{\ \alpha}$ $(i,j = 1,2,3)$ having Minkowskian data at past times. As the approach is quite different from approaches based on conformal compactification, it will be interesting to find out what the behaviour of this class of solutions is at null infinity. In any case, they provide excellent candidates for radiating sources.

4.2 Energy, momentum, angular momentum and moments

While a succession of papers from many authors has settled the "straight" positive energy conjecture (that the ADM mass and Bondi mass of isolated systems are positive), there still remains the problem of relating the Bondi mass of a black hole to its area, and of relating the Bondi mass to the ADM mass. Schmidt described an approach involving a $1/u$ expansion which, assuming convergence, showed that the Bondi mass tended to the ADM mass and revealed the need for $1/r^s.\ln r$ terms on the space-like hypersurfaces, underlining the extreme delicacy of posing correct boundary conditions for the initial value problem in asymptotically flat space-times.

The same point emerged from Chrusciel's analysis of the conditions on the data needed in order that the ADM angular momentum be unambiguously defined; these turned out to be weaker than those of Ashtekar and Hansen (1978).

There was further work by Polishchuk on Hamiltonian-based definitions of energy.

The long-standing problem of defining angular momentum at $\mathscr{I}$ has led to general analyses of momentum and angular momentum on null hypersurfaces. Swift and Vickers examined the relationship between Penrose's (1982) definition and definitions arising from positive-energy techniques. They used a propagation equation that enabled them to relate moments on a slice of $\mathscr{I}^+$ to those on a surface obtained by moving down the outgoing null geodesics. A particularly interesting feature was the ability of the technique to handle caustics in the surface.

In the stationary axisymmetric case, where there is the possibility of defining a full set of multipole moments, Koppel and Lember gave a new approach to the computation of the Thorne/Hanson moments.

4.3 Conformal Killing fields

Garfinkle showed that asymptotically flat space-times that are vacuum near infinity, asymptotically Minkowskian and with positive Bondi mass do not admit any conformal Killing fields that are not Killing fields. This effectively strengthens the part of the result of Eardley et al. in the "Mathematical Problems" section showing that, apart from Minkowsky space, asymptotically flat space-times with dominant-energy sources do not admit conformal Killing fields that are asymptotic to a non-trivial dilation.

4.4 Examples

Lessner had an interesting vacuum space-time whose distant appearance was interpreted (on the basis of the transformation properties of the integral of the energy-complex) as that of a massless extended particle.

Background References

Ashtekar, A. & Hansen, R.O. (1978). J. Math. Phys. 19, 1542-1566.
Beem, J.K. & Ehrlich, P.E. (1981). Global Lorentzian Geometry. New York: Marcel Dekker.
Christodoulou, D. (1984). Violation of cosmic censorship in the gravitational collapse of a dust cloud. Comm. Math. Phys. 93, 171-195.
Khan, K.A. & Penrose, R. (1971). Nature, 229, 185-186.
Penrose, R. (1971). Combinatorial space-time. In Quantum Theory and Beyond, ed. E.W. Bastin, pp. 151-180. Cambridge: Cambridge University Press.
Penrose, R. (1982). Proc. Roy. Soc. Lond. A381, 53.

References or affiliations for papers abstracted at GR11

Allison, D.E., Dept. of Mathematics, Southern Illinois University, Carbondale, Illinois 62901, U.S.A.
Bleyer, U. & Liebscher, D.-E. Preprints PRE-ZIAP 85-02, 86-01, Zentralinstitut für Astrophysik der Akademie der Wissenschaften der DDR, Rosa-Luxemburg-Str. 17A, DDR-1502 Potsdam-Babelsberg.
Cagnac, F. & Noutchegueme, N. Faculty of Science, P.O. Box 812, Yaounde, Cameroon.
Canarutto, D. Preprint: On the stability of b-Incompleteness in the Whitney topology on the space of connections. Istituto di Matematica Applicata "G. Sansone", Via S. Marta 3, I-50139 Firenze, Italy.
Chruściel, P.T. Institute of Mathematics, Polish Academy of Sciences, P.O. Box 137, 00-950 Warsaw, Poland.
Dray, T. & 't Hooft, G. Nuclear Physics B253, 173 (1985). Comm. Math. Phys. 99, 613 (1985). The gravitational effect of colliding planar shells of matter, Class. Quantum Grav. (in press).
Eardley, D., Isenberg, J., Marsden, J. & Moncrief, V. Preprint: Homothetic and Conformal Symmetries of Solutions of Einstein's Equations. (NSF-ITP-85-154) Institute for theoretical physics, University of California, Santa Barbara, Ca 93106, U.S.A.
Galloway, G.J. Comm. Math. Phys. 96, 423-429 (1984). Math. Proc. Camb. Phil. Soc. 99, 367-375 (1986).
(See Beem, J., Ehrlich, P., Markvorsen, S. & Galloway, G. J. Diff. Geom. 22, 29-42 (1985).)
Garfinkle, D. Asymptotically flat space-lines have no conformal Killing fields. Preprint: Department of Physics, Washington University, St. Louis, Missouri 63130, U.S.A.
Gavrilov, S.P. Kazan State University, Kazan 420008, USSR.

Gertsenshtein, M.E. & Melkumova, E. Yu. Nuclear Physics Research Institute, M.V. Lomonosov State University, Moscow 117234, USSR.
Hollier, G.P. A naked strong curvature singularity. To appear as a letter in Classical and Quantum Gravity.
Halliwell, T.M. & Konkowski, D.A. Preprint: Singularity Structure of Colliding Plane Gravitational Waves. School of Mathematical Sciences, Queen Mary College, London UK, E1 4NS.
Koppel, A. & Lember, T. Department of Theoretical Physics, Tartu State University, Tartu 202400, Estonia, USSR.
Królak, A. Dept. of applied mathematics and astronomy, University College, PO Box 78, Cardiff CF1 1XL, UK.
Kupeli, D.N. Curvature and closed trapped surfaces in 4-dimensional space-times, Gen. Rel. and Grav., to appear.
Curvature and compact spacelike surfaces in 4-dimensional space-times, Proceeding of AMS-IMS-SIAM joint summer conference at Santa Gruz (1986).
Lessner, G. Axially symmetric pp-waves and their interpretation as extended massless particles, Gen. Rel. and Grav., 18, 899 (1986).
Lisker, R. Causal Algebras. P.O. Box 1996, Berkeley, Ca. 94701, U.S.A.
Moreschi, O. Preprint: General future asymptotically flat space-times (37/86/A), SISSA, Trieste, Strada Costiera 11, Italy.
Polishchuk, R.F. State committee for standards, 9 Levinsky Prospekt, 117049 Moscow, USSR.
Rudnicki, W. Institute of Physics, Pedagogical University, Rejtana 16A, 35-311 Rzeszów, Poland.
Schleifer, N. Phys. Lett. 112A, 204 (1985).
Schmidt, B. Max-Planck-Institut für Physik und Astrophysik, 8046 Garching bei München, FRG.
Scott, S.M. & Szekeres, P. The Curzon Singularity I, II. Gen. Rel. and Grav. 18, 557-570 & 571-583 (1986).
Singh, H. Preprint: The space-time cut locus and focal locus. Department of Mathematics, M.L.K. (P.G.) College, Balrampur, Gonda (U.P.), India.
Sussman, R.A. Preprints: On spherically symmetric shear free configurations (Charged and neutral), parts I, II, III. Department of Mathematics, Queen Mary College, Mile End Road, London, UK, E1 4NS.
Swift, S.T. & Vickers, J.A.G. Preprint: Mass, momentum and angular momentum in general relativity. Department of Mathematics, University of Southampton, Southampton, UK, SO9 5NH.
Swift, S.T., d'Inverno, R.A. & Vickers, J.A.G. Preprint: Everywhere invariant spaces of metrics and isometries. Department of Mathematics, University of Southampton, Southampton, UK, SO9 5NH.
Tomozawa, Y. Preprint: Quantum corrections to the gravitational collapse, Randall Laboratory of Physics, University of Michigan, Ann Arbor, Michigan 48109, U.S.A.
Unt, Väino. Institute of Astrophysics and Atmospheric Physics, 202444 Tóravere, Tartu, Estonia, USSR.

A Brief Report on the "Early Universe" Symposium

D.V. Nanopoulos

Department of Physics, University of Wisconsin–Madison, Wisconsin 537

Nowadays, the early Universe has become the "meeting point" of Cosmology, Astrophysics, and Particle Physics. In a rather fascinating way, it becomes more and more apparent that the three aforementioned fields of interest need each other, if progress has to be made in any one of them. Clearly, understanding better the basic laws of Particle Physics is expected to shed new light on Cosmology and Astrophysics, but why do Particle Physicists have to rely on Cosmology and Astrophysics? The answer is rather straightforward: to get invaluable information, otherwise unavailable at Particle Physics laboratories. The unification programs in Particle Physics seem to demand Cosmological energies ($E \geq 10^{15} GeV$) in order to be verified and thus the early Universe may provide the only place to test these ideas. So, by studying Cosmology and/or Astrophysics we can get a lot of information about unified theories. At the same time, once we get interested, we may help to resolve some tantalizing problems both in Cosmology and Astrophysics. For a simple review on unified theories and the early Universe see Nanopoulos (1984).

Several of these connections between Particle Physics and Cosmology have been vigorously discussed in this symposium. Let us follow the cosmic time evolution and discuss along the way the different contributions to the symposium. In the framework of the standard Big-Bang Cosmology (BBC) it is usually assumed that at some time $t_o \equiv 0$ a big explosion occurred and the Universe expands and cools down since then. Following the classical equations of general relativity, it is not difficult to see that in BBC the existence of an initial singularity is unavoidable. It is possible that quantum corrections may take care of this peculiarity, but several people believe that we may be in for drastic modifications, like the one discussed by Sinha where gravity becomes repulsive above some critical temperature, T_c, see Sathyaprakash et al. (1984, 1986). Then the onset of repulsive gravity may help avert the gravitational singularity. One thing that is needed to be settled at such very early times is the number of space-time dimensions. Modern unified theories like Kaluza-Klein theories or Superstrings may need more than four dimensions, with $D = 10$ the favorable superstring number. How, then do we find ourselves living in four dimensions? That kind of tough problem was discussed by Maeda (1986) who presented stability and attractor conditions for a (four dimensional Friedman Universe) $\otimes$ (a constant internal space). He also showed that fashionable superstring compactifications (a la Calabi-Yau), interestingly enough, satisfy these stability and attractor conditions.

Another thing that is needed to occur at these very early times is inflation. Inflation, an exponential growth of the cosmic-scale factor R of the Universe for a limited but critical time period at the very beginning, seems to a lot of us an unavoidable panacea. The horizon, flatness and origin of energy density perturbation problems are resolved at once, but the basic problem of constructing a realistic inflationary scenario in particle physics, essentially remains still unresolved. Enqvist discussed different attempts, emphasizing the importance of getting inflation after compactification and presented an inflationary scenario emerging from the superstrings due to Ellis et al. (1986 a). He made it apparent, though, that these are only first attempts and a much more satisfactory picture should be expected. On the same issue of inflation, Tipler (1986) argued that the very idea of quantum cosmology contains naturally the idea of inflation: the expansion mechanism is a basic (and elementary) property of wave functions: if the wave function is initially highly localized in space, then the wave function will literally explode outward in the next instant. For certain, quantum cosmology promises a lot of surprises yet to come. As the four-dimensional, smoothed out, baby Universe evolves, we better make sure that it contains some matter in it. This is a two-fold problem: 1) The Universe, after inflation, should reheat by transforming the stored vacuum energy to radiation: photons, quarks and leptons, etc., and 2) The mechanism for the creation of the observed cosmic baryon asymmetry should operate, after inflation, otherwise all the hard-won baryon asymmetry will be diluted. Sato (1986) discussed the tight constraints that the simultaneous resolutions for problems 1 and 2 impose on inflationary models and stressed, once more, the difficulty of constructing inflationary models satisfying all desirata. Actually the cosmic baryon asymmetry, as Enqvist (1986) discussed, imposes severe limits on intermediate mass scales $M_I(\leq 10^9 GeV)$. Otherwise, during the phase transition related to M_I, too much energy is generated and as a result, the baryon asymmetry is unacceptably diluted.

Eventually, as the Universe evolves we should be able to form the different observed chemical clements. It is one of the triumphs of BBC that the primordial abundance of H, ^4He, D, ^3He, and ^7Li can be easily explained, see Yang et al. (1984). Furthermore, the success of primordial nucleosynthesis put constraints on the number of particles with substantial energy density during the nucleosynthesis era. For example, it is usually quoted that there cannot be more than four "light", left-handed, neutrinos with the usual weak strength. Enqvist argued that this number is too conservative, and that a more liberal use of astrophysical data (especially $^3He + D$) may push the allowed number of neutrinos to 5.5, as has been suggested by Ellis et al. (1986 b). This result may be of considerable importance in a class of superstring models where there are three left-handed and three right-handed neutrinos. Even if the old limit of four neutrinos is firm, as recently argued by Steigman et al. (1986), there is not necessarily a problem with the three extra right-handed neutrinos, since their energy density contribution may be negligible if their interactions are weaker than standard weak interactions. In this case as Ellis et al. (1986 b) found there is a lower limit of $0(200GeV)$ on the mass of the extra neutral gauge boson Z_E, which incidentally coincides with the lower bound coming from Particle Physics experiments obtained by Cohen et al. (1986).

Here is a good point to stop our early Universe tour, with the hope that it is apparent by now the eminent importance of unifying Particle Physics with Cosmology. My firm belief is that by the next Conference of this series, even more stunning results will come out from the study of the early Universe and thus guarantee an exciting GR-12 meeting at least like the present one, which admittedly is a hard act to follow.

REFERENCES

1. Cohen, E., Ellis, J., Enqvist, K., Nanopoulos, D.V., (1985). Phys. Lett. B161, 76.

2. Ellis, J., Enqvist, K., Nanopoulos, D.V. and Quiros, M. (1986 a). Nucl. Phys. B277, 231.

3. Ellis, J., Enqvist, K., Nanopoulos, D.V. and Sarkar, S. (1986 b). Phys. Lett. B167, 457.

4. Enqvist, K., Nanopoulos, D.V. and Quiros, M. (1986). Phys. Lett. 169, 343.

5. Maeda, K. (1986). Phys. Lett. B166, 59.

6. Nanopoulos, D.V. (1984). in Unified Field Theories and the Early Universe in Proc. 1st ESO-CERN Symposium "Large Scale Structure of the Universe, Cosmology and Particle Physics", Eds. G. Setti and L. Van Hove (ESO-CERN Publication, Geneva 1984) p. 349.

7. Sathyaprakash, B.S., Lord, E.A. and Sinha, K.P. (1984). Phys. Lett. A105, 407.

8. Sathyaprakash, B.S., Goswani, P. and Sinha, K.P. (1986). Phys. Rev. D33, 2196.

9. Sato, K., Kodama, H., Yokoyama, T. and Sato, N. (1986). Contribution #6:23 in this Conference.

10. Steigman, G., Olive, K.A., Schramm, D.N. and Turner, M.S. (1986). Phys. Lett. B176, 33.

11. Tipler, F.J. (1986). Contribution #6:28 in this Conference.

12. Yang, T., Turner, M.S., Steigman, G. Schramm, D.N. and Olive, K.A. (1984). Ap. J. 281, 493.

EXACT SOLUTIONS. REPORT OF SYMPOSIUM 07

D. Kramer
Friedrich-Schiller-Universität Jena, Sektion Physik,
Max-Wien-Platz 1, Jena, DDR-6900, G.D.R.

1 THE PROCEDURE

Exact solutions is one of the topics which attract many scientists working on General Relativity: 65 of the 621 abstracts of contributed papers included in the GR 11 abstract books (and the supplementum) were assigned to the symposium "Exact Solutions". In view of that large number of papers the symposium was allocated two successive sessions of 1.5 hours each. The first session was devoted to vacuum and electrovacuum fields whereas perfect fluids were the subject of the second one.

The general strategy was to choose some authors who should present their work verbally and to give, at appropriate places between the talks, brief summaries of other papers which seemed to me worth mentioning.

The speakers explained the methods used and the results obtained in their papers in a clear manner and kept to the time limit of 20 minutes (including discussion). The duration of the talks was no further restricted because, in my opinion, at least that time is necessary to present a reasonably amplified version of a printed abstract. Some of the talks were followed by a longer discussion.

The chairman's summaries were thematically ordered in groups. After each group the corresponding authors if present at the symposium could comment on their work and answer questions from the participants. I estimate only half of the authors mentioned in the summaries actually attended the meeting. The summaries of about 30 abstracts took almost one-third of the total time available for the symposium.

The present report roughly follows the course of the symposium. No doubt some omissions or inaccuracies will have occurred in my account, and I apologize in advance for them.

In preparing the symposium I asked some authors to send me preprints or reprints related to their abstracts. I thank all these colleagues for their support; this additional information proved to be very helpful for me. I thank especially the speakers for their excellent talks, and those participants who actively took part in the discussions, in particular Professors W.B. Bonnor and G.C. McVittie.

The reference list, which is not at all complete, contains some published work known to me on the contributions to the symposium. Most of the summarized papers exist at the moment only in the form of preprints.

2 FIRST SESSION: VACUUM AND ELECTROVACUUM SOLUTIONS

2.1 Generation methods

Originally three speakers in this group were selected (G.A. Aleksejev, N.R. Sibgatullin and M. Gürses) but none of them was able to attend GR 11. G.A. Aleksejev had shown that the linear problem for Einstein-Maxwell fields (outside the sources) admitting an Abelian Group G_2 of motions is equivalent to a singular integral equation. To obtain special classes of exact solutions one has to specify holomorphic functions on a given contour in the complex plane of the spectral parameter. For details, see Aleksejev (1985). In a second abstract Aleksejev had considered the N-soliton Einstein-Maxwell field, in particular the superposition of two hyper-extreme Kerr-Newman solutions.

Einstein-Maxwell (plus neutrino) fields admitting an Abelian G_2 were also studied by N.R. Sibgatullin and M.N. Zaripov. Starting with the Kinnersley-Chitre form of the field equations in terms of a 3 x 3 matrix, these authors gave a concise expression for the infinitesimal action of the internal Lie-Bäcklund symmetry group which corresponds to the infinite-dimensional Kinnersley-Chitre algebra. The methods applied are explained in detail in the book by Sibgatullin (1984).

M. Gürses submitted an abstract on his attempts to construct a linear problem and to find Bäcklund transformations for the vacuum field equations without prescribed space-time symmetries. Imposing certain other conditions Gürses found special types of Bäcklund transformations, e.g., within the Kerr-Schild class of solutions.

H. Salazar I introduced complex Ernst potentials and formulated a variational principle for the stationary Einstein-Born-Infeld equations. The Lagrangian contains a structure function which is determined by the type of the nonlinear electrodynamics. In the nonlinear case only trivial internal symmetries of the Lagrangian survive. Generalizing the results on Einstein-Maxwell solitons to the Projective field theory, T. Matos found a linear 3 x 3 matrix problem which implies the five-dimensional vacuum equations of stationary axisymmetric fields. Solutions can be constructed by means of the usual polynomial method. H. Lu and Ch. Feng used linear coordinate transformations in V_5 to obtain new solutions.

V. Ferrari and J. Ibañez applied the soliton techniques to vacuum fields when the Abelian G_2 acts on spacelike orbits. Starting with the Kasner metric as seed solution these authors generated and discussed a generalization of the Khan-Penrose solution which describes the collision of linearly polarized plane waves. Guided by the typical behaviour of solitons after their penetration, D.S. Chellone looked for appropriate conditions on the energy-momentum tensor to produce negligible interaction terms after the collision of plane waves. U. Yurtsever considered the collision of almost plane gravitational waves.

G. Holmes and R.A. d'Inverno investigated the internal symmetries of cylindrically symmetric vacuum fields, applied generation techniques to that case, classified the cylindrically symmetric solutions, and considered radiative thin shells (Holmes 1986).

2.2 Other methods

Z. Perjés gave a talk on the utility of Ernst coordinates in tackling the problem of stationary vacuum fields and in other situations when the complex Ernst potential exists. Its real and imaginary parts (which are assumed to be functionally independent) are identified with space-time coordinates. It turned out that the stationary vacuum fields with a conformally flat 3-space of Killing trajectories are axisymmetric and that, apart from exceptional cases which were excluded from the beginning, there are no solutions of that kind (Perjés 1986). Using Ernst coordinates Perjés improved on Simon's characterization of the Kerr solution by relaxing the asymptotic conditions. For the Kerr solution the so-called Simon tensor which is an analytic continuation of the Bach-York tensor vanishes. My own contribution concerned the rigidly rotating perfect fluids with zero Simon tensor. This property characterizes the Wahlquist solution as the generic case, and its limits.

S.B. Edgar introduced special coordinates for certain classes of Petrov type I vacuum fields. Choosing two of the Weyl tensor components (Ψ_o and Ψ_2 in Newman-Penrose notation) as two complex space-time coordinates, Edgar (1986) succeeded in reducing the field equations to one differential equation of the Monge-Ampère type. S. Bonanos generalized the Kerr-Schild ansatz by introducing a flat-space tensor which satisfies the same algebraic conditions as the energy-momentum tensor of a non-null electromagnetic field. The Einstein equations split into main equations and supplementary conditions.

2.3 Special solutions

Einstein-Maxwell fields. V.I. Khlebnikov's talk dealt with a new class of algebraically special Einstein-Maxwell fields with non-zero cosmological constant. Khlebnikov (1986) assumed that the Maxwell field is aligned and non-radiative ($\Phi_o = \Phi_2 = 0$) and that the remaining null-tetrad component Φ_1 of the electromagnetic field tensor is a constant. Arbitrary functions of the retarded time enter the final form of the space-time metric which generalizes the non-diverging type N Einstein-Maxwell fields to include the cosmological constant. The cases with spherical, pseudospherical, and plane wave surfaces were considered. Another class of Einstein-Maxwell fields with non-zero cosmological constant was given by Y.Wang and Z. Tang. These algebraically special solutions belong to the (generalized) Robinson-Trautman class (non-twisting, diverging null eigendirection of the Weyl tensor) and describe the field of an accelerated mass and electric (and magnetic) charge.

The explicit solution for the field of a massless magnetic dipole given by T.I. Gutsunajev and S.L. Elsgolts contains one real parameter (the magnetic dipole moment) and does not belong to the class of magnetostatic solutions which can be obtained from Weyl's class of electrostatic solutions by a duality rotation.

Ch.A. Kolassis and N.O. Santos investigated pure radiation (null dust) solutions admitting an Abelian G_2 which is not orthogonally transitive. The authors assumed that the preferred geodesic null vector is shear-free and arrived at a set of ordinary differential equations. The condition for the existence of a null Maxwell field was examined and special cases were identified with known solutions.

Vacuum fields. S. Persides and B. Xanthopoulos discussed the properties, in particular the asymptotic behaviour, of the class of stationary axisymmetric vacuum solutions for which the Ernst potential is separable in Weyl's canonical coordinates, and of the solutions dual to that class. In both cases the problem reduces to a Painlevé equation. Solutions with a similar mathematical structure had been discussed in the context of cylindrically symmetric gravitational waves. D.W. Hobill assumed a particular form for the metric function γ in the Bondi metric and obtained a special case of the Newman-Tamburino solution which, as a result, is now explicitly given in terms of Bondi coordinates.

S.T.C. Siklos proved the theorem that all space-times admitting a Killing spinor are conformal to pp waves and investigated the subclass of Einstein spaces (including the cosmological constant) of that type. These spaces admit a twist-free null Killing vector and can be foliated by families of totally geodesic 2-surfaces of constant negative curvature. For a detailed discussion, see Siklos (1985). J.P. de S. Lemos studied the geodesics of the Rindler spaces above and below a photonic counter-rotating disc (equal number of photons moving in each sense of rotation).

2.4 Non-empty solutions from vacuum

In his talk N. Van den Bergh reported on new results on conformally Ricci-flat space-times and posed the problem of finding the Einstein-Maxwell fields and perfect fluids which are related to vacuum fields by a conformal transformation (Van den Bergh 1986). Results on null Einstein-Maxwell fields and on shear-free or vorticity-free perfect fluids were obtained by means of the null-tetrad formalism. For instance, if the four-velocity of the (vorticity-free) fluid is orthogonal to the gradient of the conformal factor, there are perfect fluid solutions which are the conformally transformed versions of a new hypersurface-homogeneous vacuum solution of Bianchi type VI_0. The statements on the generation of perfect fluids from vacuum formed the transition to the session on perfect fluid solutions.

3 SECOND SESSION: PERFECT FLUID SOLUTIONS

3.1 Solutions with special slices

The first speaker in the second session was H. Stephani. He considered space-times admitting a foliation by timelike and conformally flat hypersurfaces with the same inner geometry and zero exterior curvature. The four-velocity lies in these slices. Assuming a special algebraic form of the Ricci tensor of these 3-spaces (at most two different eigenvalues, the four-velocity being an eigenvector), Stephani was able to determine explicitly the rotating dust and perfect fluid solutions. Their Petrov type is D and, in general, they do not admit an isometry group. The pressure is a constant, the fluid motion is shearing and expanding, and the slices are necessarily 3-spaces of constant curvature. The non-rotating solutions of the class considered by Stephani were in principle known: they are either conformally flat or belong to the Szekeres-Tomimura solutions.

T. Wolf found all perfect fluid (and vacuum) solutions when the slices are flat and the exterior curvature is covariantly constant. Several cases occur according to the distinct algebraic forms of the tensor of exterior curvature. The new solutions arising from Wolf's calculations are either of non-degenerate Petrov type I with an unspecified equation of state and an Abelian group

G_3 of motions, or of Petrov type D with a stiff equation of state and less than three Killing vectors. For all cases the kinematical characterizations were given. For details, see Wolf (1985) and Stephani & Wolf (1985).

3.2 Spherical symmetry

Non-static solutions. H. Knutsen talked on physically reasonable models of non-static gas spheres with zero pressure and mass density at the boundary. The exact solution discussed in detail was originally given by Wyman and is a member of the class of shear-free spherically symmetric perfect fluid solutions due to Kustaanheimo and Qvist. It turned out that pressure and mass density are positive everywhere inside the sphere, increase monotonely towards the centre, and decrease (increase) for an expanding (contracting) model (Knutsen 1984). The collapse of the gas sphere is stopped at a finite extension; afterwards the matter spreads out to infinity. Knutsen proposed interpreting this bouncing solution as a model for a supernova explosion.

The shear-free solutions given in the abstract by D.C. Srivastava also belong to the Kustaanheimo-Qvist class. The relation to the solution studied by Glass and Mashhoon was investigated. Y.K. Gupta and R.S. Gupta gave Petrov type D perfect fluid solutions of embedding class one which are non-static generalizations of the Kohler-Chao solutions and do not possess a finite surface of zero pressure (Gupta & Gupta 1986). These authors also submitted an abstract on a static charged fluid sphere of embedding class one which can be smoothly matched to the Reissner-Nordström solution.

In addition to spherical symmetry C.C. Dyer and L.M. Oattes assumed that there is a conformal Killing vector in the (r,t) surfaces. A subclass of the McVittie solutions satisfies this condition. The program to determine all solutions of this type is still in progress.

Static solutions. S.N. Pandey and D.V. Gupta started with a polynomial ansatz for one of the metric functions and imposed the condition that pressure and mass density vanish at the boundary. The expression for the mass density is given explicitly and the pressure results from a Riccati equation. R. Hojman prescribed a certain combination of pressure and metric functions and obtained the full metric and the equation of state (in parametric form) by quadratures, see Hojman & Santamarina (1984).

Using a retarded time coordinate, A.M. Baranov and J.N. Vlasenko found spherically symmetric static solutions which, in general, are singular at the centre and contain the interior Schwarzschild solution as a special case.

S. Haggag proved that there are no isolated static axisymmetric perfect fluid sources for which the equipressure surfaces are "spheres" r = const where the radial coordinate r defined by the metric used does not give the same physical distance in all directions.

3.3 Petrov type D solutions (with electromagnetic field)

J. Carminati reported on Petrov type D solutions of the Einstein-Maxwell equations with a perfect fluid. From some conditions (non-null electromagnetic field, aligned case, zero magnetic part of the Weyl tensor, barotropic equation of state, zero vorticity of the fluid) a series of theorems was derived using the Newman-Penrose formalism. The results lead in some cases to a considerable reduction of the field equations. There are two distinct classes, DI and DII, depending on the invariant relation between the non-zero curvature components. The known type D solutions may be classified according to this subdivision.

R.M. Misra and G.S. Dubey also considered type D solutions but started with an a priori specified form of the space-time metric which goes over into the well-known Plebański-Demiański form for type D electrovacuum solutions when no perfect fluid is present. Exact solutions were obtained for the equation of state $p + 3\mu = \text{const}$, e.g., the Wahlquist solution, and its limiting cases, endowed with a (rather limited) electromagnetic field.

3.4 Special solutions

A. Das et al. considered solutions which are conformally related to static metrics which, for instance, describe the fields of charged sources in equilibrium under their mutual gravitational attraction and electric repulsion. The additional part of the energy-momentum tensor ascribed to a cosmic fluid was defined by the conformal factor.

R.A. Sussman derived and classified spherically symmetric shear-free charged perfect fluids and found "electrified" versions of the Kustaanheimo-Qvist class.

G.J.G. Junevicus studied the Einstein-Maxwell equations for a static spherically symmetric dielectric, non-conducting and uncharged perfect fluid with r-independent mass density and dielectric constant. The field equations were reduced to an elliptic integral and the solution was matched to the Reissner-Nordström metric.

Rotating dust was considered by A. Georgiou who treated cylindrically symmetric, infinitely conducting dust in a magnetic field aligned along the rotation axis. Another paper dealt with an interior and exterior solutions for rigidly rotating charged dust with zero Lorentz force.

R. Portugal and I.D. Soares found new classes of axi-symmetric, Einstein-Maxwell fields with a conducting perfect fluid (Ohm's law). These exact solutions are of the Kantowski-Sachs type, include a non-vanishing cosmological constant, and contain the Bertotti-Robinson solution as a limiting case. The type D radiating fluid solution given by A.M. Baranov and N.N. Paklin yields a source for the Vaidya metric; the total luminosity was calculated. Radiating fluids were also considered by G. Procopiuc.

4 OTHER TOPICS

Some abstracts not mentioned above took into consideration other types of energy-momentum tensors: two-component perfect fluid (R.A. Daishev), anisotropic fluid (Y.K. Gupta and D.V. Gupta), and massless scalar field (V.S. Dorodnykh). Other abstracts dealt with such topics as, e.g., the separability of the Dirac equation in space-times with local rotational symmetry (B.R. Iyer and C.V. Vishveshwara), wave solutions in Marder's space-time

(G. Mohanty), recurrency conditions for space-times with symmetries (F. Söler), perfect fluids with homothetic motions (R.A. Daishev), and symmetries of static spherically symmetric space-times (A.H. Bokhari and A. Qadir). Z.-Q. Kuang et al. studied plane-symmetric Einstein-Maxwell fields when the Maxwell tensor does not share the full space-time symmetries (Li & Liang 1985). M.J. Rebouças applied the equivalence problem methods to homogeneous Gödel-type metrics; this work is also related to the symposium on algebraic computing.

5 COMMENTARY

New exact solutions found by different methods and assumptions were given at GR 11. However, for most of them the physical meaning has not yet been revealed and in some cases it is unlikely that they even admit a reasonable interpretation at all. On the other hand, successful attempts were made (e.g. by H. Knutsen) towards a better understanding of solutions known for a long time.

There were only a few papers on the generation techniques for vacuum and electrovacuum fields with two commuting Killing vectors, and related problems. At the moment, the possibilities of tractable and interesting applications of Bäcklund transformations and the inverse scattering method seem to be exhausted and new insights and stimulating ideas are necessary to go ahead. Some authors studied generalizations of these methods either to the case of less than two Killing vectors (M. Gürses) or to the Projective field theory (T. Matos).

Other tools to find new solutions and to derive new statements were proposed or applied, e.g., the choice of preferred space-time coordinates (Z. Perjés, S.B. Edgar), conformal transformations (N. Van den Bergh), the assumption of flat or conformally flat slices (H. Stephani, T. Wolf), an appropriate metric ansatz (R.M. Misra and G.S. Dubey), and the Newman-Penrose formalism (V.I. Khlebnikov, J. Carminati). The session talks were mainly devoted to these current developments. Some of these approaches, or combinations of them, seem to be quite promising for future research. However, completely new and exciting ideas, which might lead to a break-through in long-standing unresolved problems, were not presented at the symposium on exact solutions. No progress has been made in constructing models of realistic rotating perfect fluid sources.

Solutions of cosmological interest were assigned to the symposium on Mathematical Cosmology, in particular the work by K. Rosquist and R.T. Jantzen which, in my opinion, is one of the most important achievements on exact solutions during the last few years.

One motivation for our book on exact solutions was to avoid rediscoveries of previously known solutions. I am aware of only two GR 11 abstracts on exact solutions in which results were rederived without reference to the original papers.

Nowadays the search for exact solutions in Classical Gravity is closely related to other branches of Mathematical Physics. That contact has clearly had an impact on the studies on exact solutions. A considerable amount of work has been done since the last GR conference in Padova. Not only the large number of contributions but also the great portion of valuable papers which may stimulate further investigations is remarkable. The talks and discussions at GR 11 brought about a useful exchange of ideas.

REFERENCES

Aleksejev, G.A. (1985). Dokl. Akad. Nauk 283, 577 - 582
Edgar, S.B. (1986). Int. J. Theor. Phys. 25, 425
Gupta, Y.K. & Gupta, R.S. (1986). Gen. Rel. Grav. 18, 641 - 648
Hojman, R. & Santamarina, J. (1984). J. Math. Phys. 25, 1973 - 1974
Holmes, G. (1986). Ph. D. Thesis, Southampton University
Khlebnikov, V.I. (1986). Class. Quantum Grav. 3, 169 - 173
Knutsen, H. (1984). Gen. Rel. Grav. 16, 777 - 788
Li, J.-Z. & Liang, C.-B. (1985). Gen. Rel. Grav. 17, 1001 - 1013
Perjés, Z. (1986). Gen. Rel. Grav. 18, 531 - 547
Sibgatullin, N.R. (1984). Oscillations and Waves in Strong Gravitational and Electromagnetic Fields (in Russ.). Ch. III, Moscow: Nauka
Siklos, S.T.C. (1985). In Galaxies, Axisymmetric Systems and Relativity, ed. M.A.H. MacCallum, pp. 247 - 274, Cambridge: Cambridge University Press
Stephani, H. & Wolf, T. (1985). ibid. pp. 275 - 287
Van den Bergh, N. (1986). Gen. Rel. Grav. 18, 649 - 668
Wolf, T. (1985). Dissertation, Univ. Jena

SYMPOSIUM 18: GRAVITATIONAL WAVE DETECTORS

Reviewed by William M. Fairbank, Stanford University, USA

The following is a list of papers presented at the two contributed sessions on Gravitational Wave Detectors:

J. Hough, G. P. Newton, N. A. Robertson, H. Ward, B. Meers, R.W.P. Drever, B. F. Schutz and I. F. Corbett, United Kingdom, PLANS AND PROSPECTS FOR A LONG BASELINE GRAVITATIONAL WAVE DETECTOR IN BRITAIN

R. Schilling, K. Maischberg, A. Rudiger, L. Schnupp, D. Shoemaker and W. Winkler, Fed. Republic of Germany, AIMING AT A LARGE GRAVITATIONAL WAVE ANTENNA IN GERMANY

G. Pizzella, E. Amaldi, P. Bonifazi, P. Carelli, M. G. Castellano, G. Cavallari, E. Coccia, C. Cosmelli, V. Foglietti, R. Habel, I. Modena, G. V. Pallottino, P. Rapagnani and F. Ricci, Italy, REPORT ON THE OPERATION OF A 2300 KG CRYOGENIC GRAVITATIONAL WAVE ANTENNA WITH A DC SQUID AMPLIFIER

W. O. Hamilton, B-X Xu, A. G. Mann, G. Wang, N. Solomonson and A. Sibley, USA, PERFORMANCE OF THE LSU TUNED BAR GRAVITY WAVE DETECTOR

R. Weiss and R. Drever, USA, A LASER INTERFEROMETER GRAVITATIONAL WAVE OBSERVATORY

P. F. Michelson, M. Bassan, C. Bowdery, W. M. Fairbank, M. S. McAshan, B. Moskowitz, J. Price, T. Stevenson and R. C. Taber, USA, THE STANFORD LOW TEMPERATURE GRAVITY WAVE DETECTOR EXPERIMENT; A MECHANICALLY-MODULATED MICROWAVE CAVITY FOR DETECTION OF LOW-FREQUENCY GRAVITATIONAL RADIATION

J. Weber, USA, QUANTUM THEORY OF GRAVITATIONAL RADIATION ANTENNAS

D. G. Blair, P. J. Veitch, M. J. Buckingham, C. Edwards, J. Ferreirinho, A. J. Giles, S. K. Jones, F. J. van Kann, L. D. Mann, N. P. Prestage, D. K. Ramm and M. Zeng, Australia, PROGRESS REPORT ON THE DEVELOPMENT OF A NIOBIUM GRAVITATIONAL RADIATION ANTENNA

R. Spero, USA, THE CALTECH GRAVITATIONAL WAVE DETECTOR

P. Saulson, R. Benford, M. Burka, N. Christensen, A. Jeffries, P. Linsay, J. Livas and R. Weiss, USA, THE MIT PROTOTYPE GRAVITATIONAL WAVE ANTENNA

P. Tourrenc and J-Y Vinet, France, RESONANT RECYCLING IN GRAVITATIONAL INTERFEROMETRIC ANTENNA

K. Tsubono, M. Ohashi and H. Hirakawa, Japan, DEVELOPMENT OF PARAMETRIC TRANSDUCER FOR GRAVITATIONAL RADIATION DETECTOR

D. Hils, P. L. Bender, J. E. Faller and R. F. Webbink, USA, THE GRAVITATIONAL WAVE SKY AT MICRO-HZ FREQUENCIES

H. J. Paik, USA, WIDEBAND DETECTION OF GRAVITATIONAL RADIATION USING NON-RESONANT MECHANICAL AMPLIFICATION

G. Callegari, P. Fortini, and C. Gualdi, Italy, ON THE CRUCIAL ROLE PLAYED BY THE REFERENCE FRAME IN ELECTROMAGNETIC DETECTORS OF GRAVITATIONAL WAVES

The large number of papers submitted on gravitational wave detectors to GR 11 (42 from 24 groups) attests to the very large effort being made around the world to detect gravitational radiation. This effort is directed toward three kinds of detectors: bar detectors, laser interferometer detectors and laser detectors in space. We will summarize here the contributions on each of these three kinds of detectors.

BAR DETECTORS

A milestone was reached in gravitational wave detectors with the simultaneous operation for the first time of three low temperature bar detectors with sensitivities ranging from $h= 10^{-18}$ to $h= 5 \times 10^{-18}$. The three groups, Stanford, Louisiana State University and The University of Rome operating at CERN reported on the first initial analysis of data taken during May and June of 1986. The Stanford group had previously operated their detector for over a year with a noise temperature of 10^{-2} K and a strain sensitivity of $h = 10^{-18}$. However no detector with comparable sensitivity was on the air during this period. During the present set of simultaneous measurements the Rome group was at a noise temperature of 10 millikelvin and a sensitivity of 10^{-18}, the Stanford group at a noise temperature of 50 millikelvin and a strain sensitivity of $h = 2 \times 10^{-18}$, while the LSU group was at a sensitivity of 5×10^{-18}. The coincidence data were taken between day 121 and day 152 of 1986. A small part of this data was brought to the GR-11 conference for analysis during the conference. The data analyzed during the conference included four days of coincidence data between LSU and Stanford, six days between LSU and Rome, eight days between Rome and Stanford and three and one half days of triple coincidence between all three groups . The data was brought to the GR-11 conference and all events above a preset threshold were compared to determine any coincidences between two or all three of the groups above the preset thresholds. The Rome group set a threshhold level between .3 and .6 K, the Stanford group at .9K and the LSU group between 1 and 1.3 K. The two figures below display the preliminary coincidence data as shown by W. O. Hamilton of LSU to the GR-11 conference. The upper figure is a histogram of the number of coincidences in twenty second intervals with a time delay of -300 seconds to +300 seconds between any two of the three antennas. The time difference is considered positive if the signal occurred first in the more eastern of the two antennas, i.e., Rome vs. LSU or Rome and LSU with respect to Stanford. The lower figure shows all coincidences between any two antennas that occurred within 12 seconds of each other. There were no triple coincidences observed between the three antennas. From this preliminary analysis of the data it is seen that only one event occurred between any pair of antennas within the same second. This event occurred

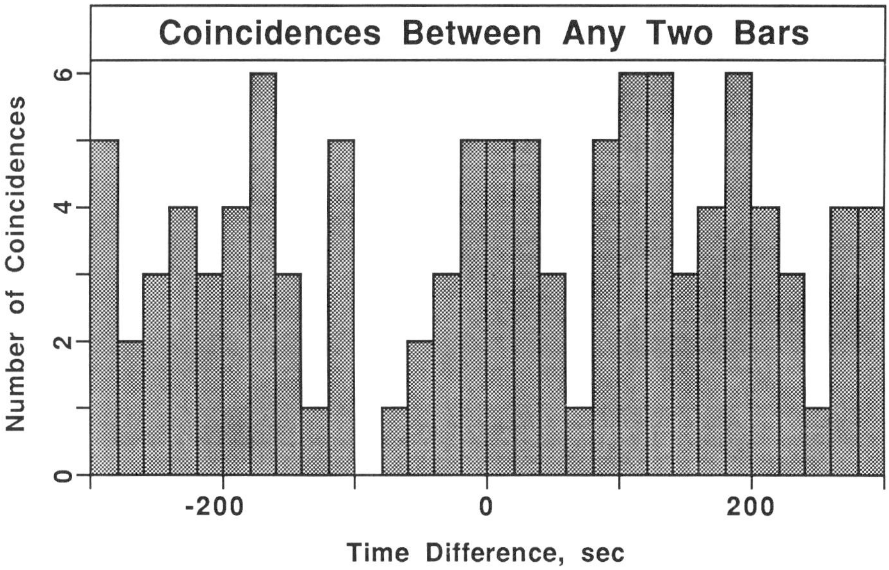

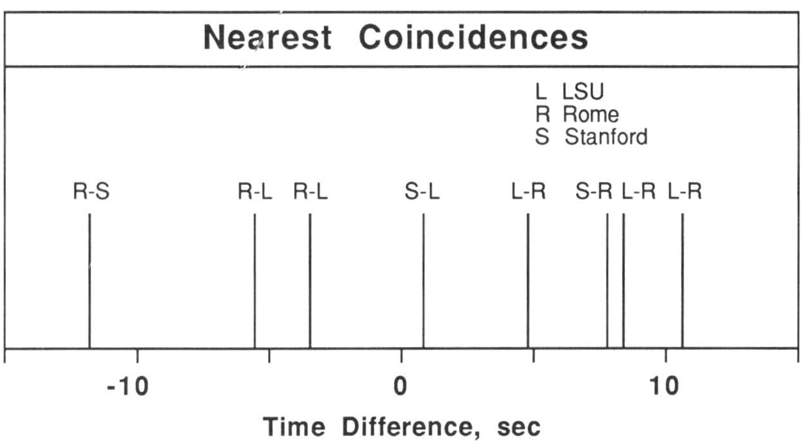

Data presented by the Rome, LSU, and Stanford gravity wave groups from coincidence run of spring and summer 1986 (see text). The histogram gives the number of coincidences versus the time delay between the detectors. The number of coincidences at T=0 is not statistically above the noise level.

between Stanford and LSU and occurred .8 seconds earlier in the LSU antenna than in the Stanford antenna. The next nearest coincidence occurred 3.5 seconds earlier in the LSU antenna than in the Rome antenna. Altogether, as shown in the lower figure, there were 8 events that occurred within 12 seconds of each other in two of the three antennas. As stated above, in no case was a triple event observed. Although the event that occurred within .8 seconds in the LSU and Stanford antennas cannot be ruled out as a gravity wave, it is statistically insignificant when compared to all the data shown in the two figures as a function of time delay between two antennas. The three groups are in the process of analyzing all of the coincidence data taken including a more complete analysis of the data brought to the GR 11 conference. All three systems have been warmed up to remove the difficulties which appeared during the cooldown of the antennas. It is planned to make a long series of coincidence measurements starting early in 1987 with all three systems expected to be at a sensitivity of $h= 10^{-18}$.

It is anticipated that within a relatively short time the University of Western Australia group and the Maryland group will also be on the air with a sensitivity of 10^{-18}, making possible five fold coincidence measurements. A sensitivity of $h=10^{-18}$ could detect the collapse of a star in the center of our galaxy if one percent of a solar mass were changed into gravitational radiation in the order of one millisecond. Since the predictions of the amount of radiation emitted during the collapse of a star vary, depending on the asymmetry of the collapse, and could be as small as 10^{-6} of a solar mass, a sensitivity of $h=10^{-20}$ is necessary to detect essentially any collapse in our galaxy. For the same sensitivity for the entire Virgo cluster the sensitivity would need to be increased to $h=10^{-23}$, although a sensitivity of $h=10^{-21}$ would see the largest predicted collapses in the entire Virgo cluster of galaxies. The objective of all five gravity wave groups with low temperature bar detectors, reporting at this conference, is to improve the transducers and to cool the bars to temperatures well below 1K to improve the sensitivity of the systems to an expected level of $h=10^{-20}$. To accomplish this objective all five groups are working on better transducers and are planning to cool the bar detectors to temperatures well below 1K. With the improvement of the SQUID to the quantum limit they may be able to reach the quantum limit of 3×10^{-21}. It is possible in principle to go beyond this limit with quantum nondemolition techniques, but this development seems to be for the future.

Michelson reported for the Stanford group on the present sensitivity of the Stanford detector and their future plans. He also reported on a promising technique to make a very sensitive 10 - 100 Hz detector capable of seeing coalescing large black holes anywhere in the universe. The detector consists of a low temperature split bar connected by a microwave cavity. Such a detector constructed from two 20-foot aluminum cylinders would have a calculated sensitivity sufficient to see the predicted radiation from the coalescence of two 1000 solar-mass black holes anywhere in the Universe if the detector could be operated at the quantum limit and isolated sufficiently from mechanical disturbances. The major difficulty of low frequency antennas from 1-100 Hz is vibration isolation. Stanford is putting considerable effort on the vibration isolation problem at low frequencies.

The present Stanford detector has been increased in bandwidth to 15 cycles by increasing the mass of the transducer to 350 grams. A refrigerator has been built into the system to allow cooling with separately pumped liquid helium to temperatures below 1K and to serve as a prototype for a He^3-He^4 dilution refrigerator capable of cooling to 10 millikelvin or below. This system at 4.2 K has operated at a higher noise level (50 millikelvin) than the former system, which ran for more than a year at a sensitivity of 10 millikelvin, $h= 10^{-18}$. It is presently being warmed up for improvements with the expectation of a long run starting in February 1987. A second system is being built to incorporate an improved transducer now being developed and a dilution refrigerator to cool the antenna to tempraratures well below 1 K. The transducer makes use of sputtered niobium coils on single crystal silicon. The number one objective is to increase the electrical Q of the transducer from 5 x 10^4 to a value well above 10^6. The number two objective is to increase the coupling between the antenna and the SQUID by decreasing the spacing between the niobium sputtered coils and the transducer mass. This would allow optimally coupled transducers of larger mass and therefore larger bandwidths to be used. Multi-mass transducers are also being studied. It is anticipated that the second system will reach a sensitivity in h of 10^{-20}. When this is accomplished the first system will be warmed up and improved to the level of the second system. Two systems will allow local coincidence measurements in addition to coincidence measurements with antennas in other parts of the world.

In November 1985 the gravitational wave antenna of the Rome group installed at CERN began operation. It consists of an aluminum cylinder 3 meters long, with mass 2270 kilograms cooled to T = 4.2 K. The antenna vibrations are detected by means of a resonant capacitive transducer working with constant electrical charge. The electrical signal from the transducer is amplified by means of a dc SQUID coupled to the transducer by a superconducting transformer. The sensitivity of the antenna to short gravitational wave bursts in terms of an effective noise temperature is of the order of 10 mK. This corresponds to a strain of h = 10^{-18}. For monochromatic waves the antenna is sensitive to h ~ 2 x 10^{-25} with three months of integration time and a bandwidth of about 1/2 Hz. An improvement of a factor of 10 in T_{eff} is envisioned, with small modifications in the mechanics of the transducer. A value of T_{eff} of 1 mK represents about the limit that can be reached with the antenna at T = 4.2 K. A new antenna with T below 0.1 K is in construction.

David Blair, representing the bar group from the University of Western Australia, presented a progress report on their low temperature niobium bar antenna. Their antenna consists of a 1.5 ton niobium Weber-type gravitational radiation antenna at a resonant frequency of 700 Hz cooled to 4 Kelvin. The bar is supported by a low-loss, high-isolation two-mode mechanical filter on which the niobium bar has an acoustic Q of 2 x 10^8 at 5 K. This Q factor is sufficient to reduce antenna thermal noise to less than the quantum limit for sampling times of about 10^{-3} seconds. To exploit this potentially low noise, a highly coupled transducer is required. For this purpose, they are developing a broadband "whip" impedance transformer with an impedance matching ratio between 10^2 and 10^4. However, in its present configuration the antenna has a resonant transformer which consists of a resonant bending flap. The effective mass of the bending flap is about .6 kg, giving a minimum antenna sampling time of about 50 msecs. A magnetically levitated, noncontacting microwave cavity serves as a parametric transducer. The ultra low phase noise pump signal is derived from a sapphire-loaded superconducting cavity. It is expected that the sensitivity of this system will be limited by the current noise in the final stage of the transducer servo loop, which can be minimized by the use of a high-Q notch filter. With an anticipated 20 dB of filtering, they expect to achieve an antenna noise number of 4 x 10^4, corresponding to a noise temperature of 1.6 mK or h of the order of 3 x 10^{-18}. The limiting servo current noise could be eliminated if the reentrant cavity were bonded to the antenna. Experiments in low acoustic loss bonding have shown that the above Q need not be degraded by bonding. Success with this, with the impedance matching and improvement of the microwave circuitry would permit a sensitivity h of the order of 10^{-19}.

Ho Jung Paik from the University of Maryland reported on a new method of obtaining impedance transformation which does not involve resonance or a continuous taper. Wideband detection of gravitational radiation by means of a resonant antenna requires large energy coupling between the antenna and the amplifier. Signal coupling near unity can be achieved by using resonant transducers, but only with some sacrifice of the available bandwidth. A tapered antenna, which could in principle provide the necessary impedance transformation without involving multiple resonances, is difficult to construct in practice. Ho Jung Paik proposed an alternate approach involving a "two-dimensional lever" structure somewhat similar to a scissor jack. He analyzed such a structure for applications to gravitational radiation detection. He showed that, unlike a linear lever, a large displacement gain (greater than 100) can be achieved in a two-dimensional structure without being limited by the inertia of the lever structure. The resonant-mass gravitational radiation detector would therefore be made a completely wideband system by coupling the transducer through this structure. Paik pointed out that the non-resonant displacement amplifier is an exact mechanical analog of an electrical transformer. He discussed practical design requirements for such a lever structure.

Joseph Weber of the University of Maryland discussed the cross section of a gravitational wave bar antenna. He pointed out that an elastic solid antenna is a large number of atoms coupled by chemical forces and satisfying Schroedinger's wave equation. He argued that there are many processes for interaction with radiation. Some have the same cross section as a continuous solid, described by classical physics. There are other kinds of energy exchange, with gravitational radiation, having cross sections orders greater than for the classical continuous structure. He argued, therefore, that the cross section of a bar gravitational wave antenna for gravitational waves is orders of magnitude larger than assumed in the classical calculations used up until now. He argued that this should be taken into account in calculating the performance of bar gravity wave detectors.

Kimio Tsubono, representing the gravitational radiation group at the University of Tokyo, reported on progress with the low frequency detector developed at the University of Tokyo to search for continuous radiation from the Crab pulsar and on a new microwave cavity parametric transducer. The gravitational antenna consists of two masses connected with a torsion bar. The newest version of the antenna, Crab III, has an expected sensitivity of $h=5 \times 10^{-22}$ with 1000 hours of observation. The mass of this antenna including the central fin is 100 kg. A planned antenna, Crab IV, of mass 1300 kg, will have an estimated sensitivity of $h=5 \times 10^{-23}$ after 3 hours of integration time. This antenna will be cooled to cryogenic temperatures. They also reported the development of a microwave parametric transducer for use with cryogenic gravitational wave antennas. This low noise parametric transducer uses a pair of superconducting microwave cavities pumped by a 9.7 GHz superconducting cavity stabilized oscillator. The cavities convert mechanical vibration into microwave sidebands which, after demodulation, reproduce the low frequency signal. For the performance test of the transducer system, the cavities were mounted on a 15-khz prototype disc antenna at 4.2 K. They obtained a transducer noise temperature of 23 mK at the antenna frequency, and hence a quantum noise number of 3.4×10^4.

LASER DETECTORS

Four groups reported on experiments with prototype laser interferometer gravitational wave detectors of 1.5 to 40 meters in length and plans to build large antennas with arms from 1 to 4 kilometers in length. These groups are located at the Max-Planck-Institut fur Quantenoptik in Garching, Germany, Glasgow University in The United Kingdom, and MIT and Cal Tech in the United States. The highest sensitivity was reported by the German group, who reported a noise equivalent strain of $h=1.1 \times 10^{-19}$ $Hz^{-1/2}$ with an interferometer arm length of 30 meters. For a 1 kilohertz pulsed source this is a sensitvity of $h=3 \times 10^{-18}$. All groups reported sensitivities near the shot noise limit. To illustrate the reported activity, success and problems with the prototype detectors at all of the locations, I will review the details presented by the German group. These represent necessary intermediate steps toward a large laser interferometer to be used as a gravitational wave antenna..

The Garching 30 meter prototype was used to gain better knowledge of the most prominent noise mechanisms. The measured displacement noise in the system is approaching the fundamental shot noise limit in an extended frequency range (500 Hz to 6 kHz) , and at an improved light power level (70 mA of photo current, corresponding to aproximately 200 mW). With N = 90 beams in the optical delay line, the noise spectral density in displacement was reduced to 3.5×10^{-18} m $Hz^{-1/2}$, giving a noise equivalent strain of $h = 1.1 \times 10^{-19} Hz^{-1/2}$. The excess noise at frequencies below 500 Hz has largely been diagnosed; corrective measures are being implemented.

To reduce the shot noise, the light circulating in the interferometer must be increased. In addition to better optical components and more powerful lasers, the idea of recycling the light looks promising and is simple and convenient to implement with a delay line system. Recent experiments have indicated the feasibility of this idea.

To achieve the desired light storage time (of the order of 1 ms), either Fabry-Perot, resonators or optical delay lines could be used in the interferometer arms. The delay line scheme, quite successful experimentally, has disadvantages associated with the very large mirrors required (at least 0.6 m in diameter). An alternate delay line scheme, using a number of smaller, individually suspended mirrors, promises a number of advantages and is currently being investigated.

Although the construction and housing of the long vacuum tubes will be a major cost factor, an arm length of as much as 3 km is proposed, in order to reduce the effects of seismic and thermal vibrations and of fluctuations in the residual gas pressure.

To detect gravitational waves of arbitrary polarization, at least two interferometers are required. A redundant antenna of three such interferometers, each with arms subtending 60°, is proposed as an optimum final configuration. Preliminary design studies and a computer assisted search for a site are now being undertaken.

The plans for the large scale antennas at the other three locations are as follows: The MIT and Cal Tech groups plan two identical antennas, one in Maine and one in California. The arms, in the form of an L, will each be 4 km long.The vacuum pipe will be 48 inches in diameter. There will be facilities for full length and half length interferometers in the vacuum pipes. Later there will be capability for multiple or specialized receivers. It is hoped that construction can start in 1988 and that the detectors can begin to operate in 1991-1992. The Glasgow group plans a 1 km baseline in the first phase of the large antenna, later to be expanded to 3 km. The geometry is also L shaped and a site has been chosen. There will be facilities for full length and half length interferometers in the vacuum pipe.

Tourrenc and Vinet discussed the possibility of building a large laser interferometer in France. They also discussed a method of recycling and its application to detectors with two Fabry-Perot cavities as arms. They pointed out that such a system would be continuously and conveniently tunable.

Schutz presented a very interesting paper on the sources of radiation and what these detectors would be able to detect. He suggested that if the destruction of a compact binary system could be observed and one could use a network of detectors to find the direction and polarization of the source, then it would be possible to deduce the absolute distance to the source. This would allow a direct determination of Hubble's constant, possibly to a precision of ten percent in one year. This illustrates the potential power of gravitational wave astronomy.

Another very interesting potential source of gravitational radiation was discussed by Peter Michelson. It has been predicted that a neutron star with a sufficiently small magnetic field will be spun up by accretion from the companion star until it becomes relativistically unstable. Such a star, it is predicted, would emit gravitational radiation and x-rays modulated at the same frequency. A proposal has been made to fly a large area x-ray antenna on the space station to detect the modulation of the x-rays from such a star. This would allow simultaneous detection of the gravitational radiation and the x-rays.

P. Bender discussed gravitational waves at very low frequencies and a proposal for a laser heterodyne gravitational wave antenna in space with 10^6 km arm lengths and 100 mW of transmitted laser power. The expected sensitivity of such a detector for detecting short pulses ranges from 10^{-22} at 10^{-4} Hz to 10^{-21} at 10^{-1} Hz. Sources of interest are: normal main-sequence binaries, compact binaries, cataclysmic variables, neutron star binaries, and close white dwarf binaries.

KALUZA-KLEIN THEORIES

D. Brill
Department of Physics
University of Maryland
College Park, Maryland 20742, USA

INTRODUCTION

Growing interest in theories of the Kaluza-Klein type was already in evidence at the Padova meeting. Since then they have been further explored, not only as possible approaches to unification in particle physics, but also from the point of view of traditional general relativity, with emphasis on their geometrical features.

The diversity of approaches being explored today was reflected in the wide range of topics represented in the 25 contributed abstracts. But if there was long-range disorder we were fortunate in having short-range order: most of the contributions fell into one of five categories, and could therefore be combined in the presentation. This remarkable cooperation between the authors enabled us to hear most of the essential results in the allotted 2 1/2 hours.

Virtually all higher-dimensional theories have in common the mechanism of dimensional reduction, whereby the size of spacetime in the "extra" dimensions is so small as to be unobservable. Quantum effects associated with these internal dimensions were the subject of our first group of contributions. This was followed by papers on cosmology and on black holes, where higher-dimensional Einstein theory can differ qualitatively from the familiar 4-dimensional case. We then heard about alternatives to Einstein theories that present themselves naturally in higher dimensions, and about other classical topics.

QUANTUM EFFECTS

The size R of the internal dimensions in Kaluza-Klein theories is thought to be due to quantum effects such as the Casimir effect. H. P. Leivo reported on work by Kunstatter and Leivo, and by Huggins and Toms, on self-consistent solutions in a $M_4 x S_1$ background, using the effective action due to gravitons to solve self-consistently for R and for the coupling constants.

Two difficulties present themselves: (1) In addition to the vacuum (Casimir) energy of gravitons, an expansion of the effective action in powers of curvature contains terms that look like the four-dimensional Einstein-Hilbert and Maxwell actions. When these are included in the self-consistent scheme, a solution for real R, Λ, and e does not exist (unlike the case when only the Casimir energy is considered). This difficulty can be removed by including the quantum effects of matter fields. (2) To test gauge invariance the calculations were carried out in a one-parameter family of gauges, and the solutions for such physical parameters as R were found to be gauge-dependent. Thus $M_4 x S_1$ appears not to be a suitable background.

T. C. Shen reported on work with B. L. Hu concerning internal spaces characterized by more than one size parameter. This leads to more than one coupling constant for the corresponding four-dimensional gauge fields. Taub space (a deformed 3-sphere) was taken as the internal space because of its symmetry group SU(2)xU(1). The electroweak coupling constant ratio was computed as a function of the amount of deformation, that is the shape of the internal space. To find a value of these parameters compactification was assumed to be due to the vacuum energy of scalar fields (rather than the previous paper's gravitons). The requirement of self-consistency then leads to a model with Minkowski external space and deformed internal space (in addition to the undeformed, internally spherical solution).

COSMOLOGY

Even in four dimensions it remains today a fact that we can write down many more cosmological models than can possibly be relevant to a representation of our actual universe. Yet one believes that a study of models is useful, particularly if generic properties emerge. This applies even more to models in higher dimensions, where we have less experience and many more possible models that might teach us something new. Thus the mathematical tractability tends to become as important as physical motivations in selecting a model. J.E.F. Skea reported on work with J. Stein-Schabes on models of this type. They were selected so that a phase plane analysis is possible, the variables being the Hubble expansion rates of the internal and external spaces respectively. The advantage of this analysis is that large classes of such models can be studied.

E. Verdaguer explored the possibilities of compactifications by classical evolution. In his work with J. Ibañez a model was chosen by demanding exact FRW behavior of the external space, a flat n-torus for the internal space, and Einstein vacuum evolution in the n+4 dimensional total space. They find that only if the FRW behavior corresponds to the "open" model do the internal dimensions at large times become vanishingly small compared to the (ever expanding) external space; however even in this model at least one of the internal dimensions must already be compactified near the cosmological singularity.

M. Henneaux, reporting on work with Demaret, Hanquin and Spindel, and by Barrow and Stein-Schabes, extended the Belinskii-Khalatnikov-Lifshitz analysis of the generic behavior near the cosmological singularity to Kaluza-Klein cosmologies. As in four dimensions, this behavior may be velocity dominated (Kasner-like, non-generic in 4 dimensions) or oscillatory (chaotic mixmaster-like, generic in 4 dimensions). The latter can be understood as a modification of the velocity dominated behavior by a spacelike curvature that acts like a time-dependent potential with which the system point collides. Because this time-dependence differs in different dimensions, the probability of such collisions also depends on dimension.

For homogeneous models they find only a finite number of collisions for spacetime dimension 5 and greater, so that the Kasner behavior becomes generic near the singularity. However, inhomogeneities apparently will cause mixmaster behavior up to dimension 10. In dimension 11 or greater a finite number of collisions leads to such Kasner exponents that space curvature can be neglected in the subsequent time development toward the singularity. (Furthermore, solutions with such Kasner exponents form an open subset of all Kasner solutions in dimension 11 and higher). Thus for these spacetime dimensions the chaotic oscillatory behavior is in general replaced by monotonic Kasner behavior.

BLACK HOLES

Nearly as bewildering as the number of different Kaluza-Klein cosmologies is the number of Kaluza-Klein black holes. However, in the 4+1- dimensional theory there is at least this simplification, that black holes with zero electromagnetic charge and angular momentum must also have vanishing scalar charge. In the paper by L. Sokolowski and B. Carr this result was extended to higher dimensions. A multi-dimensional neutral static black hole is defined to be the product of an internal Riemannian manifold (y) having nonsingular metric $e^{F(x)}h_{ij}(y)$ and an external Lorentzian 4-manifold (x), such that the total spacetime is static, solves the (vacuum) field equations, and has a curvature singularity hidden under an event horizon on which the warp factor F is regular. The field equations imply that the metric h_{ij} is Ricci flat, and in a conformally rescaled "physical" external metric the external field equations become independent of h_{ij}. The general spherically symmetric static solution of these equations can be written down explicitly: they have no curvature singularity on the horizon only if F = 0. Thus the only neutral static black holes are of the trivial type (4-dimensional Schwarzschild) x (n-dimensional Ricci flat).

ALTERNATE THEORIES

Higher-dimensional theories allow action principles, other than the standard Einstein-Hilbert action, that lead to second order field equations, and hence are on an equal footing as candidates for physical theories. Such theories were the subject of two workshop contributions.

F. Müller-Hoissen reported on his work in d dimensions where there are d/2 such independent Lagrangians, of order up to d/2 in the curvature. (In lower dimensions the field equations corresponding to these actions would be identities.) They are interesting for example because they appear in heterotic string theory, and because they allow new (e.g. spherical) classical compactifications over Minkowski space. The Einstein-Hilbert (curvature scalar) term is needed in all such theories in order to yield field equations that can predict the geometry from Cauchy initial values.

N. Deruelle in work with J. Madore showed that the spherical compactification of Minkowski space, $M_4 x S_n$ ($n \geq 4$), can be stable against linear perturbations in these theories if suitable terms up to cubic order in the curvature are included. She then considered this same, stabilizing Lagrangian but supplemented its field equations by a particular fluid stress-energy and applied it to models of the type $(FRW)_4 x S_n$. In the limit of large times t the radius of S_n becomes constant, and the external space has FRW behavior corresponding to a cosmological "constant" that decreases as $1/t^2$. Thus such models evolve toward a ground state appropriate to the present universe.

OTHER CLASSICAL TOPICS

A. H. Taub reported on a positive energy theorem in Kaluza-Klein theories. The energy is given by a d-2 dimensional surface integral on a spacelike surface, a generalized Nester's energy expression. The total space is assumed to be a principal fiber bundle with a semi-simple gauge group. (This assumption excludes the known counterexamples to energy positivity, say in n=5). The proof uses the spinor techniques introduced by Witten and, in the non-abelian case, modified covariant derivatives. In the discussion, similar work by O. Moreschi was mentioned.

The final contribution to the workshop was a paper by P. Gravel and C. Gauthier that proposed multiple connectivity as a solution to some problems associated with compactification. For example, an internal space that is highly curved to make it unobservable usually leads to a large observed cosmological constant. However, one can construct spaces of small curvature that nonetheless have small circumference in all directions. Typical examples are of the form W/Γ, where W is of constant curvature and Γ a suitable discrete subgroup of the isometry group of W. The order of Γ must be quite high. An estimate can be obtained from the observed limit on the cosmological constant and the relation of the internal space's size to coupling constants; one obtains values of the order 10^{63}. A candidate for Γ may be a discrete approximation to SU(3).

REFERENCES

This list contains the most recent publications available to me concerning to the workshop contributions. Abstracts of all contributions, together with further references, are found in the first reference below.

Astracts of contributed papers, 11th International Conference on General Relativity and Gravitation, Vol. 1, pp. 289-314, Stockholm: Akademitryck 1986

Demaret, J. & Hanquin, J.-L., Henneaux, M., Spindel, P., Taormina, A. (1986) Phys. Lett. B.

Deruelle, N. & Madore J. (1986) Phys. Lett. 114A, no. 4, 185-190.

Gravel, P. & Gauthier, C. (1986) Class. Quantum Gravity 3.

Hu, B. L. & Shen, T.C. (1986) Phys. Lett. 178B, 373-8.

Huggins, S. D. & Toms, J. D. (1986) Nucl. Phys. B263, 433-457.

Ibañez, J. & Verdaguer, E. (1986) Phys. Rev. D34, no. 4, 1202-4.

Kunstatter, G. & Leivo, H. P. (1986) Phys. Lett. 166B, no. 3, 321-4; also Nucl. Phys. B.

Moreschi, O. M. & Sparling, G. (1986) J. Math. Phys. 27, 2402-8.

Müller-Hoissen, F. (1985) Phys. Lett. 163B, 106-110; also Class. Quantum Gravity.

Skea, J. (1986) Queen Mary College Preprint.

Sokolowski, L. & Carr, B. (1986) Phys. Lett. 176B, 334-340.

Taub, A. H. (1986) J. Geom. Phys. 3.

THE GR11 SYMPOSIUM ON MATHEMATICAL COSMOLOGY

G. F. R. Ellis
Applied Mathematics, University of Cape Town, Rondebosch 7700
and
Mathematical Sciences, Queen Mary College, London E1 4NS

Twenty-one abstracts were accepted, and two oral sessions and a poster session were held (unfortunately due to pressure of time, not all the accepted papers could be presented orally). These papers showed a considerable progress in understanding exact solutions in cosmology and their relations to various possible matter fields and gravitational laws. Particular papers presented were as follows.

HOMOGENEOUS AND ISOTROPIC UNIVERSE MODELS

Some papers looked at these old friends from a new viewpoint. The initial singularity in the standard FLRW (Friedmann-Lemâitre-Robertson-Walker) homogeneous and isotropic universe models with positive spatial curvature was presented from an elegant geometrical viewpoint by E. Schücking, using the globally rigid imbedding of these models in flat 5-dimensional Minkowski space-time. This displayed the initial singularity and null-cone structure in some striking and beautiful diagrams (the method is described in S. Lauro and E. L. Schücking, *Int. Journ. Theor. Phys.* 24, 367, 1985).

The dynamics of these standard models depends on the theory of gravitation used, and recent experiments by Fischbach *et. al.* have suggested there could be a Yukawa-type component in the gravitational force. *A priori* it seems that this could have serious implications for cosmology. A Newtonian-type homogeneous and isotropic cosmological model incorporating this kind of gravitational theory has been investigated by M. P. Ryan and J. C. D'Olivo. Ryan explained the surprising result that such a Yukawa-type potential is compatible with a homogeneous and isotropic Newtonian cosmology, and leads to the same equations as in the ordinary FLRW cosmology.

HOMOGENEOUS BUT ANISOTROPIC COSMOLOGIES

The dynamics of Bianchi (spatially homogeneous) Models was discussed in an interesting pair of papers by R. T. Jantzen and K. Rosquist. Considerable progress has been made in understanding homogeneous minisuperspace dynamics through the introduction of variables which are adapted to the spatial gauge symmetry, and by compactifying the reduced Hamiltonian system. These methods have led to a more complete understanding of the qualitative behaviour of anisotropic cosmological space-times on the one hand, and to the discovery of new exact power law solutions on the other. These solutions appear as singular points of the Hamiltonian system (see R. T. Jantzen and K. Rosquist, *Classical and Quantum Gravity*, 3, 1986). This work, building on ideas of Bogoyavlensky, represents a substantial step forward towards understanding the dynamics of the family of Bianchi universes.

INHOMOGENEOUS COSMOLOGICAL MODELS

The growth and decay of inhomogeneity in different circumstances is the central issue in inhomogeneous world-models.

The decay of anisotropy and inhomogeneity

A. M. Feinstein presented work with Ch. Charach on the possibility that gravi-solitons could act as efficient mechanisms for the flattening and isotropisation of the universe. Exact solutions presented are interpreted as representing two spatially homogeneous regions with different homogeneity groups, matched across an inhomogeneous region. Kinematical quantities show a clear soliton-like dependence; as this 'gravi-soliton' passes by, the spatial curvature dies away. At late times the solution evolves towards a spatially homogeneous Bianchi I universe model. Thus this shows promise of being another mechanism towards fulfilling the Misner "Chaotic Cosmology" idea.

The growth of inhomogeneities

The opposite viewpoint is expressed by Penrose's proposal (based on entropy arguments) that the Weyl tensor C_{abcd} might act as an arrow of time in cosmology, increasing indefinitely from zero at the initial singularity. Wainwright et al have shown that the related quantity

$$P := (C^{abcd}C_{abcd})/(R^{pq}R_{pq}),$$

which measures tidal forces and the clumpiness of matter, is a more appropriate candidate in the non-vacuum case. The quantity P is zero at the Big Bang if the latter is like an FLRW singularity (an 'isotropic singularity', according to Wainwright).

W. B. Bonnor presented a paper showing that in the context of the Szekeres inhomogeneous cosmological models, if the quantity P is initially zero then the initial singularity must be universally simultaneous (i.e. is 'free of white holes'). It can then be shown that in this case P is non-decreasing along the world-line of matter and so does indeed in this case provide an arrow of time of the kind proposed by Penrose. It measures the growth of anisotropy in these exact world-models, from an initially smooth state.

A paper by S. W. Goode also considered universes that admit initial 'isotropic singularities'. He showed that within the framework of General Relativity theory with a perfect-fluid matter source, the existence of an isotropic singularity with a smooth conformal structure implies that the fluid flow must have zero vorticity at all times (a condition automatically fulfilled in the Bonnor example). This result suggests that the Penrose condition will not be fulfilled in realistic universe models (which might be expected to have non-zero vorticity).

Self-similar inhomogeneities

B. J. Carr presented work with A. Yahil looking at spherically-symmetric perfect fluid similarity solutions that are regular at the origin and asymptotically FLRW at spatial infinity, and so can be taken to model inhomogeneities in the early universe. These exact non-linear solutions have a homothetic Killing vector, and represent density fluctuations whose scale grows at the same rate as the universe. Solutions with small overdensities are permitted (too large a density contrast will result in black hole formation), as well as solutions representing regions of underdensity. This is of interest in view of the recent detection of cosmological-scale 'voids'.

Approximate homogeneity and smoothing

A. Krasinski discussed inhomogeneous solutions of Einstein's equations which contain the FLRW models as a limiting case. The family examined are space-times with conformally flat space-sections orthogonal to the fluid flow, which is taken to be shear-free. They include plane symetric, axially symmetric, and line-homogeneous as well as spherically symmetric universes, some with interesting inhomogeneity properties (A. Krasinski, Proceedings of the 4th Marcel Grossman Meeting, Academic Press). In particular one family can be regarded as containing universes which are homogeneous on a large scale but inhomogeneous in the small.

A major problem is the issue of what the effective field equations will be at a cosmological scale in a universe which is 'lumpy' on a small scale. M. Carfora presented elegant results by himself and A. Marzuoli on the problem of smoothing out space-time geometry in a cosmological context. They have extended work by Hamilton to define a deformation map from any universe with closed space-sections of positive Ricci curvature into a FLRW universe with the same average (volume) behaviour. This finite deformation map has some remarkable properties, such as preserving all symmetries. They are then able to show that Einstein's equations hold at the smoothed out scale, but with extra contributions to the effective energy density and pressure due to the averaged effect of small-scale inhomogeneities (M. Carfora and A. Marzuoli, *Phys. Rev. Lett.* 53, 2445, 1984). The physical implications of this interesting paper have still to be explored.

OBSERVATIONAL RELATIONS

Two papers dealt with the relation between cosmological models and observations.

Recent theoretical work has emphasized our uncertainty as to the gravitational field equations applicable at a cosmological scale. J. Ehlers and W. Rindler have returned to the issue of what aspects of a FLRW geometry can be determined direct from possible astronomical observations when the gravitational field equations are unknown.

It is known that in this case, even ideal measurements of both source luminosity and number counts as a function of redshift do not determine the curvature of a Robertson-Walker metric. Rindler pointed out that these relations do determine the metric if L(z) and n(z), the intrinsic luminosity of a galaxy and number density of galaxies in the past, are provided by an independent theory. In that case an elegant geometric formula gives the present spatial curvature K from the limiting relation between the volume and area of a geodesic sphere around the observer; and these quantities can be determined from number count and luminosity measurements.

The observational verification of whether or not the universe is spatially homogeneous is difficult. W. Bonnor and G. Ellis have proposed a new approach in which a universe is said to be 'observationally homogeneous' if galaxies in distant regions look very similar to galaxies nearby. The implication is that similar physical conditions must have occurred there, in order to result in the occurrence of similar objects; more precisely, thermal histories must have been the same. The question then arises: does this in turn imply that the universe itself is spatially homogeneous ? (It is plausible that this should be so, because of the restrictions implied on initial data on a spacelike hypersurface).

Ellis reported a simple counter-example, showing that exactly uniform thermal histories do not always imply that space-time is spatially homogeneous (W. Bonnor and G. Ellis, *Mon. Not. Roy. Ast. Soc.* 218, 605, 1986). However the geometry involved is rather special (it is spherically symmetric); it still seems plausible that uniform thermal histories almost always do imply spatial homogeneity.

FURTHER WORK

Other contributions of interest considered the dynamics of FLRW universes (with conformally invariant scalar fields, or Einstein-Liouville solutions); spin dynamics: the non-minimal coupling of a vector field and gravity: magnetic fields, particle flows, and homoclinic phenomena in Bianchi universes; and the distribution of the spatial orientation of galaxy axes in a universe model. The subject is thus alive and well at present, with interesting work being done on a variety of frontiers.

MATHEMATICAL PROBLEMS

Y. Choquet-Bruhat
Institut de Mécanique, Université Paris VI, France

About 60 papers were attributed to this section. Its general heading allowed for many different subjects. Due to the lack of time only 8 abstracts have been presented orally. I was guided in the choice by the desire that a variety of topics be touched upon, but that all be directly related to General Relativity, with specific results. Each speaker had 10 minutes.

This session as well as the general lecture by S.T. Yau about his and Christodoulou's work proved that remarkable progress has been made recently on old problems. The theorems which have been obtained use the full complexity of Einstein equations. It will certainly also be the case for the new results about the global mathematical problems in General Relativity which can be hoped for in the near future.

A number of abstracts were concerned with the applications of the 3+1 Regge calculus. Ruth Williams gave a very clear and substantial account not only of her own work, but also of results on the same topic included in the abstracts of Dubal, Friedman and Jack or Barrett.

A few papers were concerned with General Relativity or Yang Mills fields in euclidean signature, in particular K.B. Marathe's "Geometry of generalized gravitational instantons" and F. Flaherty's "Yang Mills equations on the multiplicative 4-torus".

Sorkin gave a flavor of what purely topological methods can offer to General Relativity, in arbitrary dimensions, by examining the conditions under which a cobordism mediating between two closed space like hypersurfaces, with different topologies, admits a lorentzian metric. The results suggest that Lorentz cobordism is not an appropriate mechanism for changing topology (for instance creating a pair of monopoles) except if the space time has an even dimensionality.

Bombelli presented work done in collaboration with Ashtekar

and Koul on a Hamiltonian formulation of General Relativity without a 3+1 splitting : these authors define a symplectic structure on the space of all asymptotically flat (in a suitable sense) solutions of Einstein equations. They obtain the asymptotic generators of the spi group and the B.M.S. group, and the corresponding conserved quantities in a unified way.

Several interesting abstracts, extending previous results of Choquet-Bruhat - Deser, Fisher - Marsden, Brill, York, concerned linearisation stability : Einstein equations coupled with a perfect fluid (Bao and Marsden), local results, i.e. on a subset with boundary (Brill, Reula and Schmidt) - Brill and Vishveshwara sketched a general study of linearization instabilities occurring in a joint set of equations (for instance Einstein equations supplemented by symmetry conditions) which are both separately linearization stable.

The previous results rely essentially on the linearized constraints. The theorem presented orally by B.S. Kay under the title "linearization stability of spherically symmetric black holes" has a different meaning, since it concerned the long time behaviour of the solution of the Klein-Gordon equation in the exterior Schwarzchild metric; he presented improvements on previous rigorous results (in particular by Wald) on the linearization stability of spherically symmetric black holes.

General properties of the curvature of lorentzian metrics were included, in particular, in the abstracts of G.S. Hall (problem of determination of a metric by its sectional curvature); C.D. Collinson (null tetrad approach to the infinitesimal holonomy group), while groups of invariance were the central object in the abstracts of Swift, d'Inverno and Vickers (invariant spaces of metrics), Gotay, Isenberg, Marsden, Montgomery, Sniatycki and Yasskin (constraints and momentum mappings in Relativistic field theories) and, with a more applied point of view, Maartens (symmetries in Relativistic kinetic theory).

Mc Lenaghan gave a short oral presentation of his last result, obtained in collaboration with Carminati, on Huyghens' principle: they have shown that there exist no Petrov type D space-times on which the conformally invariant scalar wave equation, or Maxwell's equation or Weyl's neutrino equation satisfies Huygens' principle. For Petrov type III space-times, it is shown that the validity of Huygens' principle for any of these equations implies that the space-time is conformally related to one in which the repeated principal null vector field of the Weyl tensor is recurrent.

The paper by Isenberg, Eardley, Marsden and Moncrief, presented orally by Isenberg, was a fine study of conformal Killing fields for solutions of Einstein equations. The result is that such fields are in general full Killing fields, generators of isometries. More explicitly the authors prove the following theorem (a proper conformal Killing field ξ is such that $L_\xi g = \phi g$, with ϕ a non constant function; if ϕ is a non zero constant ξ is called "homothetic" Killing field) :

Theorem :

1) The only vacuum solutions of Einstein's equations which contain proper conformal Killing fields are those which are flat, or certain plane wave space-times (this result was already given in Defrise-Carter's thesis).

2) If a globally hyperbolic space-time satisfies the Einstein or Einstein-Yang-Mills equations (with or without matter which obeys certain energy conditions), admits a compact constant mean curvature space-like hypersurface, and contains a homothetic Killing field, then either the space-time is flat or that homothetic Killing field is actually a Killing field.

3) If a globally hyperbolic space-time solution of the vacuum Einstein equation admits a conformal Killing field (and also a compact constant mean curvature hypersurface) then either the space-time is flat or ξ is a Killing field.

4) Minkowski space-time is the only asymptotically flat solution of the Einstein equations (with sources satisfying the dominant energy condition) that has a conformal Killing field which is asymptotic to the dilatation Killing field .

The proof of the different sections of this theorem requires various tools. The first result is local : the proof relies on the integrability conditions of Ricc(g) = 0 and $L_\xi g = \phi g$.

The second is much more difficult to prove : the authors use a 3+1 decomposition of Einstein equations, together with the homothetic Killing field equations and elaborate arguments of analysis. Section 3 is essentially a corollary of 1 and 2.

To prove 4 the authors use the positive mass theorem, together with order by order examination of the fall off rates of the components of the metric and the conformal Killing field.

Moncrief presented his program and results for the reduction of the vacuum Einstein (or Einstein Maxwell) equations with a space like

Killing field to a hyperbolic system for a pair of scalar fields, where the coefficients are functionals of these fields. He sketched the realization of this reduction program for the case of Lorentz metrics defined on S^1-bundles over the base manifold $S^2 \times \mathbb{R}$. The metrics were each assumed to have a U(1)-isometry group generated by a spacelike Killing field tangent to the fibers of the bundle. By imposing suitable coordinate conditions (existence of a section with constant mean curvature) and solving the elliptic system of constraint equations he showed how the three metric is determined as a functional of the two scalar fields. The reduced dynamical system deduced from a reduced hamiltonian takes the form of a pair of coupled hyperbolic equations defined on $S^2 \times \mathbb{R}$ harmonic map with target the hyperbolic 2-space. It is essentially independent of the particular S^1-bundle from which the system descended. Only the discrete value of a certain conserved quantity (which must coincide with the Chern class of the appropriate bundle) serves to distinguish solutions corresponding to one bundle from those of another. Moncrief then used the $SL(2,\mathbb{R})$ invariance of the one-Killing-field Einstein equations to transform solutions corresponding to one bundle to solutions corresponding to an inequivalent one. This phenomenon is most interesting for the case of transferring solutions from the trivial bundle to a non-trivial one or vice versa. Only certain (infinite dimensional) proper subsets can be so transferred, often with striking changes in their qualitative properties. For example the so-called generalized Taub-NUT (GTN) solutions which he defined on $S^3 \times \mathbb{R}$ can all be transformed to solutions on the trivial bundle in such a way that the non-singular Cauchy horizons of the GTN solutions are mapped to curvature singular boundaries of the image space-times.

One aim of the reduction, which is still an open problem, is to prove that every local solution extends uniquely to a global one covering the interval $t = \tau \in (-\infty, \infty)$, exhausting the maximal Cauchy development.

H. Friedrich gave a short oral summary of the remarkable results he has obtained in recent years concerning the global existence and asymptotic structure of solutions of Eisntein's equations, with or without cosmological constant. All these results are obtained through the use of what Friedrich calls "Regular conformal equations" : these equations appear as a first order system for a lorentz frame e_k (which determines the non physical metric g) and its riemannian connection $\Gamma_{i\ j}^{\ k}$ on the one hand, a scalar Ω (which will, when positive, determine the physical metric $\tilde{g} = \Omega^{-2} g$), its gradient Σ_i, its Laplace operator $s = \frac{1}{4}\nabla^\mu \nabla_\mu \Omega$. Other auxiliary unknowns are the trace free 2-tensor $s_{ih} = \frac{1}{2}(R_{ij} - \frac{1}{4}R\eta_{ij})$ and the rescaled Weyl tensor $d_{ijk} = \Omega^{-1} c_{ijk}$. The system results, when $\Omega > 0$, from the usual Einstein equations

$$\mathrm{Ricc}(\Omega^{-2} g) = \Lambda\, \Omega^{-2} g$$

it is regular, even for $\Omega \leqq 0$, and invariant under rescaling $(g,\Omega) \to (\theta^2 g, \theta\Omega)$. The R.C.E enable Friedrich to transform global problems on Minkowski or De Sitter space time into global problems on the Einstein cylinder (procedure used, with no need of such involved reasoning, for Yang-Mills equations by Y. Choquet-Bruhat and D. Christodoulou). The main results are as follows :

(i) If h is a Riemannian metric on an orientable compact 3-dimensional manifold S, d a symmetric trace free tensor field of valence two with vanishing divergence on (S,h), and Λ a positive "cosmological constant", then the (sufficiently smooth) triple (S,h,d) and Λ determine a unique past asymptotically simple solution of Einstein's equations $\mathrm{Ric}(g) = \Lambda g$, for which (S,h,d) may be identified with the "**asymptotic initial data** set" implied on past conformal infinity. All past asymptotically simple solutions of $\mathrm{Ric}(g) = \Lambda g$ with compact Cauchy surface are obtained in this way.

(ii) For asymptotic initial data (S,h,d) with $S = S^3$, which are sufficiently "close" to asymptotic De Sitter data, the resulting solution of $\mathrm{Ric}(g) = \ g$ is asymptotically simple in the future as well as in the past. These solutions are "global" in the sense that they are null-geodesically complete.

(iii) In the "hyperboloidal initial value problem" for Einstein's equations $\mathrm{Ric}(g) = 0$ data are prescribed on a surface which is thought to intersect future null infinity in a space-like 2-sphere. "Hyperboloidal initial data" sufficiently close to Minkowskian hyperboloidal initial data evolves into a solution of $\mathrm{Ric}(g) = 0$ which is null geodesically future complete and in fact asymptotically simple in the future. The conformal extension can be chosen such that future null infinity is obtained as the past null cone of a point i^+ which represents future time-like infinity of the solution.

The full generality of the solutions obtained by Friedrich is still under discussion.

NUMERICAL RELATIVITY

Takashi NAKAMURA
Department of Physics, Kyoto University, Kyoto, Japan

1. TEN YEARS OLD NUMERICAL RELATIVITY

Ten years have passed since the birth of Numerical Relativity in 1976. Here the birth means the work by Smarr (1977) on two black hole collision as it gave the first numerically generated non-spherical space times with asymptotic flatness. Since then there have been many works on the general relativistic collapse problem. The status of the field before 1983 was reviewed by Smarr (1984) in GR10. The present status including characteristic initial value problems and Regge Calculus was also reviewed by Piran (1986) in GR11. Here I mainly concentrate on the summary of the symposium and discuss the past and the future development in the (3+1) approach which seems to be the most advanced.

At the stage of GR10 there were two 2D (axially symmetric) Einstein/hydro codes by a US group (Evans 1984; Evans, Smarr and Wilson 1984) and the Kyoto group (Nakamura 1981; Nakamura and Sato 1981 a,b, 1982). In GR11 we can add the third 2D code by UK-Israel Group (Stark and Piran 1985). By using the latter two codes, the general relativistic collapse of rotating stars has been simulated for various cases. Relevant questions are;

1) What is the final structure of space time after the collapse of rotating stars?
2) What kind of information can we extract from the gravitational radiation emitted during the collapse?
   By observing gravitational waves directly what kind of physics in strong gravity can we know?

Concerning the first question, numerical relativity seems to have an answer for 2D cases. Let us define non-dimensional angular momentum of the system q by $q=cJ/M^2G$ where J and M are the total angular momentum of the system and the gravitational mass, respectively. q corresponds to a/m of the Kerr black hole. Then the critical values of $q(=q_c)$ for the formation of black holes (formation of an apparent horizon) are;

Kyoto Group (Nakamura 1981; Nakamura and Sato 1981 a,b)

	q_c	remarks
Model A	1.05	supermassive stars(almost rigid rotation)
Model B	0.92	supermassive stars(differential rotation)
Model C	0.86	stars of mass $10M_0$(effect of strong interaction is considered)

Stark and Piran (1985) suggest q_c=1.2±0.2 or 0.8±0.05 for γ=2 polytrope depending on the extraction of the internal energy at the initial time.

The above results suggest that q_c seems to be almost independent of equations of state, rotation laws, initial density distribution, codes, coordinate conditions and numerical methods. If we consider numerical relativity as an experiment, an experimental value of q_c up to now is 1.0±0.1, which suggests the cosmic censorship hypothesis is relevant under plausible initial conditions like those adopted in these numerical simulations.

One of the largest advances in Stark and Piran's code is the ability to estimate the gravitational radiation emitted by a collapsing rotating star. This is due to their gauge condition (Bardeen and Piran 1983) in which the metric outside the star tends quickly to the Schwarzschild metric. They found that the energy radiated for $q \leqq q_c$ is proportional to q^4 and the wave pattern has a similar shape irrespective of q. The waves show the dominance of the quasi-normal mode of a black hole. The efficiency of the gravitational radiation is found to be small (at most $10^{-3}mc^2$) partly due to the phase cancellation effects for 2D cases (Nakamura 1985). (See also Piran's review in this volume for details.)

If q is greater than unity, final results depend on the initial conditions (Nakamura 1981; Nakamura and Sato 1981 a,b; Nakamura 1983; Stark and Piran 1985). For example in general relativistic collapse of accreting neutron stars with rotation (Nakamura 1983), depending on initial rotation laws, oblate shape or disk-like or ring-like rapidly rotating neutron stars (not black holes) are formed in 2D calculations for q>1. The shapes of these stars suggest they may be unstable to fragmentation. As early type stars which are responsible for the formation of neutron stars and black holes are rotating fast, the value of q of collapsing cores is expected to be greater than unity (cf. q for the sun is 0.18). Then the collapse, pursuit and plunge scenario will be relevant (MTW 1971), which is essentially three dimensional (3D). As Nakamura

(1985) pointed out by extrapolating the results of the perturbative calculations, an efficiency of gravitational radiation up to 9% is expected in such a fully 3D process while only 0.1% is expected for 2D cases.

3D processes should be necessary for a strong emitter of gravitational radiation. In fact as shown by Miyama, Nagasawa and Nakamura (1986), in 3D processes such as fragmentation, the efficiency of emission of gravitational radiation increases by at least a factor 10 even though they used the results of Newtonian collapse calculations and the Landau-Lifshitz formula to estimate the energy. It is now desirable to construct a fully general relativistic 3D code to answer the questions shown in the beginning of this section. Fortunately the recent speed of super computers is fast enough to construct such a code. As a first step to 3D code in numerical relativity, two efforts have been dedicated recently. Marck and Piran 1986 are now constructing a 3D relativistic hydrodynamic code around a Kerr black hole. They are now expanding every quantity by Discrete Fourier Transform in ϕ-direction and Chebychev polynomials in r and θ directions. On the other hand Nakamura 1986 has been constructing 3D codes for pure gravitational waves by which the propagation of a localized wave packet can be simulated. In his code every quantity is expanded by using spherical harmonics, and a part of the numerical results was shown in the symposium.

Another important advance in numerical relativity has been made by Shapiro and Teukolsky (1985 a,b,c). They performed fully general relativistic calculations of the gravitational collapse of collisionless systems. In their calculations the metric is assumed to be spherically symmetric but the orbit of each particle is non-spherical. They found that if the central gravitational redshift of the star cluster is greater than 0.4, a black hole is finally formed. This may explain the origin of the central black holes in Quasars and AGNs. In the symposium, evolutions of dense stellar systems including the formation of black holes were reviewed by Kandrup.

2. SUMMARY OF SYMPOSIUM

In the (3+1)-formalism of the Einstein equations, basic equations are divided into constraint equations and evolution equations of the metric tensor and the extrinsic curvature. In numerical relativity, we first determine initial data by solving constraint equations for a given distribution of matter. To evolve initial data, equations for matter are usually solved. So only a part of the constraint equations and the evolution equations of the metric tensor and the extrinsic curvature is needed and the other part is left. For example, it is enough to solve only evolution equations, which is called a free evolution. In this scheme there has been a question concerning a consistency of the accuracy of numerical schemes (Piran 1983). The constraint equations include the first and the second derivatives of the metric and the extrinsic curvature. Therefore even if the deviations of the numerical solutions of the evolution equations from the exact ones are $O(h^2)$, where h is a mesh size, the deviations in the constraint equations seem to be O(1) or O(h) because the finite difference versions of the first derivative and the second derivative are $O(1/h)$ and $O(1/h^2)$, respectively.

Choptuik gave arguments against the above claim in the symposium. He first referred to Richardson's statement that " The errors of the integral (solution of the difference equations) and of any differential equations derived from it due to central differences instead of differential coefficients are of the form,

$$\text{Error}=h^2 f_2(x,t)+h^4 f_4(x,t)+\ldots\ldots\ldots\ldots \tag{1}$$

where f_2, f_4,... are <u>independent</u> of the mesh spacing". If this statement is accepted for our system, it is easy to show the errors of the constraint equations should be $O(h^2)$. As f_2, f_4,... are essentially analytic functions, the derivatives of the errors should be as

$$\text{Derivative of Error}=h^2 \partial/\partial x f_2(x,t)+h^4 \partial/\partial x f_4(x,t)+\ldots\ldots\ldots\ldots$$

which clearly shows errors are $O(h^2)$.

However there is no rigorous proof of the above statement for the Einstein equations. So Choptuik considered spherically symmetric space time with scalar fields. For this system, he used a free evolution and checked the errors of the constraint equations by using numerical results with various mesh size h. As a result he showed errors are proportional to h^2. This resolves a consistency problem at least for one example.

In numerical relativity due to the finiteness of memories and the speed of the computers, we solve the Einstein equations only in a limited region of space. Thus for the exterior region where we have no grid points, how to impose proper boundary conditions is a major problem in numerical schemes. In the symposium Anderson and Hobill discussed this important problem. They considered a model problem consisting of a spherically symmetric scalar field $\psi(r,t)$ in a Schwarzschild-like background coupled to a nonrelativistic harmonic oscillator having an amplitude $Y(t)$. The system in many ways parallels an isolated general relativistic system in the quadrupole approximation with damping due to the radiative loss of energy. The model is defined by equations as;

$$((1+k(r))^2\partial_t^2-\nabla^2)\psi(r,t)=\lambda\rho(r)Y(t), \qquad (2)$$

$$\partial_t^2Y(t)+\omega^2Y(t)=4\pi\lambda\ \rho(r)\psi(r,t)r^2dr, \qquad (3)$$

$$k(r)=\begin{cases} kr & r\leqq 1 \\ k/r & r\geqq 1, \end{cases} \qquad (4)$$

$$\rho(r)=\begin{cases} 1 & r\leqq 1 \\ 0 & r\geqq 1. \end{cases} \qquad (5)$$

In the outer zone the field equation is written as,

$$((1+k/r)^2\partial_t^2-\nabla^2)\psi(r,t)=0. \qquad (6)$$

From the characteristic equation the retarded null coordinate becomes

$$u=t-r-k\ln(r). \qquad (7)$$

Introducing $\Phi(r,t)\equiv r\psi(r,t)$, we have a solution Φ in successive powers of k as,

$$\Phi=\Sigma\ k^n\Phi_n, \qquad (8)$$

where

$$\Phi_0(u,r)=\Phi_0(u),\ \Phi_1(u,r)=-\int_{-\infty}^{u}\Phi_0{}'(s)ds/(u-s+2r),\ldots \qquad (9)$$

Anderson and Hobill solved Eqs.2) and 3) using the analytic solutions (Eqs. 8 and 9) as an outer boundary condition. They found that in general matching at a distance of at least one wave length gave an accuracy of less than 1% provided that the background curvature was sufficiently small. They also found the accuracy depends on the order of approximation in Eq.8) and a matching radius. For example in the second order approximation, a fifth matching radius is enough to have 2% accuracy compared with that in the first order approximation. This will open the possibility of saving memories and computing time in numerical relativity.

Nakamura presented some numerical results of 3D time evolution of pure gravitational waves. Adopting the (3+1)-formalism of the Einstein equations with α(lapse)=1 and β^i(shift)=0, he first writes down the basic equations in (x,y,z) coordinates. Then γ_{xx}, γ_{yy},etc are the basic variables. Let us assume every quantity Q in (x,y,z) has an Taylor expansion as,

$$Q = \sum_{p,q,r}^{\infty} a_{pqr}(t)x^p y^q z^r/p!q!r!. \tag{10}$$

Then it is very easy to show that Eq.(10) can be reexpressed as,

$$Q = \sum_{l,m} r^l Q_{lm}(r^2,t)Y_{lm}(\theta,\phi). \tag{11}$$

We need $(\partial/\partial z)Q$,$(\partial^2/\partial^2 z)Q$ etc to calculate the Ricci tensor R_{ij}. As $(\partial/\partial z)Q$,$(\partial^2/\partial^2 z)Q$ etc should be expressed like Eq.(10), they have also the form in Eq.(11), that is, for example ,

$$(\partial/\partial z)Q = \sum_{l,m} r^l (\partial/\partial z)Q_{lm}(r^2,t)Y_{lm}(\theta,\phi). \tag{12}$$

It is easy to determine the relation between $(\partial/\partial z)Q_{lm}(r^2)$ and $Q_{lm}(r^2)$ etc.

For the time evolutions, K_{ij}(extrinsic curvature) and γ_{ij}(metric) are expanded as Eq.(12). Then $(\partial/\partial t)\gamma_{ij}$ has a simple form for (l m) component while $(\partial/\partial t)K_{ij}$ is solved as

$$(\partial/\partial t)K_{ij}{}^{lm}(r^2,t) = \int ({}^{(3)}R_{ij} + K\, K_{ij} - 2K_i{}^n K_{nj})\, Y_{lm}d\Omega/r^l, \tag{13}$$

In Eq.(13) 41 points Gauss quadrature in θ-direction and 16 points Discrete Fourier Transform for ϕ-direction are used. The number of grid points in the r-direction is 100 with $r_{max}=7r_0$ where r_0 is an initial size of a localized wave packet. For each r, m ranges from -8 to 8 and l from $|m|$ to $|m|+10$. Computing time for a typical evolution takes three hours for 700 time steps by using FACOM VP200 which is a Japanese super computer with 400MFLPS.

For the weak amplitude case (amplitude of the wave is 1%) the numerical solution should essentially agree with the linearized one, which is guaranteed within an error of a few percent. ADM energy flux at $r=5r_0$ is estimated from numerical data. It is found that the relative error(or difference) between the computed one and that of the linearized solution is at most a few percent up to the time when the wave leaves the numerical boundary. As the nonlinear terms are taken into account in the numerical calculations, a few percent difference (error) is expected in this case. This suggests to us a possibility of estimating the energy of the gravitational waves from the numerical results with an error of a few percent even in 3D numerical relativity.

Slagter discussed an approximate wave solution of the Einstein equations coupled to a Higgs field by using the third order multi-scale method. In the method the metric and Higgs field are expanded in powers of a small parameter ε,

$$g_{\mu\nu}(x^\sigma,\xi) = \underline{g}_{\mu\nu}(x^\sigma)+\varepsilon h_{\mu\nu}(x^\sigma,\xi)+\varepsilon^2 k_{\mu\nu}(x^\sigma,\xi)+......$$

and

$$\phi(x^\sigma,\xi) = \underline{\phi}(x^\sigma)+\varepsilon\psi(x^\sigma,\xi)+\varepsilon^2\chi(x^\sigma,\xi)+......$$

with

$$\xi\equiv\varepsilon^{-1}\Pi(x^\sigma)$$

where Π is a scalar function(phase).

Slagter especially considered some numerical solutions on an Einstein-Rosen background metric and found that depending on initial conditions solitary wave like solutions or the formation of a trapped surface is seen. His calculation is related to the fluctuation threshold for primordial black hole formation. However more numerical work is necessary in models inhomogeneous in all directions, which is one of the author's conclusions.

Two papers were presented in the poster session. Lanza considered a method for determining self-gravitating toroidal figures of equilibrium around a rotating black hole by using a multigrid method. Dubal applied a

(3+1) formalism for inhomogeneous Regge Calculus to static polytropic spheres and spherically symmetric dynamical collapse.

3. NUMERICAL RELATIVITY AFTER TEN YEARS

In the symposium on Algebraic Computing in GR11 the power of algebraic computing software such as SHEEP, MACSYMA, MAPLE and REDUCE was demonstrated (Frick 1986). In this symposium on Numerical Relativity these packages were shown to be useful to construct the numerical codes in numerical relativity. Hobill and Anderson (1986) used MACSYMA for their analysis of matching of the numerical solutions with analytic solutions. Nakamura (1986) used REDUCE to construct his 3D code for time evolution of pure gravitational waves. After looking at their long expressions of the source list of FORTRAN, it seems that the use of the algebraic computing software is indispensable in constructing codes as well as checking and analyzing numerical results. So one of the trends shown in the symposium was the use of the full power of the computer not only for numbers but also for symbols. In this sense the present status of the numerical relativity can be called CAR(Computer Aided Relativity).

At the same time, we can say the stage of this symposium is unique in relation to the observation of gravitational radiation from the collapse. As shown in the symposium on Gravitational Wave Detectors, the development in the field is remarkable. By the Bar type detectors a burst source in our Galaxy can be detected. As for the laser interferometer detectors, plans to make kilometer size interferometers were shown. If everything proceeds along their plans, it is certain burst sources in Virgo cluster of galaxies can be detected in early 1990s.(Fairbank 1986) It is certain that 2D numerical relativity has grown up and will give definite information to experimenters, while 3D numerical relativity may be in a baby phase at present. However we believe it will grow up by the time when the gravitational wave will be detected and play an important role in gravitational astronomy.

In conclusion in GR12 in Brazil, we can expect much more developments in numerical relativity or CAR.

REFERENCES

Papers presented to the GR11 Conference and published in:
Abstracts of Contributed Papers
11TH INTERNATIONAL CONFERENCE ON
GENERAL RELATIVITY AND GRAVITATION

Bakhankov, A.A. "The numerical solution method for the Einstein equations", p. 401.

Choptuik, M. "Consistency in numerical relativity", p. 402.

Anderson, J.L. and Hobill, D.W. "Matched analytic-numerical methods for isolated systems", p. 403.

Nakamura, T. "Three dimensional time evolution of pure gravitational waves", p. 404.

Piran, T. "Numerical relativity: from cosmology to gravitational collapse", P. 405.

Slagter, R.J. "On approximate and asymptotic wave solutions in gravity theory coupled to a Higgs field", P. 406.

Kandrup, H.E. "Relativistic stellar dynamics", p. 511.

Dubal, M.R. "3+1 Regge calculus:formalism and application to spherical collapse", p.354.

Lanza, A. "Self-gravitating toroidal figures of equilibrium around black holes", 11:07.

OTHER REFERENCES

Bardeen, J.M.(1983) and Piran, T. "General relativistic axisymmetric rotating systems; coordinates and equations". Physics Reports, 96, 205.

Evans, C.(1984) "A method for numerical simulation of gravitational collapse and gravitational radiation generation", in J.Centrella, J.Leblanc and R.Bowers (eds) "Numerical Astrophysics", Jones and Bartlett Publishers,Inc. Boston, pp. 216- 256.

Fairbank, W. "Gravitational wave detectors", this volume.

Frick, I. "Algebraic computing", this volume.

Marck, J.A.(1986) and Piran, T. "3D relativistic hydrodynamics around a Kerr black hole" ,in H.Sato and T. Nakamura(eds) "Proceedings of the

14-th Yamada Conference on gravitational collapse and relativity" World Scientific Publishing Co, Singapore.
Misner, C.W., Thorne, K.S. and Wheeler, J.A.(1973) "Gravitation", Freeman, San Francisco
Miyama, S.M.(1986) "Gravitational radiation in the 3D processes", Nagasawa, M. and Nakamura, T. in H.Sato and T. Nakamura(eds) "Proceedings of the 14-th Yamada Conference on gravitational collapse and relativity" World Scientific Publishing Co, Singapore.
Nakamura, T.(1981),"General relativistic collapse of axially symmetric stars leading to the formation of neutron stars and black holes" Prog.Theor.Phys. 65, 1876.
Nakamura, T.(1981a),"General relativistic collapse of rotating and Sato, H. supermassive stars", Prog.Theor.Phys. 66, 2038.
(1981b),"General relativistic collapse of rotating stars", Phys.Lett. 86A, 318.
(1982),"General relativistic collapse of non-rotating, axisymmetric stars", Prog.Theor.Phys. 67, 1396.
Nakamura, T.(1983),"General relativistic collapse of accreting neutron stars with rotation", Prog.Theor.Phys. 70, 1144.
Nakamura, T.(1985) "Gravitational radiation from Kerr black holes", in R.Ruffini(ed) "Proceedings of the fourth Marcel Grossman Meeting".
Nakamura, T.(1986) "Gravitational radiation and 3D numerical relativity", in H.Sato and T. Nakamura(eds) "Proceedings of the 14-th Yamada Conference on gravitational collapse and relativity" World Scientific Publishing Co, Singapore.
Piran, T.(1986) "Numerical relativity: from cosmology to gravitational collapse", this volume.
Shapiro, S.L.(1985a) "The collapse of dense star clusters to massive black and Teukolsky, S.A. holes: the origin of quasars and AGN", Ap.J 292, 41.
Shapiro, S.L.(1985b) "Relativistic stellar dynamics on the computer. I and Teukolsky, S.A. Motivation and numerical method", Ap.J 298, 34.
Shapiro, S.L.(1985c) "Relativistic stellar dynamics on the computer. II and Teukolsky, S.A. Physical applications", Ap.J 298, 58.
Stark, R.F.(1985) " Gravitational wave emission from rotating collapse", and Piran, T, Phys.Rev.Lett. 55 ,891.

Stark, R.F.(1986) and Piran, T. "Gravitational radiation from rotating gravitational collapse", in H.Sato and T. Nakamura(eds) "Proceedings of the 14-th Yamada Conference on gravitational collapse and relativity" World Scientific Publishing Co, Singapore.

Smarr, L.(1977) "Space-times generated by computers, black holes with gravitational radiation", Ann.N.Y.Acad.Sci. 302, 569.

Smarr, L.(1984) "Computational relativity: numerical and algebraic approaches" in B.Bertotti, F. de Felice and A. Pascolini (eds) "General Relativity and Gravitation (GR10)", D.Reidel, Dordrecht, pp. 163-183.

Symposium 20: Particle Motion and Continua in Relativity.

Chairman: J. N. Goldberg, Syracuse University,
Syracuse, New York 13244-1130 USA.

There were 27 abstracts selected for this symposium. In subject matter they fell into five general groups: continuum problems including thermodynamics and polytropic structure; equations of motion for test particles including gravitational radiation; conservation laws which are observer dependent or do not arise from symmetries in the Lagrangian; explicit calculations using known solutions to examine specific physical phenomena; and four contributions involving tachyons. I decided to invite people to talk on topics which were central to the abstracts submitted and also to select from among the abstracts those which I thought could be most easily understood in a short presentation. Unfortunately, in the latter group my selection picked out many who did not actually attend GR 11. Therefore, during the session only five talks were given.

1. Henry Kandrup of the University of Maryland opened the session with a talk on Recent Developments in Relativistic Statistical Mechanics and Kinetic Theory. The idea was to focus on continuum problems from a fundamental point of view. Therefore, the object of the report was to describe the development of a systematic and rigorous theory of nonequilibrium statistical mechanics for self-gravitating systems in the context of general relativity. A principal application of this work has been the study of the structure and evolution of systems of stars such as a galaxy or a globular cluster.

The basic philosophy of the program was influenced

heavily by recent work in Newtonian statistical mechanics which showed that ordinary kinetic theory can be viewed as a well-defined limiting case of a more complete many-body theory. For example, from that point of view, the fundamental object of a statistical description of N identical point mass stars is an N-particle distribution function μ, defined in a 6N-dimensional phase space, whose evolution is governed by an N-particle Liouville equation. For practical purposes, however, the physically relevant observable is instead a reduced one-particle distribution f(x,v,t), which yields the probability density for finding a given star at the point x with velocity v at time t. If there were no interactions amongst the stars, each would evolve independently, and one would conclude that the evolution of f is governed by a linear and local equation. However, stars do interact, and the evolution of the one-particle f involves the two-particle g in a fashion embodied in the ordinary BBGKY hierarchy. The key achievement of recent work in statistical mechanics has been the observation that, by introducing the notion of a "subdynamics," one can project out from this hierarchy an exact nonlocal and nonlinear equation for the evolution of f which contains no explicit reference to interstellar correlations buried in such quantities as g. Thus, in a first approximation, the evolution of f will be governed by the collisionless Boltzmann equation, but it will be affected by correlations among the stars. If, in a first approximation, one supposes that the stars follow trajectories determined solely by the average gravitational field, these correlational corrections lead to the collisional Boltzmann (or Fokker-Planck) equation.

In order to generalize this picture in a relativistic framework, one must identify a meaningful notion of "gravitational force." The key idea is to view the deviations from geodesic flow in a smoothed out "background" spacetime as being the manifestion of the desired "forces."

To the extent that the two spacetimes are "nearly the same," these forces may be derived from linearized equations and the physical picture is analogous to the evolution of an electromagnetically interacting system in a given background spacetime.

With this identification of gravitational force, one can try to formulate a covariant statistical description. The "obvious" arena of physics is then an 8N-dimensional phase space, defined as the direct product of N one-particle cotangent bundles. The imposition of mass shell constraints restricts the fundamental physics to a 7N-dimensional hypersurface in that space, but a further restriction to a (6N+1)-dimensional surface, which could make direct contact with the ordinary Newtonian theory, is impossible unless one breaks manifest covariance by specifying some "gauge" condition relating the various "times." This is simply a result of the fact that there exists no ordinary notion of dynamics in general relativity unless one introduces a preferred splitting into space plus time. Another difficulty with this approach is that this general approach to dynamics does not admit a clean initial value formulation. Even if one breaks manifest covariance in this fashion, one is faced with the problem that the evolution of the system is not specified uniquely by stipulating the coordinates and velocities of each of the stars on a given spacelike hypersurface: one must also specify the values of the accelerations and all their derivatives.

The approach to relativistic statistical mechanics developed by Israel and Kandrup is based on the assumption that the dynamics can be approximated as local, so that the specification of coordinates and momenta at a given instant do suffice to determine the subsequent evolution. One can then generalize the Newtionian theory to define an N-particle distribution function μ, and, in terms of an arbitrary "gauge" condition connecting the various times, derive an exact nonlinear equation for the evolution of a reduced one-particle f. A collisionless Boltzmann equation then follows in the limit that the background spacetime is chosen self-consistently as a solution to an "average" Einstein equation with a stress-energy $T^{\mu}{}_{\nu}$ generated from f. A collisional Boltzmann equation follows by evaluating corrections to this mean field description as if the stars in question interact via retarded potentials appropriate for geodesic flows in the average spacetime.

A more rigorous formulation of relativistic statistical mechanics requires that particle and field degrees of freedom be treated on an equal footing. Unfortunately, one does not know how to approach this problem covariantly. Consequently, work to date has been formulated in the context of a specified 3 + 1 spliting, with the field being viewed as an infinite collection of oscillators. The fundamental object of the theory is then a generalized distribution function $\tilde{\mu}$ which is a function of 6N particle variables, an infinite number of oscillator variables, and a single time t. Standard Newtonian techniques then lead to coupled nonlinear equations for reduced one particle f's and one-oscillator g's. The collisionless Boltzmann equation of general relativity follows rigorously if one assumes i) that the gravitational forces are derived from linearized equations and ii) that correlations between particles and oscillators may be ignored. By neglecting correlations, but allowing for nonlinearities in the equations of motion, one obtains a more complicated description which sheds insight into the "averaging ambiguities" recently considered by Ellis. And, by restricting attention to linearized equations, but allowing perturbatively for the effects of correlations, one recovers, in an appropriate limit, the collisional Boltzmann equation first derived by Israel and Kandrup.

2. For more than twenty-five years there has been considerable controversy over the validity of the quadrupole formula for gravitational radiation. Over the past six

years there has been a coming together of results from several groups using different technical approaches to the problem of motion and gravitational radiation. Therefore, it seeemed appropriate to ask James Anderson of the Stevens Institute of Technology to speak on the Current Status of the Quadrupole Formula.

He began by pointing out that there are different versions of the formula which divide into two basic classes. One is a far zone formula which involves the calculation of gravitational flux. Typically they have the form $L = \langle I_{ij}\ I_{ij} \rangle /5$, where I_{ij} is the reduced mass quadrupole of the source and the brackets indicate an average over a period of the system. L is interpreted as the energy flux at infinity. As it stands such a flux formula is of little value since the total energy flux is not measurable. Furthermore, such a measurement cannot be used to determine the period change of the binary pulsar. The second class is the near zone quadrupole formula or balance equation: $E = -\langle I_{ij}\ I_{ij} \rangle /5$, where E is the Newtonian energy of the system. Of course, E must be given in terms of source coordinates for the formula to be useful. Similar statements to the above can be made for the angular momentum. One needs the energy balance equation to obtain the change of period while one also needs the angular momentum balance equation to obtain the eccentricity in the binary system.

In order to derive either flux or balance equations, one needs a source model, appropriate boundary conditions, coordinate conditions, and a method of approximation in order to construct an approximate solution to the Einstein equations. Delta-function sources are unsatisfactory. They do not model black holes, cannot relate to extended sources in the near zone, and lead to renormalization problems in approximation methods. There are extended source models with appropriate equations of state which are applicable to compact bodies. While one might like to state boundary conditions on I^+ or I^- current efforts do not do so. Rather they employ initial value conditions. For example, Schutz and Futamase set free initial data to zero on a space-like hypersurface. They argue that so doing corresponds to averaging over initial conditions. Isaacson, Winicour and Welling set data on an initial null surface. They rule out incoming radiation by the requirement that the initial data be Newtonian in an appropriate sense. Damour uses an extended source model for the source in four dimensions. He assumes that incoming radiation is ruled out by the use of the retarded Green's function. Anderson sets arbitrary data on a space-like surface in the near zone and zero data on a null cone in the far zone. He then argues that the long time behavior is insensitive to the initial data. This latter argument is similar to that of Futamase and Schutz. The detailed relationships among these approaches is not yet clear.

There are problems connected with the use of harmonic coordinate conditions and it is not clear how the results depend on their use. The need to introduce logarithmic terms in the higher orders of all the approximation methods may be related to this problem.

Most analytical studies deal with slowly moving systems for which the ratio of light travel time across the system to the characteristic time (e.g. orbit period) of the system can be an expansion parameter. The expansion should not lead to obvious inconsistencies such as divergent integrals and secular terms. The divergent integrals which were pointed out by Ehlers, Havas, Goldberg, and Rosenblum have been shown to be due to the non-analyticity of the expansion and are now no longer a problem. Anderson himself has used a multiple time formalism to avoid secular terms. In general, the most efficient method for treating the slow motion problem has been the method of matched asymptotic expansions introduced by Thorne and Burke and applied by Kates and Anderson. While it would be sufficient to prove that the expansions are asymptotic to an actual solution, neither that nor the convergence of any of the approximation schemes has in fact been proven.

If one has a satisfactory approximate solution, then the flux equations are relatively easy to obtain. One merely needs to evaluate whatever expressions one adopts for the definition of the flux. However, once again, such expressions have little practical value. It is more difficult to derive balance equations. Very few satisfactory derivations exist for gravitationally bound systems such as the binary pulsar. One approach is to derive a radiation reaction force term in the source equations of motion by using the covariant conservation law for matter $T^{ab}{}_{;b} = 0$. It is then a straightforward procedure to obtain the balance equations. There are two problems with this approach: It cannot be applied to sources with black hole components and, as Schutz has pointed out, the expression for the radiation reaction force is coordinate dependent. In the harmonic gauge, for example, it does not have the usually accepted form. In linearized theory it is possible to find a gauge transformation which yields this form, but it is not clear that in the non-linear theory one can find such a coordinate transformation. According to Anderson, this is a serious problem which remains open.

All derivations of the quadrupole formula which purport to be more than a formal expansion are applicable only to slow motion. In the case of the binary pulsar, the slowness parameter is of the order of 10^{-9}. Therefore, it is valid to use the balance equation to calculate the change in the orbital period as a consequence of general relativity.

3. Recently James Hartle and Kip Thorne have begun a program to study the equations of motion for compact bodies in a manner which would focus on general relativistic effects which may become observable in neutron star systems. That is, they consider the post-Newtonian motion of compact

bodies including the precession of the spin axes. I found their approach quite exciting and Kip Thorne of the California Institute of Technology agreed to speak (for the two of them) on The Motion and Precession of Black Holes and Other Compact Bodies: Curvature Coupling Effects Derived by Surface Integrals.

The idea is based on earlier publications of P. D'Eath. In the usual calculations one starts with a background Minkowski space and builds up a series expansion in a parameter proportional to v/c or M/R (R is a mean interparticle distance). D'Eath's proposal breaks up the space-time into a region near the compact body and a far region. In the near region the field is dominated by the compact body. As a zeroth order metric one can choose the Schwarzschild or Kerr solution. The far field represents the external universe, which may or may not be asymptotically flat, but it is assumed to be smooth at the location of the compact body. Furthermore, there is assumed to be an intermediate buffer region in which both solutions are weak and where the solutions are to be matched.

These ideas are taken over by Hartle and Thorne. Because their motivation comes explicitly from the expectation that observations of compact bodies - particularly black holes and neutron stars - will become sufficiently accurate that the precession of their spin axes as well as their motion in binary systems will become observable tests of general relativity, their approximation method is explicitly related to the various lengths constructed out of the compact body and the external curvature. The internal (near) field depends on the lengths M,L, T which characterize mass, dimensions, and the time associated with any change of the compact body. Similarly, the external (far) field is characterized by $\mathcal{R}$, $\mathcal{L}$, $\mathcal{T}$ which determine a mean curvature, the distance defining inhomogeneities, and the time for external changes to become important.

Separate coordinate systems are used in the internal and external schemes. The internal metric is asymptotically flat so they choose the internal coordinates to be tied to a non-rotating Minkowski frame. The external metric is to be smooth at the location of the compact body. Therefore, they introduce non-rotating Riemannian coordinates with the origin at the body center. The matching in a buffer region is carried out in these coordinates. Since the coordinates are fixed only up to some order, there remains considerable gauge freedom which is used to simplify the calculations. If one had an exact solution of the Einstein equations corresponding to a compact body immersed in an external field, one would obtain the laws of motion as follows: The energy- momentum and angular momentum would be defined in terms of surface integrals in the buffer zone using the Landau-Lifshitz super-potential. Higher moments are defined in accordance with a prescription given by Thorne using "asymptotically Cartesian and mass centered" coordinates. The rates of change of the energy- momentum and angular momentum are then determined from the conservation laws and are expressed in terms of integrals involving the pseudo- tensor. The formal results obtained in this way relate the changes in energy-momentum and angular momentum to the electric and magnetic parts of the Weyl tensor:

$$dP_j/dt = - \mathcal{B}_{jk}\, J^k - 1/2\, D_{(m}\, \mathcal{E}_{jk)}\, \mathcal{I}^{km} ,$$

$$dJ_j/dt = - e_{jkm}\, \mathcal{I}^{kn}\, \mathcal{E}_n{}^m - 4/3\, e_{jkm}\, \mathcal{S}^{kn}\, \mathcal{B}_n{}^m .$$

These are the laws of motion. Actually exhibiting the results in terms of the functional form of the metric field itself leads to the equations of motion. The actual calculations are carried out only in an approximation method using an expansion in the parameters indicated above. This has been done for a many body problem. To define the external field, Hartle and Thorne start from the 1.5 post-

Newtonian solution given by EIH to which they graft terms due to spins and quadrupole moments. To focus on the motion of a particular body, the external field is defined by omitting those terms which become infinite at the body's center. The internal field is taken to be the Kerr solution. To the 1.5 post-Newtonian order there is no gravitational radiation and the center of mass remains on the central world line. Therefore, the equations of motion for the Kth body take the schematic form

$$M_K\, d^2x_K{}^i/dt^2 = \text{(EIH)} + \text{(Quadrupole forces)}$$

$$+ \text{(Spin-orbit forces)} + \text{(Spin-spin forces)},$$

$$dJ_K{}^i/dt = \text{(Geodetic precession)} + \text{(Gravito-magnetic precession)}$$

$$+ \text{(Tidally torqued precession)}.$$

What makes this approach so interesting is its focus on the measurable quantities and the orders to which they are indeed measurable.

4. Neil Ashby of the University of Colorado at Boulder presented an interesting paper describing the Construction of Normal Fermi Coordinates on a Test Body in Elliptical Orbit in a Schwarzschild Field. He began by giving the exact solution for the elliptical geodesics in terms of the Jacobian elliptic funtions sn, cn, and dn. Then one constructs Fermi normal coordinates along the geodesic. To do so one introduces an orthonormal tetrad attached to the geodesic such that the time-like vector is tangent to the geodesic. All the tetrad vectors are parallel propagated along the curve. New coordinates are introduced whose spatial coordinates have their origin on the geodesic and the time-like coordinate is the proper time along the geodesic. The spatial coordinate curves are extended as

space-like geodesics whose tangent vectors at the origin are the tetrad vectors. These are the local inertial coordinates in the freely falling frame. Since each point on the geodesic may be the origin of Riemannian coordinates, the metric tensor to second order in the local inertial coordinates takes the well known simple form:

$$g_{oo} = -1 + R_{omon}\, x^m\, x^n,$$

$$g_{ok} = 2/3\; R_{omkn}\, x^m\, x^n,$$

$$g_{jk} = \delta_{jk} + 1/3\; R_{jmkn}\, x^m\, x^n.$$

Ashby gave the physical components of the Riemann tensor explicitly in terms of the Jacobi elliptic functions. This calculation gives a model of the tidal forces acting on a body in motion around a compact object.

5. The final speaker was E. P. Esteban of the University of Puerto Rico who spoke on the Geodesics of a Compact Body in a Constant External Magnetic Field. He presented a calculation of geodesics in a magnetized Kerr metric which represents a rotating black hole in a constant magnetic field. He finds that the presence of the magnetic field has little effect on the geodesics unless the black hole is very massive, $>10^8\ M_0$ and the magnetic field is $>10^{10}$ gauss. This is true for both time-like and null geodesics. In these cases the magnetic field decreases the size of the ergo region as well as the stability threshold of stable circular orbits.

QUANTUM FIELD THEORY ON CURVED SPACE

P. Hajicek
Institute for Theoretical Physics, University of Bern
Sidlerstrasse 5, CH-3012 Bern, Switzerland

THE MAIN STIMULI IN THE DEVELOPMENT OF THE THEORY

The theory of quantum fields on a fixed classical background spacetime has a long history (see, e.g. Fulling (1984)). However, a real interest for this theory came only after the discovery of the so-called "black hole evaporation" (Hawking (1975)). Since then, a great deal of work has been done on particular examples of backgrounds and quantum states as well as on general foundations of the theory. This evolution was stimulated and influenced by further historical coincidences, of which at least the following two should be mentioned. First, the success of semiclassical approximation in the theory of Yang-Mills fields, the notions of soliton and instanton and the discovery of an important role of topology (cf. Coleman (1977)) brought a current of new ideas. Second, a great interest in big bang cosmology among physicists and the problems encountered in this field led to a number of interesting research programs; from these we must quote at least the idea of the inflationary scenario (Guth (1981)).

Maybe, most relativists have been motivated towards research in this area by the vague hope of obtaining some hints concerning the old great problem - how to construct a quantum theory of gravity. Indeed, there are attempts in this direction, a quite spectacular one being the change in the axiomatic structure of the quantum theory (Hawking (1982)), to allow for the loss of quantum coherence in quantum processes. This is an extrapolation of a feature of the quantum field theory on curved spacetimes to a full quantum theory of gravity. Even if such attempts are controversial, they at least broaden the basis of discussion.

SEMICLASSICAL APPROXIMATION

The quantum field theory on curved space cannot be considered as a complete and exact theory which just generalizes the "ordinary" quantum field theory in Minkowski spacetime, because it does not account for the influence on the geometry of spacetime which the quantum fields must have - the so-called back reaction. Most people consider it, therefore, as the first stage in the so-called semiclassical (or loop-) expansion. It seems that some such kind of approximation must have some validity, because the classical Einstein theory is a very good description of our experience.

The nature of the semiclassical approximation in quantum field theory is explained, for example, in Coleman (1977). There, a self-interacting scalar field ϕ on the two-dimensional flat background is taken as a model. The interaction term in the Lagrangian contains a dimensionless coupling constant λ. The usual expansion of the theory in powers of λ around the exact classical solution $\phi = 0$ is renormalizable and defines a sensible quantum theory. However, there are other exact solutions; the corresponding values of ϕ_{cl} are proportional to λ^{-1} and their total energy to λ^{-2}. One can expand the scattering theory of the small disturbances $\delta\phi$ which are considered as a quantum field around ϕ_{cl} and study their interaction with the background and with each other. The expansion in powers of the Planck constant h coincides with partially summed expansion in powers of λ; it is the expansion in the number of loops.

This scheme can work only if the classical solution ϕ_{cl} is classically and quantum stable so that all quantum corrections to it remain small in the whole space and for all times. Only then are the asymtotic states of $\delta\phi$, necessary for the scattering theory, are well-defined. In particular, the corrections to the total energy of ϕ_{cl} must be negligible with respect to the classical value.

In the context of the cosmological particle creation at the big bang, or of the black hole evaporation, a sort of semiclassical approximation is often used, which is slightly different from the above one (which we shall call "lst kind"). It is based on the equation

(1) $$G_{\mu\nu} = 8\pi G \langle T_{\mu\nu} \rangle,$$

(see, e.g. Birrell-Davies (1982)), where $G_{\mu\nu}$ is the Einstein tensor of a background classical metric $g_{\mu\nu}$ (which does not need to satisfy the classical Einstein equations) and $<T_{\mu\nu}>$ is the expectation value of the energy-momentum tensor operator of the small disturbances around $g_{\mu\nu}$ in some quantum state. One attempts to solve the equation (1) for a given metric $g_{\mu\nu}$ together with a given quantum state of the disturbances at some initial (or final) time (or time interval) similarly as the classical equations are solved; that is, one is going to solve a sort of initial value problem as opposed to the scattering problem of the 1st kind of semiclassical approximation (cf. Jordan (1986)). This is, in fact necessary, because the corresponding solution of the classical equations, $g^{cl}_{\mu\nu}$, does not remain a good approximation to $g_{\mu\nu}$ for the whole time, but must change gradually by quantum corrections to something completely different. at large times. The scattering theory around $g^{cl}_{\mu\nu}$ does not make much sense: for example, in the black hole evaporation case, the corrections to the total energy of the classical black hole will not only not be small - they will diverge as the retarded time, u, grows to infinity.

According to my knowledge, nobody has undertaken any attempt to justify this 2nd kind of semiclassical approximation by showing, e.g. that it works for some simplified model (cf. Fulling (1984)). In fact, it has rather been shown that the equation (1) exhibits some, speaking in a very mild way, surprising features even if one expands around the flat empty spacetime (Horowitz (1981)): in some cases, exponential run-aways from the classical solution, in other cases tachyonic waves are obtained. This may be interpreted in different ways (see the loc. cit.), one possibility being that any kind of semiclassical approximation breaks down for gravity coupled to other fields, another being that only the semiclassical approximation based on the equation (1) is inappropriate, etc. In any case, the applicability of any semiclassical approximation to situations in which the global changes by quantum corrections are large remains an open problem.

The semiclassical approximation has been critized from different standpoints as well (e.g. Duff (1981)). Also, one could feel that an approximation cannot be very well founded, if the exact theory which is going to be approximated is not known. Aware of all these difficulties, we will still assume that the quantum field theory on curved space is some sort of effective semiclassical approximation in the present review. Thus, we will use the language of the semiclassical approximation to describe the contributions, and, in fact, we will have in it a basis of a "Linnéan" classification scheme of all activities in the field.

THE TREE STAGE

Most of the work done in quantum field theory on curved space is at the tree stage. Here, no regularization or renormalization is necessary, and everything can be obtained in a unique way from some properties of the exact solutions to the full or to the linearized classical equations. One can distinguish three main groups of problems.

Exact classical solutions

Here, one is looking for new classical solutions of both the Lorentzian or Euclidean Einstein equations. One obtains background spacetimes as new candidates for vacuum (say, anti-de Sitter spacetime, or $M^4 \times K^d$ - solutions for 4+d-dimensional Kaluza-Klein theory, etc.), new candidates for solitons (say, some extremal black holes), or for instantons. We had only "solitonic" contributions.

F. Degen. New extreme black hole solutions to the Einstein-Yang-Mills-Higgs equations are searched for.

C. Teitelboim. Analogons to Dirac monopole are found in a new type of field theory - the so-called "p-form electrodynamics".

R. Sorkin. The definitions of spin and statistics are generalized to solitons of a non-linear σ-model, and the well-known spin statistics relation is shown.

Linear theory of small disturbances

After an exact classical solution is chosen, the classical dynamical equation is linearized around it. The corresponding solutions, the so-called small disturbances, are considered as a quantum field.

There is no interaction except for the influence of the external spacetime on the small disturbances. One often considers incomplete models, for example a scalar field or a Dirac spinor field on a curved background.

The first step is to find some complete system of solutions to the linearized equation, the so-called modes. If the spacetime is asymptotically flat or static, then there is a well-defined spectrum of these modes. Interesting problems are, for example, the existence of zero modes, their relation to the topology of background or to the degrees of freedom of the soliton, the question of classical stability (exponentially growing modes), etc.

The corresponding contributions are the following.

V.M. Mostepanenko. Certain modes in cosmological backgrounds like Bianchi type I model are shown to be unstable.

G.V. Shishkin and J.E. Andrushkevich. Variables are separated in the Dirac equation on a class of background metrics and some complete systems of modes are found on the De Sitter spacetime.

V.D. Ivashchuk. Stationary solutions to the Dirac equation are found for the background combined from the open Einstein universe and a Dirac monopole electromagnetic field.

V.G. Bagrov and A.V. Shapovalov. Properties of the Dirac equation in the Yang-Mills external field are studied.

O. Cristea. The Dirac equation in an infinitesimal gravitational field of a rotating body is examined.

D.V. Gal'tsov and A.A. Ershov. The Klein-Gordon and Dirac equations are studied on the electrovac type D classical background, in particular on the Kerr-Newman black hole.

The next step is to consider the small disturbances as a quantum field and study the various possible quantum states of the field on the given background. There will, in general, be no analogue of the ground state of the flat background, but there will be states with interesting symmetries, asymptotical or other boundary properties, and special relations to some families of detectors or observers. Here, several papers have been submitted.

B.S. Kay and R.M. Wald. The authors consider the Schwarzschild spacetime as a background and a massive or massless scalar as a quantum field. The emphasis is on completely rigorous proofs and methods. The results can be described as follows. By assuming the Hadamard from (regularity in a certain sense) of the state everywhere in the Kruskal manifold (i.e. in particular, on the horizon), symmetry with respect to the time translation of the Schwarzschild background (staticity), and some technical points, the authors show the uniqueness of the so-called Hartle-Hawking state (and its extension beyond the horizon). For the observers outside the black hole, it is a thermal state corresponding to the Hawking temperature.

This is an important negative result in the following sense: it shows that there are no reasonable states in the Schwarzschild spacetime including both the horizon and the infinity. If we use the Hartle-Hawking state in the equation (1), then the Schwarzschild metric is not even an approximate solution to it. Even for large black holes, when the Hawking temperature is extremely low and the energy density is very small everywhere, it nevertheless integrates to an infinite contribution to the total energy of the hole.

M. Castagnino and R. Ferraro. An interpretation of various states of quantum fields on curved classical backgrounds is attempted by associating each of them with some privileged family of observers.

S. Hacyan. Energy spectrum as seen in a uniformly accelerated frame is calculated for the flat spacetime vacuum of a massive scalar and spin-1/2 field.

O. Cristea. Effects which can be observed in a rotating frame on a one-electron state in the flat spacetime are studied.

The third and last step is to describe the dynamics of linear quantum fields on curved spaces. For that, various methods can be chosen (e.g. C*-algebra approach, Fock space approach, etc.). Mostly, a Hilbert space of states is chosen, whose Fock space structure depends on the momentary geometry. Then, one can calculate, for example, the particle production by the background, etc. We have one contribution.

A. Anderson. The Green's function for the Schrödinger equation on an arbitrary symmetric spacetime is calculated from the Green's function on its maximal torus. This exact result can be used, via the proper time formalism, to calculate the Feynman propagator for particular Fock space structures on the symmetric spacetime background.

Tree scattering of small disturbances off each other

Here, we have already interacting fields in an external spacetime. However, the calculations are limited to those processes whose Feynman diagrams do not contain any loops. Thus, there is no need for renormalization. One can study the effects of topology and curvature on the well-known scattering amplitudes.

J. Audretsch. The Compton effect and the particle decay are studied using two component scalar models on an expanding Robertson-Walker background.

H.E. Kandrup. Some relations between entropy generation and particle production are studied. It is shown that under very general conditions the total entropy will be conserved in the quantum field theory on a background. The new concept of the so-called reduced entropy is introduced; this entropy is associated with a measurement which does not probe a complete set of observables. Generation of such a reduced entropy is possible and can be associated with a particle creation. This is shown for a model of a self-interacting field theory.

THE ONE LOOP STAGE

By calculating up to the one-loop graphs inclusive, two main groups of problems can be attacked. First, one can try to obtain the back reaction of the created particles and vacuum polarization on the external potential which created them. The calculations of the quantum corrections to the mass of a soliton, or to the transition probability associated with an instanton belong to the same category. Second, one can obtain further effects of the curvature or topology on the quantum fields.

The back reaction

The changes in the metric due to quantum corrections are governed by effective equations like (1). These equations can be obtained from effective actions, which can have the in-out, or the in-in form. In this way, the $\langle T_{\mu\nu}\rangle$ "industry" comes into being. The most important application seems to be the Hawking effect. The problematics of the equation (1) have been explained above. There are, however, more unsolved problems in this particular case. Consider, for example, a spherically symmetric model of a star collapse. It is extremely likely that $\langle T_{\mu\nu}\rangle$ remains very small in some neighbourhood U of the surface of the star (if we exclude the final stages with high curvature). This could very well imply that the changes in the local geometry in U are small. However, this does not justify the belief that the event horizon will be shifted only negligibly. The event horizon is a global concept and its position and existence are sensitive to the whole future development of the space-time (only "approximate" event horizons are sufficiently local, cf. York (1984), but their position has nothing to do with that of the absolute event horizon).

We have the following contributions.

A.C. Ottewill and M.R. Brown. In this interesting paper, the so-called Brown-Ottewill-Page approximation method for calculation of $\langle T_{\mu\nu}\rangle$ is studied. The method is based on a conformal transformation law for the renormalized one-loop effective action of a conformally invariant quantum field on arbitrary background. The conformal transformation relates the results on different backgrounds, and the calculation of $\langle T_{\mu\nu}\rangle$ can, therefore, be performed only in the most simple (symmetric) conformally related background. There are, however, problems with this approximation, if the quantum field has spin 1 (electrodynamics). Some paradoxical results can be mended by choosing another value for the conformal anomaly, but the approximation remains still less accurate than in the case of scalar field. The reasons for this failure are traced.

V.P. Frolov and A.I. Zelnikov. $<T_{\mu\nu}>$ of the electromagnetic quantum field near the horizon of the Kerr black hole background is calculated.

K.A. Bronnikov and E.M. Saminski. Some regularization methods of $<T_{\mu\nu}>$ on anisotropic cosmological backgrounds are criticized.

P. Candelas and B.P. Jensen. The Hartle-Hawking state is extended beyond the horizon and $<T_{\mu\nu}>$ is calculated near r = 0 on the Kruskal background.

R.W. John. The problem of existence of the De Witt expansion and its relation to the Hadamard one is addressed by using a simple spacetime background.

A. Krolak and W. Biernacki. $<T_{\mu\nu}>$ is calculated on two-dimensional spacetimes.

Effects on small disturbances

The way in which the background metric or other external field and the corresponding topology enters the one-loop corrections to the effective action is not important only for the back reaction. It leads to new properties of the quantum fields themselves. There are two interesting contributions.

B.L. Hu and D.J. O'Connor. As a model, a self-interacting, N-component scalar field ϕ on various fixed backgrounds (De Sitter space, $R^d \times S^{D-d}$, etc.) is considered. The dominant one-loop contributions to the effective potential V at $\phi = 0$ (the metastable state of the inflation theory) come from the lowest modes (infrared divergence). The infrared behaviour of the system is in turn determined by the open dimensions (R^d) and the phase transition behaves effectively as in a d-dimensional background (d = 0 for De Sitter spacetime). In such a way, the topology of the spacetime can have a strong influence on the phase transition of the inflation theory and the calculations based on the flat spacetime effective potential need not be always correct. Similar effects are known in the physics of surfaces, where they are called "finite size effects".

D. Lohiya. The behaviour of the fermion quantum field on the black hole background combined with a dyon-like configuration of the Yang-Mills field is studied. The chiral anomaly leads to a spontaneous chiral flux similar to the case of the pure Yang-Mills Julia-Zee background in spite of the different boundary condtions - regularity at the horizon in the former and at the center in the latter case.

HIGHER LOOP CORRECTIONS

Here, there is only one paper.

R. Tarrach and E. Elizalde. The self-interacting quantum scalar field is considered on some fixed backgrounds (Robertson-Walker, Schwarzschild). No back reaction to the spacetime metric is considered, but the interacting quantum field is studied by exact methods.

CONCLUSIONS

The quantum field theory on curved space has changed its face: the stormy period of development in the seventies, when it was not even clear what are the questions which may sensibly be asked, is over. A number of real problems remain, they are, however, difficult and the progress is slow.

At the end, I should like to apologize to those whose work has been misinterpreted by me as well as to those whose ideas have not matched my classification scheme and remained unmentioned (I have not attempted to classify the contributions according to their possible usefulness anyway).

REFERENCES

Birrell, N.D. and Davies, P.C.W. (1982). Quantum Fields in Curved Space. Cambridge: Cambridge University Press.

Coleman, S. (1977). Classical Lumps and Their Quantum Descendants. In New Phenomena in Subnuclear Physics, ed. by A. Zichichi, pp. 297-421. New York: Plenum.

Duff, M.J. (1981). Inconsistency of Quantum Field Theory in Curved Space-time. In Quantum Gravity 2: A Second Oxford Symposium, ed. by C.J. Isham, R. Penrose and D.W. Sciama, pp. 81-105. Oxford: Oxford University Press.

Fulling, S.A. (1984) What Have We Learned from Quantum Field Theory in Curved Space-Time ? In Quantum Theory of Gravity. Essays in Honour of the 60th Birthday of Bryce S. DeWitt, ed. by S.M. Christensen, pp. 42-52. Bristol: A. Hilger.

Guth, A.H. (1981). Phys.Rev. D23, 347.

Hawking, S.W. (1975). Commun.Math.Phys. 43 , 199.

Hawking, S.W. (1982). Commun.Math.Phys. 87, 395.

Horowitz, G.T. (1981). Is Flat Space-Time Unstable ? In Quantum Gravity 2: A Second Oxford Symposium, ed. by C.J. Isham, R. Penrose and D.W. Sciama, pp. 106-130. Oxford: Oxford University Press.

Jordan, R.D. (1986). Phys.Rev. D33. 444.

York, J.W., Jr. (1984). What Happens to the Horizon When a Black Hole Radiates ? In Quantum Theory of Gravity. Essays in Honour of the 60th Birthday of Bryce S. DeWitt, ed. by S.M. Christensen, pp. 135-147. Bristol: A. Hilger.

SYMPOSIUM 14: RADIATIVE SPACETIMES AND APPROXIMATION METHODS

B.F. Schutz
Department of Applied Mathematics and Astronomy,
University College Cardiff CF1 1XL, Great Britain.

Abstract. Talks presented at the symposium fell into three categories. Talks on the *quadrupole formula* strongly reinforced the consensus that the Einstein-Landau-Lifshitz-Chandrasekhar quadrupole approximation is robust and useful, and they clearly showed some new directions that research in the field is taking. Speakers about *gravitational radiation* widened the discussion to spacetimes with fully nonlinear radiation from fully relativistic systems. A closing section on *approximation methods* contained two papers on other aspects of weak field and/or slow motion space-times.

INTRODUCTION

Thirty-one abstracts were submitted for this session and twelve talks were presented. The field is clearly a healthy one. The controversies of previous GR meetings concerning the validity of the 'quadrupole formula' have largely been resolved, and research on that subject is taking new directions that promise a deeper understanding of radiating systems. I will summarize the three sections separately below.

It goes without saying that the division of subjects into different symposia is often of necessity arbitrary. Readers who want a good overview of the subject should refer also to the reports on Symposium 20, in which J. Anderson gave a review of work on the quadrupole formula and K.S. Thorne described an approach to equations of motion for compact bodies; on Symposium 11, where D. Hobill discussed the matching of numerical source calculations to analytic wave expressions; and to the plenary lectures of J. Ehlers and T. Piran.

THE QUADRUPOLE FORMULA

J.Winicour (MPI Astrophysik, Garching) The quadrupole formula

E.M.Kirk (U.Southampton, UK) An approach to a quadrupole formula error bound via Newtonian gravity on the null cone

T.Futamase (Washington U., St. Louis) The strong field point particle limit and the equations of motion for the binary pulsar system

J.L.Anderson (Stevens Inst. Tech., N.J.) Gravitational radiation damping of systems with compact sources

F.I.Cooperstock (U.Victoria, BC) Applications of the new forumula for gravitational radiation energy loss

G.Schäfer (Meudon) Gravitational radiation and generalized Lie-series

At GR10 it appeared that a consensus was emerging that the quadrupole approximation was a good one for nearly Newtonian systems, both for the near-zone radiation reaction and for the far-zone radiation amplitude. Work in the intervening three years has clearly strengthened that consensus. We have a new and very different null cone derivation of the far-zone formula (Winicour, Kirk), and we have new demonstrations that the formulae even extend to compact bodies like neutron stars and black holes, if they move in external nearly-Newtonian fields (Futamase, Anderson). The calculations here are still on the level of 'formal' methods, in the terminology of Ehlers' plenary lecture, although error estimates may soon be available (Kirk). Rigorous theorems in this subject still seem far away. Nevertheless, the quadrupole formula is impressively robust, not only because it is derivable in a variety of different ways, but also because it seems to give good results in some situations where it might be expected to fail.

The null-cone approach to the Newtonian limit described by Winicour is one of the most original ideas to have emerged in this subject recently. By adopting a null-hypersurface initial-value formulation before going to the limit $c \to \infty$, some connection between the Newtonian source and asymptotic null infinity seems to be maintained. The result is that a calculation of the Bondi news function in the limit $r \to \infty$ gives the standard quadrupole formula. Kirk, reporting work done jointly with J.A. Vickers, carries the Winicour formulation a step further. By using the Newman-Penrose equations, he obtains a simpler null-cone system, with the result that it seems possible to generate useful error bounds.

It is clear that these methods will stimulate much more work in the future. In this context, therefore, I think it will be important to pay attention to a subtle difference between the calculations of Winicour and Kirk. Winicour sets up his post-Newtonian expansion by taking the limit $\lambda \to 0$ ($\lambda \sim c^{-1}$) at fixed r, while Kirk allows for a limit at fixed λr. In the language of matching, Winicour's seems to be a near-zone limit, while Kirk can choose either the near or the far zone, depending on how he treats r. It is not surprising to see the quadrupole formula emerge from Kirk's far-zone limit, but how do we interpret it in the Winicour calculation?

It is my view that this is related to a point emphasized by Piran in his plenary lecture, that in his numerical calculations of gravitational collapse, he can see the gravitational wave amplitude taking shape in the near zone, well inside one gravitational wavelength. Piran's result applies to his particular gauge, but I have elsewhere (Schutz 1986a) shown that this happens in any gauge in the Newtonian limit. In this view, then the Winicour calculation gives us the most rigorous demonstration so far that by going well out in the near zone (Winicour's $r \to \infty$ limit) one can pick out the gravitational wave amplitude. As Piran remarks, the usual view is that the gravitational wave amplitude cannot reliably be identified except outside a few radiation wavelengths. Winicour provides a framework for revising this view, at least for nearly Newtonian systems.

Turning now to the motion of compact bodies, Futamase outlined an approximation method which shows that the motions of N bodies with arbitrarily strong internal gravity are the same as that of N weak-field nearly-Newtonian bodies, provided that the interior of each body is nearly stationary and the bodies are well separated. This is closely related to the work of Hartle & Thorne reported in Symposium 20, and it grows out of earlier work by D'Eath and Kates, but there are some differences. Where Hartle & Thorne solve the field equations in different regions and then match the solutions to each other, Futamase writes down a single implicit integral solution valid everywhere, which he then finds approximations to in various regions. Hartle & Thorne calculate the precession and deviations from geodesic motion caused by coupling of a body's multipole moments to external curvature; Futamase derives the Newtonian motion, the near-zone quadrupole radiation reaction, and the far-zone quadrupole radiation field. In all cases, the 'body-zone' ADM masses replace the Newtonian masses in the standard formulae. This shows clearly that the strong equivalence principle embraces 'gravitational potential energy' even in the generation of gravitational waves by bodies with strong internal fields.

Also closely related is the work reported by Anderson, in which he uses the EIH method to derive the result that compact bodies radiate energy and angular momentum by the 'quadrupole' formula if they interact weakly. The method uses integrals on surfaces around the bodies, for the evaluation of which it uses asymptotic expressions for the interior metric. In this respect it is only slightly less complete than the Futamase and Hartle & Thorne calculations. In compensation, however, the method is computationally efficient, gives finite integrals, and can be applied to black holes (which Futamase's method cannot be).

The talk by F.I. Cooperstock went some way toward resolving the differences between the general calculations of the quadrupole formula and his specific calculations of radiation from particular systems which were evident at GR10. He and R.H. Lim obtain the quadrupole formula in a particular limit and are investigating corrections in cases where the size of the source may not be small compared to the size of the "near zone".

G. Schäfer introduced a novel approach to the full problem of radiation generation in general relativity, using a canonical approach based on Lie series with retardations. This is an outgrowth of his earlier work on the ADM n-body problem.

In summing up this section, I would like to remark that the recent work is beginning to reveal new features of radiation in the Newtonian limit, going beyond the efforts at previous GR meetings simply to verify the quadrupole formula. Besides the work on wave formation in the near zone referred to above, there is the beginning of a systematic attempt to go beyond the quadrupole formula. If the next step may be called the 'octupole formula' - even though it contains more than just octupole radiation, such as the post-Newtonian corrections to the sources of the quadrupole radiation - then I believe its predictions may become testable by the end of the century. Observations of coalescing binaries with long-baseline interferometric gravitational radiation detectors promise an accuracy at the 1% level, which is where the post-Newtonian corrections to the orbit and to the radiation become important for these systems (Schutz 1986b). Achieving the octupole formula means, among other things, dealing with possible logarithmic terms in the post-Newtonian expansions, as discussed by Anderson in Symposium 20. Finally, as another direction for 'new' work, I should mention the work of Nahmad-Achar (1986) for which there was insufficient time at this meeting, in which he shows that angular momentum in gravitational radiation is free of supertranslation ambiguity in the Newtonian limit.

GRAVITATIONAL RADIATION

J.Bičák (Charles U., Prague) Radiative spacetimes with boost-rotation symmetry

J.L.Friedman (U.Wisconsin, Milwaukee) Gravitational radiation from rapidly rotating netron stars

E. Nahmad-Achar (U.Cambridge) Angular Momentum in General Relatity

A.M.Abrahams (U.Illinois) Mode-mode coupling in the implosion of Brill waves holes (which Futamase's method cannot be).

The talk, by F.I. Cooperstock went some way toward resolving the differences between the general calculations of the quadrupole formula and his specific calculations of radiation from particular systems which were evident at GR10. He and R.H. Lim obtain the quadrupole formula in a particular limit and are investigating corrections in cases where the size of the source may not be small compared to the size of the "near zone".

G. Schäfer introduced a novel approach to the full problem of radiation generation in general relativity, using a canonical approach based on Lie series with retardations. This is an outgrowth of his earlier work on the ADM n-body problem.

In summing up this section, I would like to remark that the recent work is beginning to reveal new features of radiation in the Newtonian limit, going beyond the efforts at previous GR meetings simply to verify the quadrupole formula. Besides the work on wave formation in the near zone referred to above, there is the beginning of a systematic attempt to go beyond the quadrupole formula. If the next step may be called the 'octupole formula' - even though it contains more than just octupole radiation, such as the post-Newtonian corrections to the sources of the quadrupole radiation - then I believe its predictions may become testable by the end of the century. Observations of coalescing binaries with long-baseline interferometric gravitational radiation detectors promise an accuracy at the 1% level, which is where the post-Newtonian corrections to the orbit and to the radiation become important for these systems (Schutz 1986b). Achieving the octupole formula means, among other things, dealing with possible logarithmic terms in the post-Newtonian expansions, as discussed by Anderson in Symposium 20. Finally, as another direction for 'new' work, I should mention the work of Nahmad-Achar (1986) for which there was insufficient time at this meeting, in which he shows that angular momentum in gravitational radiation is free of supertranslation ambiguity in the Newtonian limit.

GRAVITATIONAL RADIATION

J.Bičák (Charles U., Prague) Radiative spacetimes with boost-rotation symmetry

J.L.Friedman (U.Wisconsin, Milwaukee) Gravitational radiation from rapidly rotating netron stars

E. Nahmad-Achar (U.Cambridge) Angular Momentum in General Relatity

A.M.Abrahams (U.Illinois) Mode-mode coupling in the implosion of Brill waves

This group of talks spans a wide variety of investigations of radiation outside the post-Newtonian framework.

J. Bičak emphasized that the spacetimes with axial and boost symmetry are the only known exact solutions containing gravitational waves and sources. An infinite number of explicit solutions describing various "uniformly accelerated" multipoles is available at present, and they can be used for testing approximation methods and numerical codes. For example, an 'octupole formula' (with vanishing quadrupole terms) can be derived from exact solutions. There are always two incomplete generators of null infinity, and the ADM mass vanishes; but null infinity is topologically $S^2 \times R$, the radiative field peels appropriately, and these are liable to be the best analytic solutions with source and radiation that we will have for some time.

J.L. Friedman reminded us that rotating stars are subject to a gravitational-wave instability, and in particular that pulsars with millisecond periods may be spinning at the critical angular velocity for such an instability. This angular velocity depends on the neutron star equation of state, and he presented results for the critical angular velocity for a variety of equations of state and levels of viscosity, which tends to damp out the instability. The 1.6 millisecond pulsar (see the plenary talk by J. Taylor) is in the right range, although one example like this is not sufficient to establish that this effect is indeed the one that determines the angular velocity. A new pulsar with a 1.16 ms period may have been found recently, but the detection is not yet certain.

E. Nahmad-Achar presented new variational principles based upon angular momentum. After finding pseudotensorial measures of energy and angular momentum which can be evaluated in any co-ordinates (not just asymptotically rectangular ones), he examined what solutions of Einstein's equations extremize the angular momentum on a spacelike hypersurface or the angular momentum across a timelike hypersurface. In the latter case, a new method for calculating radiating axisymmetric gravitational collapse emerges: a metric plus perfect fluid is an axisymmetric solution of Einstein's equations if and only if it extremizes (under mild constraints) the angular momentum flux across the timelike hypersurface Φ = const, where $\partial/\partial\Phi$ is the axial Killing vector.

A.M. Abrahams showed how nonlinear effects in gravitational waves generate new waves. Reporting analytic and numerical calculations performed with C.R. Evans, he showed that the pure $\ell = 2$ linear Brill wave (Teukolsky wave) travelling in toward the origin emerges later with $\ell = 4,6,\ldots$ components. The numerical calculation was illustrated with a videotape of the results.

Related to this was the abstract submitted by C.D. Ciubotariu (Poly.Inst.Jassy, Romania). He discussed wave generation by wave-wave interaction in general relativity on arbitrary backgrounds, in the short-wavelength limit. This is closely related to the calculation of the Isaacson effective energy-momentum tensor, which may be regarded as the special case in which the two interacting waves generate a secondary wave of infinite wavelength. Unfortunately, Ciubotariu's abstract overstates the size of the wave-generation effect: it cannot significantly reduce the amplitude of the waves which are being looked for in gravity-wave experiments.

In summary, gravitational radiation research seems very healthy. The existence of variational principles that include radiating spacetimes may open up new analytic possibilities. Mixed numerical-analytic calculations may become more common (see also the report on Symposium 11). And Friedman's talk highlights an area of research which has new urgency: the calculation of linear normal modes of rotating relativistic stars. These are needed for two reasons: first as test-bed examples for the fully nonlinear, three-dimensional, fully relativistic hydrodynamical codes now under development; and second because developments in X-ray astronomy and in gravitational wave detectors may make the sorts of unstable stars Friedman discussed observable in the next decade.

APPROXIMATION METHODS

H.Pfister (U.Tübingen) Analysis of some characteristic physical effects in general relativity through perturbation of the Schwarzschild mass shell.

H.Okamura (Kogakuin U., Tokyo) Post-post-Newtonian approximation in general relativity

This group of talks uses approximations that do not involve radiation. Pfister discussed three approximate calculations, all of which perturb Schwarzschild mass shells. The first was 'gravitational shielding', in which the effect of external gravitational fields is reduced inside a spherical shell. The second was whether a rotating shell can be made to produce inside itself not only the Coriolis force of frame-dragging but also a centrufugal force of appropriate magnitude. The answer is that it can, and the demand uniquely determines the shell's shape. The third result was a calculation of the gravimagnetic repulsion of two nearby, even interpenetrating, rotating shells. It can balance their attraction if the shells are superimposed and relativistically rotating.

Okamura spoke on the derivation he and colleagues had given several years ago of the post-post-Newtonian equations of motion and Lagrangian for many-body systems. In particular he felt that recent work by Schäfer and T. Damour had taken insufficient account of their contributions. Schäfer responded to this during the discussion, saying that he and Damour had corrected errors and shortcomings in Okamura's work and had correctly referred to it (Damour and Schäfer 1985). A paper presenting the view of Okamura and colleagues will soon be published in Prog.Theor.Physics (Kyoto), written by T. Ohta and T. Kimura.

REFERENCES

Damour, T., and Schäfer, G. (1985), *Gen.Rel.Gravit.* 17, 879-905

Nahmad-Achar, E. (1986). Ph.D. thesis, University of Cambridge, Cambridge England

Schutz, B.F. (1986a) *In* Dynamical Spacetimes and Numerical Relativity, ed. J. Centrella. Cambridge: Cambridge University Press

Schutz, B.F. (1986b). *Nature* to be published.

SUPERSTRINGS

Lars Brink
Institute of Theoretical Physics
Chalmers University of Technology
Göteborg, SWEDEN
and
Institute for Theoretical Physics
University of California
Santa Barbara, California

1. INTRODUCTION

The search for the fundamental principle or principles of Nature has led us to consider physics at smaller and smaller distances. At scales of the order of 10^{-8} cm it was found at the turn of the century that we have to introduce a new framework: quantum physics. This meant that we have to understand how the classical theories based on large-scale physics cope with the quantum theory. In the case of Maxwell's electrodynamics, this merging was simple. Even at quantum physics scales Maxwell's action is true and can be directly taken over to Quantum ElectroDynamics (QED). The checking that this is really a consistent theory was, however, a formidable task and led to the theory of renormalization. This theory was subsequently enlarged to the Glashow-Weinberg-Salam model unifying electrodynamics with the weak nuclear interactions. Similarly the strong nuclear interactions were understood in terms of a quantum field theory, Quantum ChromoDynamics (QCD). There is also a great deal of evidence that at very small length scales the two theories will be unified to one single interaction.

Where does gravity enter in this picture? It is clear from the superposition principle underlying quantum mechanics that also the gravity force at small enough length scales must be quantized. It is also easy to see from the knowledge of Newton's coupling constant, Planck's constant and the velocity of light, the three basic numerical constants that enter, that the relevant length scale is the Planck length 10^{-33} cm. Following the success of QED we could simply try to directly quantize the Einstein-Hilbert action for gravity. However, this leads immediately to problems. There is a thumb-rule in quantum physics which says that "everything that is possible happens." Based on general relativity the Einstein-Hilbert action is just the simplest of an infinity of possibilities, the rest of which give negligible contributions at macroscopic scales and are hence never considered. However, at Planck scales all terms are roughly of equal size! Quantizing the Einstein-Hilbert action will generate all these terms according to our thumb-rule and we are left with a theory in which an infinity of normalizations (of the various terms) have to be found. The first explicit loop calculation showing the need for an infinity of terms was done by Goroff and Sagnotti (1985). This seems an awkward way out of the dilemma and has been rejected by most people working in quantum gravity. However, no one has been able to circumvent this argument within ordinary quantum gravity so far. (In quantum field theoretic language the above property is referred to as the Einstein-Hilbert action being non-renormalizable.)

From our knowledge about the unification of the three other basic forces in Nature, it is natural to seek a unification also involving gravity. Supergravity, which is a natural extension of Einstein's gravity (Freedman, van Nieuwenhuizen and Ferrara 1976 and Deser and Zumino 1976), is such an attempt. It does provide a basis for unification since it can have two coupling constants and enough fundamental particles to include gravitons, photons, W- and Z-bosons and gluons as well as matter particles. Supergravity theories also carry huge symmetries, which could forbid an infinity of possible terms in the actions and hence prohibit quantum breakdown. However, no supergravity theory is able to solve this problem (Howe and Lindström 1981, Kallosh 1981), and we must content ourselves by regarding them as a step in the right direction, but not an ultimate one.

To find a theory involving gravity with a unique action, one has to go outside the whole framework that the above-mentioned theories have been based on, namely the assumption that the basic constituents are point-like objects. If we instead consider one-dimensional fundamental quanta, strings, we can seemingly solve our problems! String theory grew out of attempts to find models for the hadrons, the strongly interacting theories, but was eventually realized to be a possible candidate for a quantum theory unifying gravity with the rest of the fundamental interactions. (For a review of string theory, see the text-book by Green, Schwarz and Witten 1986.) Let me not discuss here the details of string theory (see the talk of Green), but instead concentrate on the problems ahead, what has to be done to really convince ourselves that we have indeed found the correct unified theory for all interactions.

2. PROBLEMS WHICH SHOULD BE SOLVED

2.1 Geometry

Superstring theories have so far mainly been constructed in the light-cone gauge, i.e. the formalism is non-covariant and written in a special frame. To compare with Einstein's theory we are lacking a geometric interpretation, a generalized equivalence principle. The problem has many faces. The equivalence principle is based on macroscopic observations. For superstrings we are to build a fundamental theory at truly microscopic length scales, where our human intuition is absent. What we must do is somehow to extrapolate our mathematical understanding of general relativity.

A further aspect to keep in mind is that the fundamental string interaction (as seen in the light-cone gauge) is the three-string interaction, which is quite different from point-particle theories where a graviton can split into any number of gravitons.

A huge effort has been put into the problem of finding a covariant field theory. The initial observation was made by Siegel (1984), who realized that the relevant property of covariant quantization that should be translated into the language of second-quantized field theory is the BRST invariance (Kato and Ogawa 1983, Hwang 1983). From this observation it is, in fact, quite straightforward to construct a covariant free field theory. In a string theory with only first class constraints, the first quantized theory is completely specified by demanding (Marnelius 1981)

$$Q|phys>=0, \tag{1}$$

where Q is the BRST charge constructed out of phase space operators; coordinates, momenta and ghost coordinates. It is natural to go over to a wave functional formulation by hitting the equation (1) with the bra $< \phi|$.

Now $< \phi|\ x(\sigma), \cdots, >= \phi[x(\sigma), \cdots]$ is a functional of all coordinates (including ghost ones) and in this representation the operator Q is as above with the momenta written as functional derivatives of the coordinates. The second quantized theory is then obtained by letting ϕ become a field satisfying a canonical commutation relation and a free action is constructed as

$$S = \int Dx(\sigma) \cdots \phi Q \phi \tag{2}$$

This part has been straightforward and follows exactly the procedures used for light-cone field theories. In order to construct interaction terms, however, one needs new ideas. Witten (1986a) has proposed a three-string interaction for open strings based on the mid-points playing a special role. His full action is a Chern-Simon term with Q as an exterior derivative. This action is furthermore unique and quantization cannot give rise to new local counterterms. Another approach originates from Siegel (1984)(See also Hata, Itoh, Kugo, Kunitomo and Ogawa 1986 and Neveu and West 1986),who uses the techniques of the light-cone gauge with interactions at the end-points of the string.

Witten's approach has so far not been generalized to closed strings. This is, of course, a very important issue since this sector contains the gravitons which determine the space-time geometry.

The program sketched above is also applicable to superstrings in the Ramond-Neveu-Schwarz formulation (Witten 1986b) although the technical details are more formidable here.

One may ask if this is the ultimate formulation. The Chern-Simon term is certainly a geometric quantity and perhaps one can formulate a geometric principle that underlies it. However, the BRST charge is not completely frame independent. To remedy this fact, an action with only a three-string coupling has been proposed (Friedan 1985, Witten 1986a, Yoneya 1986), which at least formally gives Witten's action (Horowitz, Lykken, Rohm, Strominger 1986) when a background solution is substituted in.

There is so far no formulation with covariant supersymmetry. One would believe that supersymmetry plays a vital role already at the level of the underlying principle, and at least I am convinced that an ultimate formulation should rest heavily upon it.

All of these problems are, of course, most vital to solve and with the effort invested in them now, we can expect rapid progress.

I must point out here again that we have implicitly assumed field theory to be the correct framework. Perhaps the fundamental description uses a different one!

2.2 Consistent Perturbation Expansion

A signal of the breakdown of the quantum theory based on the Einstein-Hilbert action is seen if one constructs perturbation expansions of scattering amplitudes. These will in general diverge and to renormalize the theory one will need infinitely many counterterms, hence making the theory void of predictions. In superstring theory the hope is that all amplitudes are finite and well-behaved. The finiteness of the one-loop graphs was established by Green and Schwarz (1982). To establish finiteness to an arbitrary order might seem horrendously difficult, but it need not be. Mandelstam (1973) has devised a method based on functional integrals over the world-sheet. Here there is just one diagram at each loop order (at least for closed strings) and the problem is transferred to understanding Green's functions on Riemann surfaces with different topology. This is a mathematical subject which is very well developed and we can expect an answer to the question of finiteness in the near future. Mandelstam uses the light-cone gauge but covariant methods based on Polyakov's (1981) approach are also considered.

It has been proven that the light-cone field theory actions are unique if one insists on local terms in the actions (Bengtsson, Brink, Cederwall and Ögren 1985). This is a very strong indication of at least renormalizability. To complete a finiteness proof along this line one needs to further prove that no non-local terms are possible. The light-cone gauge approach is based on finding representations of the super-Poincaré group. In the quantum case one really needs to check that the quantum corrections to the generators still satisfy the algebra. A breaking here signals an anomaly and would kill the model. This method has so far proved to be a complicated way to find anomalies and other methods are probably preferrable.

As said above the hope is that one can prove that each term in the perturbation expansion is finite in superstring theories. Such a result will then also prove that there are no anomalies in the theories. However, all of these results are perturbative. Each term is a very complicated function of the momenta involved. Furthermore from Regge theory we know that at high energies the loop corrections will dominate (since the leading trajectory has intercept 2). It seems a really arduous task to prove that the perturbation expansion is an asymptotic series. There are indications that for ordinary non-abelian gauge theories this is not the case, casting doubts on these theories as truly fundamental, but for string theory I think it must be the case, if we believe in them being fundamental theories.

2.3 Non-Perturbative Effects.

As mentioned in the previous subsection, the perturbation expansions albeit most probably finite order by order have an enormously rich structure. It is then natural to assume non-perturbative effects to be present. Polyakov (1986) has given one example which illuminates the problem. In any string theory there is a Kalb-Ramond field $B_{\mu\nu}$. For a four-dimensional theory such a field can have an instanton solution. Due to Debye screening the $B_{\mu\nu}$-field can acquire a non-perturbative mass and one gets some kind of string confinement. It is not clear that this will occur in a supersymmetric theory, but one must settle such issues. One must be careful when drawing conclusions based on properties of specific states. The string contains an infinite number of states and the summing over all states for a certain process can completely change the outcome of it. This is, for example, why the one-loop graphs are finite.

2.4 Compactifications to Four Dimensions.

It was shown by Candelas, Horowitz, Strominger and Witten (1985), that if one assumes a background solution of the form $M_4 \times K_6$, where M_4 is four-dimensional Minkowski space and K_6 a six-dimensional internal space, and an $N = 1$ supersymmetry, a possible solution for K_6 is Calabi-Yau spaces (Calabi 1955 and Yau 1977), which are three-dimensional complex spaces with Ricci flat Kähler metrics and holonomy group SU(3). This has led to promising scenarios for phenomenology (Witten 1985). At the classical level there seem, however, to be tens of thousands of solutions. A dream here would, of course, be to find a unique solution. However, eventually we must seek solutions to the full quantum theory and in view of what was said above, this seems also to be an arduous task. Here we probably have to content ourselves with approximate solutions, at least to start with, and it will be important to find such methods that can single out solutions.

This subject will eventually, once we understand the models better, be a huge one. It is, of course, mandatory that we find a connection between our proposed fundamental model and low-energy phenomenology.

I also like to stress that we should be open-minded enough to continue a search for four-dimensional string theories. We cannot exclude such models now although we believe we might have the correct framework in terms of ten-dimensional models.

2.5 Gravitational Collapse and Singularities.

Suppose we consider processes where the external states are massless. We can then integrate out all massive states and we are left with an action just in terms of the massless ones. This is an infinite series of terms with, for example, the Riemann-Christoffel tensor appearing to arbitrary order. The coefficients in front of these terms must be so cleverly arranged that the theory is still finite. This corresponds hence, in a sense, to the correct extension of the Einstein-Hilbert action (if superstrings passes all tests). Macroscopically, the new terms contribute negligibly as said above, but microscopically they do change the theory. This means that one has to reconsider the small-scale structure of gravity. What happens to the Schwarzschild solution; what happens to the singularity theorems and to gravitational collapse? (Boulware and Deser 1985)

Again I must stress that the fact there is an infinity of terms can circumvent general arguments based on a finite number of terms and one should preferably consider the questions above in the full string theory and not in any truncated version.

To find the complete action in terms of massless fields seems again horrendous, but it might be possible to give it in an implicit form. There are powerful ways of computing individual terms either based on studies of string amplitudes or by non-linear σ-model techniques.

2.6 Planck-Time Cosmology.

The standard model of cosmology works very well and superstring theory should not destroy this success. It will, however, have important consequences for Planck-time cosmology which so far has been unreachable with conventional models. Here a problem arises immediately. Suppose we can consider the universe to first approximation as an ideal gas of strings. Since the string has an infinity of states one can prove that this leads to a limiting temperature (Frautschi 1971). For open strings this is a true such one in the sense that there is no way of reaching this temperature. Heating up the system only produces more strings. However, for closed strings the limiting temperature rather signals a phase transition (Sundborg 1985, Bowick and Wijewardhana 1985). Heating up the system would at the critical temperature take the system over into another phase where essentially all the energy is carried by a small number of strings. The only remaining states apart from these few ultramassive ones are the massless ones. Here we should probably use a point-particle description where the massive strings define the geometry.

These considerations are very preliminary since they depend on the lowest order in a perturbation expansion. It is unlikely that higher order terms will change the picture. However, as emphasized above, we really need the non-perturbative result before we can make definite conclusions.

This is a most profound question to answer. We believe strongly that string theory is the correct framework to describe Planck-energy physics. However, we eventually must answer the question of what physics is like at higher energies. To conclude that string theory works at all energies (if it does) one must resolve the puzzle above. Perhaps there is an even more fundamental theory above Planck-energies.

3. CONCLUSIONS

The problems discussed above were the problems discussed in the mini-symposium. It shows the broad activity and vitality in the field. For the first time we have a candidate theory that could solve the quantum gravity problem and the unification problem. Perhaps we are on the way to a unique theory for Nature. However, I hope this talk has shown that the way is long and that we have a lot of work in front of us. With the efforts invested in the field, though, I am sure there will be a lot of new achievements to report on at the next GR-meeting.

This research was supported in part by the National Science Foundation under Grant No. PHY82-17853, supplemented by funds from the National Aeronautics and Space Administration, at the University of California at Santa Barbara.

REFERENCES

1. Bengtsson, A.K.H., Brink, L., Cederwall, M. and Ögren, M. (1985). Nucl. Phys. B254, 625.
2. Boulware, D.G. and Deser, S. (1985). Phys. Rev. Lett. 55, 2656.
3. Bowick, M.J. and Wijewardhana, L.C.R. (1985). Phys. Rev. Lett. 54, 2485.
4. Calabi, E. (1955). A Symposium in Honor of S. Lefschetz. In Algebraic Geometry and Topology. p. 78. Princeton: Princeton University Press.
5. Candelas, P., Horowitz, G., Strominger, A. and Witten E. (1985). Nucl. Phys. B258, 45.
6. Deser, S. and Zumino, B. (1976). Phys. Lett. 62B, 335.
7. Frautschi, S. (1971). Phys. Rev. D3, 2821.
8. Freedman, D. Z., van Nieuwenhuizen, P. and Ferrara, S. (1976). Phys. Rev. D13, 3214.
9. Friedan, D. (1985). EFI preprint.
10. Goroff, M. and Sagnotti, A. (1985). Phys. Lett. 160B, 81.
11. Green, M.B. and Schwarz, J.H. (1982). Nucl. Phys. B198, 441.
12. Green, M.B., Schwarz, J.H. and Witten, E. (1986). Superstring Theory I and II. Cambridge: Cambridge University Press.
13. Hata, H., Itoh, K., Kugo, T., Kunimoto, H. and Ogawa, K. (1986). Phys. Lett. 172B, 186, 195.
14. Horowitz, G.T., Lykken,J., Rohm, R. and Strominger, A. (1986). ITP Santa Barbara, 86-61.
15. Howe, P.S. and Lindstrōm, U. (1981). Nucl. Phys. B181, 467.
16. Hwang, S. (1983). Phys. Rev. D28, 2614.
17. Kallosh, R.E. (1981). Phys. Lett. 99B, 122.

18. Kato, M. and Ogawa, K. (1983). Nucl. Phys. B212, 443.

19. Mandelstam, S. (1973). Nucl. Phys. B64, 205.

20. Marnelius, R. (1981). Phys. Lett. 99B, 467.

21. Neveu, A. and West, P.C. (1986). Phys. Lett. 168B, 192.

22. Polyakov, A.M. (1981). Phys. Lett. 103B, 207, 211.

23. Polyakov, A.M. (1986). Lecture submitted to Second Nobel Symposium on Elementary Particles, Marstrand, Sweden.

24. Siegel, W. (1984). Phys. Lett. 151B, 391, 396.

25. Sundborg, B. (1985). Nucl. Phys. B254, 583.

26. Witten, E. (1985). Nucl. Phys. B258, 75.

27. Witten, E. (1986). Nucl. Phys. B268, 253.

28. Witten, E. (1986). Nucl. Phys. in press.

29. Yau, S.T. (1977). Proc. Natl. Acad. Sci. 74, 1798.

30. Yoneya, T. (1986). Proc. of the 7th Workshop on Grand Unification, Toyama, Japan.

TERRESTRIAL EXPERIMENTS TO TEST THEORIES OF GRAVITATION

H.J. Paik
Department of Physics and Astronomy, University of Maryland
College Park, MD 20742, U.S.A.

Abstract. The proceedings of Symposium 19, Terrestrial Experiments, are summarized. In response to the recently rekindled interest in the hypothetical "Fifth Force", about half of the contributed papers dealt with experimental tests of the Equivalence Principle and the Inverse Square Law for which violations have been reported and interpreted as evidence for the existence of a new force. Although some old experiments are being refined and many new experiments are under preparation, no new experimental data have been presented at this conference. Other experiments that were discussed in Symposium 19 include torsion balance experiments to detect effects of intrinsic spin, atomic physics experiments to test Local Lorentz Invariance, and superconductor experiments to detect gravitational and velocity effects.

INTRODUCTION

The Equivalence Principle, the Inverse Square Law and the Local Lorentz Invariance have been the long cherished fundamental principles upon which Einstein's General Relativity was constructed. There have been steady theoretical attempts in the past two decades, however, to modify the Inverse Square Law of Gravitation in the Newtonian limit. These attempts arose from at least five different schools of theoretical physics: (1) higher derivative theories of gravity (Pechlaner & Sexl 1966), (2) generalized scalar-tensor theories of gravity (Wagoner 1970; O'Hanlon 1972; Acharya & Hogan 1973), (3) spontaneous breakdown of scale invariance (Fujii 1971; Fujii 1974; Zee 1979), (4) supergravity (Deser & Zumino 1977; Scherk 1979; Fayet 1980), and (5) grand unified theory (Weinberg 1978; Wilczek 1978; Kim 1979; Moody & Wilczek 1983; Chang *et al.* 1985). Implications of these theories can be put into a common form of the modified scalar potential:

$$\phi(r) = -\frac{GM}{r}\left[1 + \alpha e^{-r/\lambda}\right], \tag{1}$$

where α and λ are the dimensionless coupling strength and the range of an additional short-range force, respectively.

The early theoretical work on the modification of General

Relativity and its implication on gravity experiments had gone largely unnoticed until 1976. At that time, Long (1976) reported an experiment in which he claimed a violation of the Inverse Square Law by (0.37 ± 0.07)% between 30 cm and 4.5 cm. This claim awakened the field to the fact that the gravitational constant G is the most poorly constrained constant in Nature and that, perhaps, an interesting new physics could be discovered by examining the law of gravitation more carefully. Empirical limits on α and λ have been examined by various authors (Long 1974; Mikkelsen & Newman 1977; Gibbons & Whiting 1981). A flurry of experimental activities to check the Inverse Square Law at laboratory distances as well as at geological-scale distances followed (Yu et al. 1979; Panov & Frontov 1979; Hirakawa et al. 1980; Spero et al. 1980; Stacey & Tuck 1981; Chan et al. 1982, Ogawa et al. 1982; Holding & Tuck 1984; Chen et al. 1984; Kuroda & Hirakawa 1985; Hoskins et al. 1985). Long's original claim was contradicted in all ranges by two higher resolution experiments (Spero et al. 1980; Chen et al. 1984). Although the validity of the Inverse Square Law has been established by these experiments to $\alpha \lesssim 10^{-4}$ at $\lambda =$ 0.1 m and $\alpha \lesssim 10^{-2}$ at $\lambda = 1$ m, noticeable exception has been provided by the mine experiments of Stacey & Tuck (1981), and Holding & Tuck (1984), in which a consistently higher value of G, corresponding to $\alpha = -(7.2 \pm 3.6) \times 10^{-3}$, was reported.

Another dimension was added to the field in 1986 when Fischbach et al. (1986 a) came out claiming that a reanalysis of the original Eötvös experiment (Eötvös et al. 1922) indicates a violation of the Equivalence Principle. In particular, Fischbach et al. have found a strong correlation between the accelerations of various materials to the Earth and their baryon numbers or hypercharges. They have further concluded that, since such an anomaly has not been observed in the experiments by Roll, Krotkov & Dicke (1964), and by Braginsky & Panov (1972), in which the accelerations of the test bodies to the sun are compared, a composition-dependent new force with a range smaller than the Earth's radius must be postulated to explain the results. This would again imply violation of the Inverse Square Law of the type given by Eq. (1), except that α is now composition-dependent. The paper by Fischbach et al. (1986 a) has drawn many criticisms (Thieberger 1986; Bizetti 1986; Milgrom 1986; Neufeld 1986; Thodberg 1986; Keyser et al. 1986; Chu & Dicke 1986; Eckhardt 1986; Kim 1986), but it has also stimulated the experimental physics community to accelerate ongoing experiments on the Equivalence

Principle and to plan various old and new versions of Eötvös and Galileo experiments. A reasonably up-to-date, but non-exhaustive, list of some 20 experiments is found in Fischbach et al. (1986 b).

In order to review the status of the "Fifth Force", we devoted one of the two oral sessions of Symposium 19 (5 papers) to the topic of the Equivalence Principle and the Inverse Square Law. The second oral session (6 papers) covered other tests of General and Special Relativity. Due to the limited time in the oral sessions, many papers covering the topics for the two oral sessions as well as others (14 papers) had to be scheduled in a poster session. As experiments to test another important basis for General Relativity, the Local Lorentz Invariance, two papers discussed the merit and approach of the new precise atomic physics tests. Also noticeable were parallel efforts by two independent groups to test the role of intrinsic spin in gravitation. The relevance of polarized body experiments in the context of various equivalence principles was previously discussed by Ni (1983).

EQUIVALENCE PRINCIPLE

Fischbach (Paper 19:07) led off the discussion on the "Fifth Force" by reviewing the status of the reanalysis of the Eötvös experiment. Three major criticisms of the paper by Fischbach et al. (1986 a) were acknowledged: (1) omission and misrepresentation of some data points in the reanalysis of the Eötvös experiment (Keyser et al. 1986; Kim 1986), (2) a sign error for the new force (Thodberg 1986; Hayaski & Shirafuji 1986), and (3) estimation of the strength of the new force without knowing local mass anomalies (Fischbach et al. 1986 b; Neufeld 1986; Thiebeger 1986; Bizetti 1986; Milgrom 1986; Eckhardt 1986). Fischbach conceded that errors of the latter two kinds had been committed in their original work. It is noteworthy that, while others argued that the dominant effect of the "Fifth Force" comes from the local horizontal mass anomalies, according to Eckhardt (1986) they are the only source that can couple to the torsion balance experiments. Interestingly, this solves two problems encountered in the original paper by Fischbach et al.: the inconsistency in the sign and the magnitude of the new force between the Eötvös experiment and the mine experiment, since the local mass anomaly near the site of the Eötvös experiment is not well known. Fischbach also addressed the first criticism on the data selection. Data points for snakewood-platinum, tallow-copper and dissolution of copper sulfate, which were left out originally because of uncertainty in the baryon-number-to-mass ratio, have now been added. They too fall right onto the straight line determined from other data points with the exception of tallow-copper which is off by about one standard deviation. Therefore, the main assertion that the original Eötvös data does show a strong correlation with the baryon number appears to have survived the criticisms. The apparent contradiction with the anomaly reported in the geophysical experiment in sign and magnitude has also vanished. So it appears that the issues raised by these experiments will be settled only by additional independent experiments with higher resolution.

Faller (Paper 19:06) reported progress on the Eötvös experiment using the fluid-suspended torsion balance and also discussed plans for a Galileo experiment in which two of his absolute gravimeters are employed. Because of the effect of the local horizontal mass anomaly, the sensitivity of the Eötvös experiment can be greatly enhanced by conducting the experiment near a cliff where the horizontal mass anomaly is a maximum (Milgrom 1986; Thieberger 1986). The Galileo experiment in which vertical accelerations of two test masses are compared, however, does not rely on the horizontal mass anomaly. Even for an ellipsoidal Earth, a hemispherical mass of an appropriate radius underneath the apparatus will exert the "Fifth Force" on the two falling objects. Faller believes that his free-fall gravimeters have enough resolution to check the claims of Fischbach et al.

Nieto (Paper 19:14) presented an alternative explanation for possible violations of the Weak Equivalence Principle and the Inverse Square Law. Quantum gravity theories would predict additional Yukawa-type interaction potentials with approximately gravitational strength. An experiment to investigate gravitational acceleration on antimatter at LEAR was discussed. A related paper by Holzscheiter et al. (Paper 19:10), presented as a poster, discussed the actual experimental scheme for the LEAR experiment. Antiprotons are extracted from the accelerator, slowed down, and detected. Gravitational acceleration on the antiparticles is computed from the time-of-flight spectrum. The experimental approach seems to be very similar to the electron and positron free-fall experiments by Fairbank et al. (1974).

Many other experiments on the Equivalence Principle are under way or under preparation as listed in Fischbach et al. (1986 b). But the six months, between the publicity on the "Fifth Force" and the Conference, was too short a time for most experimentalists. It appears that a lot of activities and results will be reported in a year or so.

INVERSE SQUARE LAW

Stacey (Paper 19:21) reviewed the status of several geophysical experiments which are designed to test the Inverse Square Law at geological-scale distances. The gravitational constant G measured in mines continues to be higher than the accepted laboratory value. A possible systematic error due to density uncertainty has been reduced in the more recent measurements in the Hilton mine (Holding & Tuck 1984) by the use of available density data, from extensive coring of the area, and knowledge of geological structure. The result stands at

$$G_{geophys} = (6.730 \pm 0.003) \times 10^{-11} m^3 kg^{-1} s^{-2}, \quad (2)$$

whereas the best laboratory value for G (Luther & Towler 1982) is

$$G_{lab} = (6.6726 \pm 0.0005) \times 10^{-11} m^3 kg^{-1} s^{-2}, \quad (3)$$

indicating

$$\alpha = -(7.2 \pm 3.6) \times 10^{-3}. \quad (4)$$

The range λ is not constrained well by the mine experiments. Stacey has also reported a tentative result from a beam balance experiment conducted in a hydro-electric lake. A preliminary analysis of this experiment gives a null result. This result, combined with limits obtained from satellite and surface geodesy, confines the value of λ for $|\alpha|$ as large as 1% to the range between 10 m and 10^3 m (or below 1 cm). The geophysical measurements of the gravitational constant are reviewed in an upcoming article by Stacey et al.(1986). Their treatment of mass profile and density uncertainty has been challenged (Kim et al. 1986).

Chan et al. (Paper 19:03) reported progress in the null test of the Inverse Square Law in which a superconducting gravity gradiometer is employed (Chan et al. 1982). Unlike Cavendish-type experiments, this experiment tests the Poisson equation for the gravitational potential, a differential form of the Inverse Square Law. In the mass-free region, the additional term in Eq. (1) modifies the Laplace equation into

$$\nabla^2\phi(r) = -\frac{GM}{r}\frac{\alpha}{\lambda^2}\exp(-\frac{r}{\lambda}). \qquad (5)$$

By using a localized source with a certain symmetry or an extended source with known geometry and density such as water, one could perform a double null experiment: a null source and a null detector. Suppression of many error sources and high sensitivity expected in the superconducting detector may yield resolution in excess of 10^{-4} in the desired range, 10 m to 10^3 m. Unfortunately, it is difficult to expect an early result from this experiment inasmuch as the gradiometer instrument is still at the development stage.

Several other papers suggesting various types of the Inverse Square Law test were scheduled in the poster session. Piso et al. (Paper 19:15) discussed a dynamic gravity experiment, similar to the experiment of Ogawa et al. (1982), in which gravitational interaction between a rotating generator and a mass quadrupole detector is investigated. Shoemaker & Robel (Paper 19:19) proposed another dynamic test of Newton's law by use of a dynamic beam balance. Simaciu (Paper 19:20) suggested an experiment in which both the attracting mass (source) and the pendulum (detector) are put in the same vacuum to eliminate effects of molecular screening forces from the vacuum enclosure. Zaitsev & Kolosnitsyn (Paper 19:24) suggested another dynamical test of the Inverse Square Law in which a laser interferometer is employed to detect the dynamic gravitational field of a rotating dumbbell. Belyaletdinov & Frontov (Paper 19:25) proposed a torsion pendulum experiment in which a resonance frequency shift due to gravitational force is explored to obtain dG/dr at small distances. Long presented two papers: one (Paper 19:27) reporting progress in the refinement of his early experiment (Long 1976), and the other (Paper 19:28) discussing feasibility of a gravimeter experiment near a navigation lock on a river. It appears that the next few years will bring the excitement of reaping the results of the hard work on the part of these experimentalists.

OTHER EXPERIMENTS

Although this year's momentum and enthusiasm were definitely in the areas of the Equivalence Principle and the Inverse Square Law, interest in many other tests of General Relativity and space-time structure was evident. Three papers represented work on the polarized-body torsion balance experiments. Ni (Paper 19:13) discussed the motivation and theoretical aspects of such experiments in the light of their implication on the Einstein Equivalence Principle and the long-range spin-spin interactions proposed in the particle physics community. Ritter & Ni (Paper 19:18) presented an accompanying paper discussing an experimental approach to search for anomalous intrinsic spin. Graham & Newman (Paper 19:08) reported initial results from a cryogenic torsion balance experiment. An anomalous torque is inferred from the observed angular deflection of a hanging spin-polarized sample caused by reversing the spin polarization of the fixed sample. A superconducting shield attenuates the magnetic field by a factor of 10^9. The measured anomalous torque is less than a few parts in 10^{10} of the magnetic torque expected between unshielded samples.

On another front, improved tests of the Local Lorentz Invariance is sought by new precise atomic physics tests. Haugan (Paper 19:09) argued that, when a right theoretical framework is used, atomic and nuclear systems can provide the most precise tests of the Local Lorentz Invariance. Chupp (Paper 19:04) proposed one such test using polarized ^{3}He and ^{21}Ne. By taking advantage of the very long relaxation times of those polarized atoms (2 and 24 hours, respectively), an improvement of 10^6 over the original experiments of Hughes et al. (1960) and Drever (1961) is expected. Although Local Lorentz Invariance and the space-time metric in Special Relativity are widely accepted concepts in modern physics, a number of papers challenged various notions of Special Relativity. Torr & Chang (Paper 19:23) claimed experimental verification of ten-dimensional space. Recami (Paper 19:16) argued that muon neutrinos are tachyons traveling faster than light. Marinov (Paper 19:11) claimed that he had experimentally proved violations of the principles of relativity, equivalence, and energy conservation. (If this last paper is right, we must have been living in the wrong universe!)

Two papers concerned use of microwave cavities in the relativity experiments. Melissinos (Paper 19:12) reported results of a search for anomalous long-range interactions at highly relativistic velocities. A stationary detector consisting of a superconducting microwave cavity was placed near the path of the circulating beam of 800 GeV protons at FERMILAB. Unfortunately, inadequate shielding of the electromagnetic interaction resulted in an observed signal much stronger than gravitation.

This experiment puts an upper limit on the coupling constant of a possible velocity-dependent new long range force at 10^{19} times G, the Newtonian gravitational constant. Nazarov et al. (Paper 19:26) proposed use of high-Q dielectric (sapphire) microwave cavities as detectors of angular velocity. The operating principle of the device appears to be similar to that of a ring laser gyro in which a phase shift between two counter propagating electromagnetic waves is measured to detect the angular velocity of the platform. Dobrinic et al. (Paper 19:05) discussed measurement of the transverse Doppler effect in accelerated systems.

Last but not least. Anandan (Paper 19:01) discussed relativistic gravitational effects on superconductors. He was able to show rigorously (Anandan 1984) that the phase of the order parameter in a superconductor is red-shifted in a gravitational field just as light waves. This may allow a test of the Equivalence Principle in a new physical system although the resolution of such an experiment will be much poorer than those of other experiments.

CONCLUSION

It is clear that the recent announcement of the possible discovery of a "Fifth Force" (Fischbach et al. 1986 a) stimulated great interest on the part of experimentalists to resume, improve and accelerate old experiments, as well as to plan new experiments. After the storm of criticisms, the essential claim of Fischbach et al. that the original Eötvös data show a strong correlation with chemical composition seems to be intact. Whether this represents a new physics or is an artifact of statistical fluctuation, only time will tell. It is hard to understand, however, how this experiment has been interpreted as a null test of the Equivalence Principle and has been accepted so widely and so long by the physics community as a corner stone for General Relativity.

Even if the argument of hypercharge coupling does not hold in the face of new evidence provided by future experiments, an important suggestion has been made by Fischbach et al. in that the Eötvös and Galileo experiments can be used to explore new intermediate-range forces. It is useful to note, however, that only composition-dependent forces can be detected in such experiments because of the differential nature of measurement. In addition, small differences in the baryon-number-to-mass ratios between materials tend to attenuate the signal arising from the new force by a factor 10^3. Direct tests of the Inverse Square Law, which employ single test masses, do not encounter these problems. In such experiments, the density inhomogeneity of the source is a more important error source. Highest resolution experiments may therefore require a null detector, such as the Laplacian detector (Paik 1979), or a null source, such as the ocean or lake water (Stacey 1978).

It appears that we are entering into the most dynamic period in the history of Experimental Relativity. Technology is maturing. Theoretical advances continue to present challenges. Many of the ongoing and planned experiments will yield results in the next few years. We are therefore looking forward to GR 12 with great expectations.

REFERENCES

Acharya, R. & Hogan, P.A. (1973). Nuovo Cimento Lett., 6, 668

Anandan, J. (1984). Phys. Lett., 105A, 280.

Bizetti, A. (1986). Nuovo Cimento, 94B, 80.

Braginsky, V.B. & Panov, V.I. (1972). Sov. Phys. JETP, 34, 463.

Chan, H.A., Moody, M.V. & Paik, H.J. (1982). Phys. Rev. Lett., 49, 1745.

Chang, D., Mohapatra, R.N. & Nussinov, S. (1985). Phys. Rev. Lett., 55, 2835.

Chen, Y.T., Cook, A.H. & Metherell, A.J.F. (1984). Proc. R. Soc. London, A394, 47.

Chu, S.Y. & Dicke, R.H. (1986). Preprint.

Deser, S. & Zumino, B. (1977). Phys. Rev. Lett., 38, 1433.

Drever, R.W.P. (1961). Phil. Mag., 6, 683.

Eckhardt, D.H., (1986). Phys. Rev. Lett., 57, 2868.

Eötvös, R.V., Pekár, D. & Fekete, E. (1922). Ann. Phys. (Leipzig), 68, 11.

Fairbank, W.M., Witteborn, F.C., Madey, J.M.J. and Lockhart, J.M. (1974). In Experimental Gravitation, ed. B. Bertotti, p. 310, Academic, New York.

Fayet, P. (1980). Phys. Lett., 95B, 285.

Fischbach, E., Sudarsky, D., Szafer, A., Talmadge, C. & Aronson, S.H. (1986 a). Phys Rev. Lett., 56, 3.

Fischbach, E., Sudarsky, D., Szafer, A., Talmadge, C. & Aronson, S.H. (1986 b). In Progress in Electroweak Interactions, ed. J. Tran Thahn Van, Vol. 1, p. 229, Editions Frontieres, Gif Sur Yvette, France.

Fujii, Y. (1971). Nature Phys. Sci., 234, 5.

Fujii, Y. (1974). Phys. Rev., D9, 874.

Gibbons, G.W. & Whiting, B.F. (1981). Nature, 291, 636.

Hayashi, H. & Shirafuji, T. (1986). Kitasato University Report No. 86-2.

Hirakawa, H., Tsubono, K. & Oide, K. (1980). Nature, 283, 184.

Holding, S.C. & Tuck, G.J. (1984). Nature, 307, 714.

Hoskins, J.K., Newman, R.D., Spero, R. & Schultz, J. (1985). Phys. Rev., D32, 3084.

Hughes, V.W., Robinson, H.G. & Beltran-Lopez, V. (1960). Phys. Rev. Lett., 4, 342.

Keyser, P.T., Niebauer, T. & Faller, J.E. (1986). Phys. Rev. Lett., 56, 2425.

Kim. J.E. (1979). Phys. Rev. Lett., 43, 103.

Kim, Y.E. (1986). Phys. Lett., B177, 225.

Kim. Y.E., Klepacki, D.J. & Hinze, W.J. (1986). Preprint. To be published in Nature.
Kuroda, K. & Hirakawa, H. (1985). Phys. Rev., D32, 342.
Long, D. (1974). Phys. Rev., D9, 850.
Long, D. (1976). Nature, 260, 417.
Luther, G.G. & Towler, W.R. (1982). Phys. Rev. Lett., 48, 121.
Mikkelsen, D.R. & Newman, M.J. (1977). Phys. Rev., D16, 919.
Milgrom, M. (1986). Nucl. Phys., B277, 509.
Moody, J. & Wilczek, F. (1984). Phys. Rev., D30, 130.
Neufeld, D.A. (1986). Phys. Rev. Lett., 56, 2344.
Ni, W.T. (1983). In Proc. 1983 International School and Symposium on Precision Measurement and Gravity Experiment, ed. W.T. Ni, p. 531, National Tsing Hua Univeristy, Taiwan.
Ogawa, Y., Tsubono, K. & Hirakawa, H. (1982). Phys. Rev., D26, 729.
O'Hanlon, J. (1972). Phys. Rev. Lett., 29, 137.
Paik, H.J. (1979). Phys. Rev., D19, 2320.
Panov, V.I. & Frontov, V.N. (1979). Sov. Phys. JETP, 50, 852.
Pechlaner, E. & Sexl, R. (1966). Commun. Math. Phys., 2, 165.
Roll, P.G., Krotkov, R. & Dicke, R.H. (1964). Ann. Phys. (N.Y.), 26, 442.
Scherk, J. (1979). Phys. Lett., 88B, 265.
Spero, R., Hoskins, J.K., Newman, R.D., Pellam, J. & Schultz, J. (1980). Phys. Rev. Lett., 44, 1645.
Stacey, F.D. (1978). Geophys. Res. Lett., 5, 377.
Stacey, F.D. & Tuck, G.J. (1981). Nature, 292, 230.
Stacey, F.D., Tuck, G.J., Moore, G.I., Holding, S.C., Goodwin, B.D. & Zhou, R. (1986). Preprint. To be published in Rev. Mod. Phys.
Thieberger, P. (1986). Phys. Rev. Lett., 56, 2347.
Thodberg, H.H. (1986). Phys. Rev. Lett., 56, 2423.
Wagoner, R.V. (1970). Phys. Rev. D1, 3209.
Weinberg, S. (1978). Phys. Rev. Lett., 40, 233.
Wilczek, F. (1978). Phys. Rev. Lett., 40, 279.
Yu, H.T., Ni, W.T., Hu, C.C., Liu, F.H., Yang, C.H. & Liu, W.N. (1979). Phys. Rev., D20, 1813.
Zee, A. (1979). Phys. Rev. Lett., 42, 417.

SYMPOSIUM 16: SPINORS, TWISTORS AND COMPLEX METHODS

K.P. Tod
Mathematical Institute, Oxford

Seven abstracts were selected for presentation in this session. These were

C.R. Lebrun — *Ambi-twistors and Einstein's Equations*

L.J. Mason — *Twistor Theory and the Einstein Vacuum Equations*

R.J. Baston and L.J. Mason — *Conformal Gravity, Einstein Spaces and Twistors*

B.P. Jeffryes — *The Newtonian Limit of Penrose's Quasi-Local Mass*

E.T. Newman and S. Chakravarty — *The Riemann-Hilbert Problem, Twistors and the Reduction of Self-dual Gauge Theories to Quadratures*

H.J. de Vega — *Inverse Scattering Approach to Self-dual Yang-Mills Fields and Spectral Twistors*

P. Topiwala — *A Twistor Approach to the Einstein Metric on K3.*

All seven abstracts deal with twistors. In the event, C.R. LeBrun was unfortunately unable to attend the conference and some account of his work was included by R.J. Baston in the second talk.

The first three abstracts are concerned with solving the "real" Einstein vacuum equations i.e. with finding real, Lorentz-signature four-dimensional Ricci-flat manifolds. They provide three separate characterisations in twistor terms of the vacuum equations.

In order to summarize them, I begin by recalling an earlier theorem of Lebrun (1980, 1983): given a space-time $\mathcal{M}$ with metric which can be either real-analytic, four-dimensional and Lorentzian, in which case you may complexify it at once, or complex four-dimensional ab initio, let $\mathcal{N}$ be the space of (unscaled) null geodesics in $\mathcal{M}$; then $\mathcal{N}$ is naturally a five-dimensional complex manifold and it is possible to recover $\mathcal{M}$ and the conformal structure on $\mathcal{M}$ just from $\mathcal{N}$. (Note that since the concept of unscaled null geodesic is conformally invariant, $\mathcal{N}$ could not be expected to fix the scale of the metric on $\mathcal{M}$.)

To prove this, one begins by observing that a point p in $\mathcal{M}$ is represented by its light cone N_p in $\mathcal{N}$, that is by the set of null geodesics through p, which define a complex submanifold in $\mathcal{N}$. The submanifolds N_p can be recognized in $\mathcal{N}$ just by their complex analytic properties, which enables $\mathcal{M}$ to be recovered from $\mathcal{N}$. Infinitesimal considerations then lead to recovering the conformal metric.

In his abstract to this session, *Lebrun* showed what extra structure on $\mathcal{N}$ was equivalent to the Einstein equations on $\mathcal{M}$, or more accurately, was equivalent to the possibility of choosing a metric in the conformal class of metrics determined by $\mathcal{N}$ such that the Einstein vacuum equations were satisfied by this metric. Given $\mathcal{N}$ and the corresponding $\mathcal{M}$, he chooses a representative metric in the conformal class. The aim is to define a rank-two vector bundle E over $\mathcal{N}$, and the fibre above a particular null geodesic $\gamma \in \mathcal{N}$ is defined by using parallel propagation along γ in $\mathcal{M}$ with the chosen representative metric. The details have appeared in (Lebrun 1985). The conclusion is that the conformal scale of $\mathcal{M}$ may be chosen such that $\mathcal{M}$ is vacuum iff E has a global section.

As I remarked above, this is a *characterisation* of the vacuum equations but without, for example, some independent way of recognising E, which Lebrun called the *Einstein bundle*, one is no nearer solving the vacuum equations.

The work of *R.J. Baston* and *L.J. Mason* (1985), which draws on earlier work of *M.G. Eastwood* and *C.R. Lebrun*, and was described by Baston in the second talk, also involves $\mathcal{N}$. When $\mathcal{M}$ is flat or conformally flat, $\mathcal{N}$ is naturally the diagonal in the product of projective twistor space and dual projective twistor space, $\mathbb{PT} \times \mathbb{PT}^*$ (for this and other twistor matters see e.g. Penrose and Rindler 1985, 1986 or Huggett and Tod 1985). When $\mathcal{M}$ has conformal curvature, the idea is to mimic this embedding as an expansion off of $\mathcal{N}$ to various powers in a normal coordinate. (Something similar was carried through by Isenberg, Yasskin and Green (1978) and by Witten (1978) to characterise the real Yang-Mills equations in flat space).

The conclusion is that, if the extension is possible to *fifth* order, then $\mathfrak{M}$ has vanishing Bach tensor and if the extension is possible to *sixth* order, a further conformally invariant tensor on $\mathfrak{M}$ vanishes. Provided the Weyl tensor of $\mathfrak{M}$ has enough non-zero invariants, these conditions in turn imply that $\mathfrak{M}$ is conformal to an Einstein space.

The work of *Mason* (1985), which he described in the first talk, involves *hypersurface twistors*. These are twistors defined relative to a choice of hypersurface Σ in a space-time and the aim is to characterise the Cauchy problem with data on Σ in terms of hypersurface twistors for Σ.

Rather than define hypersurface twistors directly (for which see e.g. Penrose and Rindler 1986) I shall describe them via another observation of Lebrun's (1980): given a hypersurface Σ with first and second fundamental forms, there is an evolution of Σ, unique up to conformal rescalings, into a (complex) space-time $\mathfrak{M}$ with anti-self-dual Weyl curvature. This $\mathfrak{M}$ will contain *α-planes*, that is totally null two-planes with self-dual normal bivector, and the intersection of an α-plane with Σ is a hypersurface twistor. Naturally, they can be identified just at Σ with a knowledge of the first and second fundamental forms of Σ, and they turn out to define a complex three-dimensional manifold, $\mathfrak{J}(\Sigma)$.

It has long been thought that evolution of Σ according to the vacuum equations could be given an elegant transcription into evolution of the complex structure of $\mathfrak{J}(\Sigma)$ (see e.g. Penrose 1975). This is what Mason has done. He introduces two holomorphic two-forms on $\mathfrak{J}(\Sigma)$: one of them is holomorphically closed iff the vacuum constraint equations are satisfied at Σ; the other contains the information of the location of the hypersurface in space-time and the "lapse function"; and both are involved in the evolution.

In his plenary lecture, Roger Penrose reminded us about his quasi-local mass construction (Penrose 1982). This construction is tailored to be correct in linear theory i.e. in the limit as $G \to 0$. While it has some problems for arbitrary two-surfaces in arbitrary space-time, it has given a number of satisfactory answers in particular cases.

In his talk to this session, *B.P. Jeffryes* described his calculation (Jeffryes 1986) of quasi-local mass in another limit, the Newtonian limit as $c \to \infty$. He uses the formalism of Futamase and Schutz (1983) to deal with perfect fluid space-times with a Newtonian limit and considers a two-surface S which, in the limit, lies in a three-surface Σ of constant Newtonian time. The Penrose mass M_P associated with S, which is supposed to be a measure of the total mass-energy within S, becomes in the limit the (conserved) total mass M_T. However, keeping the next term in the expansion in the parameter ϵ of Futamase and Schutz

$$M_P = M_T + E\epsilon^4 + \text{higher order}$$

Jeffryes finds

$$E = \int_\Sigma \rho(\tfrac{1}{2} v_c^2 + \underline{r}.\nabla U - \Pi)dV$$

where ρ is the fluid density

$\underline{v}_c$ is the fluid 3-velocity w.r.t. the centre of momentum within S

$\underline{r}$ is position w.r.t. the centre of mass within S

$-\Pi$ is the specific internal energy of the fluid

U is the Newtonian gravitational potential.

Note that E involves the gravitational field ∇U rather than just the potential, although if S surrounds all the sources of a bounded system then the divergence theorem implies

$$E = \int_\Sigma \rho(\tfrac{1}{2} v_c^2 - \tfrac{1}{2}U - \Pi)dV.$$

Jeffryes also considers the additivity properties of this energy. Specifically, given two sources A and B in relative motion he finds

$$E(A+B) - E(A) - E(B) = \frac{M_A M_B}{M_A + M_B}(\tfrac{1}{2} V^2 - gh)$$

where E(A), E(B), E(A+B) are the energies associated with two-surfaces surrounding respectively source A, source B and both sources,

M_A, M_B are the respective masses

V is the relative velocity of the centres of mass

h is the separation of the centres of mass

g is the relative acceleration of the centres of mass

These expressions are a pleasing mixture of one's Newtonian expectations with non-local or quasi-local features.

The remaining three abstracts deal with applications of twistors to the solution of self-dual field equations.

Ted Newman described the first of these. With various co-workers, most recently S. Chakravarty, he has reduced the characteristic initial value problem for self-dual Yang-Mills fields when data is given on $\mathscr{I}^+$ to quadratures plus a Riccati equation. The data consist of a matrix $A(u,\zeta,\bar{\zeta})$ of functions on $\mathscr{I}^+$, the matrix to lie in the Lie algebra of the gauge group, where u is the usual coordinate along the generators of $\mathscr{I}^+$ and ζ is the usual stereographic coordinate on the space of generators.

This matrix is the $O(r^{-1})$ part of one component of the Yang-Mills potential $\gamma_a(x^b)$ where x^b are coordinates on Minkowski space $\mathbb{M}$. To recover the potential from A, one first restricts A to the intersection with $\mathscr{I}^+$ of the future light cone of a particular point $p \in \mathbb{M}$. This entails setting

$$u = x^a \ell_a(\zeta,\bar{\zeta})$$

in A where $\ell^a(\zeta,\bar{\zeta})$ is the usual parametrisation of future-pointing null vectors by the stereographic coordinate ζ and x^a are the coordinates of p. Now one must solve what Newman calls the *Sparling equation*

$$\eth G = -GA$$

for a square matrix $G(x^a,\zeta,\bar{\zeta})$, regular on the Riemann sphere of ζ.

From G one recovers $\gamma_a(x^b)$ by a simple process of differentiation and manipulation (Newman 1980, Kozameh and Newman 1985). The hard part in this process is solving the Sparling equation. Newman and Chakravarty first observe that this is actually easy and indeed just a matter of quadratures if A is upper (or lower) triangular. They then consider the gauge freedom in the Sparling equation. This is of the form

$$G \to G' = Gg$$
$$A \to A' = g^{-1}(Ag - \eth g)$$

for suitable g. Now they observe that there is sufficient freedom here to transform A to triangular form, at least for the group $GL(2,\mathbb{C})$.

What they find is a Riccati equation for a ratio of components in the transforming matrix g. Using existence theorems for the Riccati equation they are then able to assert the existence of a suitable g.

Newman illustrated the technique with the example which leads to (the complexification of) the single instanton solution. He also observed that, given the connections which R.S. Ward has established between many different integrable systems and the self-dual Yang-Mills equations (Ward 1985), these techniques could be of wide application in integrable systems.

H.J. de Vega described work of himself and collaborators on the *Inverse Scattering Approach to Self-dual Yang-Mills Fields* (Avan, de Vega and Maillet 1986, Avan and de Vega 1986). We recall that the Ward construction for self-dual Yang-Mills fields (Ward 1977) begins with a consideration of the equation

$$\pi^{A'} D_{AA'}\Psi = 0 \qquad (1)$$

where $\pi^{A'}$ is a constant spinor field on Minkowski space and $D_{AA'}$ is the Yang-Mills covariant derivative. In the Ward construction, the solutions of (1) on a particular α-plane define the fibre over the corresponding twistor of a vector bundle over twistor space. The integrability condition for (1) is the self-duality condition on the cconnection, and this observation is also the starting point for the Belavin-Zakharov (1978) inverse scattering approach.

De Vega and his associates remark that (1) does not have an invariant form under the conformal group. This is of no concern for the Ward construction but affects the inverse scattering discussion. They therefore propose a different linear system to replace (1). I have taken the liberty of translating this into 2-component spinors when I believe it to be

$$\Omega^{A'} \nabla_{AA'}\Psi = 0 \qquad (2)$$

where now $\Omega^{A'}$ is a (dual) twistor i.e. a spinor field satisfying

$$\nabla^{A(A'}\Omega^{B')} = 0.$$

The integrability condition is as before.

Given a square matrix $\Psi(x;W_\alpha)$ of solutions to (2), parametrised by the dual twistor W_α, de Vega constructs what he calls "infinitesimal Bäcklund transformations" for new solutions near to the given one, or equivalently, solutions of the self-dual Yang-Mills equations linearised about the given solution. In a 2-component notation again these seem to be

$$\delta A_{AA'} = \Phi_{A'}{}^{B'} Q_{AB'}$$

where $\Phi_{A'B'}$ is an arbitrary constant spinor, and Q_a is a pure-gauge potential constructed from Ψ. He goes on to find more complicated Backlund transformations and to investigate the algebra which they generate.

It has been known for some time (see e.g. Hitchin 1975) that the only simply-connected real four-dimensional compact manifold which could possibly carry a Riemannian metric with self-dual Riemann tensor (and therefore with vanishing Ricci tensor) was the so-called *K3 surface*. This is actually a complex two-dimensional manifold and while there are many different complex structures on a K3 surface, the underlying real manifold is the same in every case. The simplest example is a non-singular complex surface of degree 4 in $\mathbb{C}\,\mathbb{P}^3$.

Another example, relevant below, may be described as follows: take a two-dimensional complex torus T^2, i.e. $\mathbb{C}^2$ modulo the identifications

$$(z_1,z_2) \sim (z_1+m_1\alpha_1+n_1\beta_1, z_2+m_2\alpha_2+n_2\beta_2) \text{ for } \forall\, m_i,n_i \in \mathbb{Z}$$

and some fixed choice of α_i, $\beta_i \in \mathbb{C}$;
now identify

$$(z_1,z_2) \sim (-z_1,-z_2);$$

finally, resolve the $2^4 = 16$ singular points which arise from the antipodal identifications.

It is a consequence of Yau's proof of the Calabi conjecture (Yau 1977) that the K3 surface actually *does* carry a self-dual metric. This means that there must be a twistor space for K3 from which the self-dual metric can be obtained by Penrose's non-linear graviton construction (Penrose 1976).

In his talk to this session, *P. Topiwala* described his proof that the K3 surface carries a self-dual metric which proceeds by directly proving the existence of the twistor space. We recall that, for these purposes, a twistor space is a complex 3-manifold $\mathcal{J}$ with holomorphic projections p : $\mathcal{J} \to \mathcal{M}$ to a complex 2-manifold $\mathcal{M}$ and $\Pi : \mathcal{J} \to \mathbb{C}\mathbb{P}^1$. The fibres of p are copies of $\mathbb{C}\mathbb{P}^1$ and the fibres of Π are copies of $\mathcal{M}$. $\mathcal{J}$ needs some other structure which I won't dwell on except to remark that as a real 4-manifold $\mathcal{J}$ is diffeomorphic to $\mathcal{M} \times S^2$. When $\mathcal{M}$ is $\mathbb{C}^2$, $\mathcal{J}$ is $\mathbb{C}\mathbb{P}^3$ with a $\mathbb{C}\mathbb{P}^1$ removed. The aim here is to prove that a suitable $\mathcal{J}$ exists when $\mathcal{M}$ is K3. Topiwala uses the second model of a K3 surface described above. The various identifications on $\mathbb{C}^2$ lead to similar identifications on the twistor space for $\mathbb{C}^2$. Thus much was anticipated some while ago (Hitchin 1984). The subtlety,which allows the construction to go through, is in the treatment of the fibres of $\mathcal{J}$ above the singular points in the identified T^2, which involves *relative deformation theory*.

One ends up with an existence theorem for some self-dual metrics on K3, which one had already with the work of Yau. The extra that one has is the novelty of approach, which could presumably be adapted to give conformally self-dual metrics on other manifolds, and some local information about the metric. In particular, this confirms Page's picture of the K3 metric as a superposition of 16 Eguchi-Hansen metrics (Page 1978).

There were five posters in the associated poster session that I wish to mention.

The first, by *H. Pedersen*, described his paper *Einstein Metrics, Spinning Top Motions and Monopoles* (1986). This is based on another construction of Lebrun (1982): given a real analytic 3-manifold $\mathcal{S}$ with conformal structure there exists a unique real 4-manifold M with Einsteinian metric and cosmological constant -1 (i.e. $R_{ab} = -g_{ab}$) which has $\mathcal{S}$ as its conformal infinity. The simplest example takes $\mathcal{S}$ to be S^3 with the usual metric when M turns out to be $\mathbf{H}^4$, the hyperbolic space. The procedure is to take the space Z of null geodesics in the complexification of $\mathcal{S}$ (Z is actually a twistor space). Then the points of M can be recognised as certain curves in Z.

Pedersen applies this construction to the left-invariant metric

$$ds^2 = \sigma_1^2 + \sigma_2^2 + \lambda^2\sigma_3^2$$

on S^3, where $\lambda = 1$ gives the usual metric. The space of null geodesics, following Arnold (1978), may be regarded as the motions of a spinning top with moments of inertia $(1,1,\lambda)$ but with *zero* energy (because null). M is still hard to find but can be guessed, and the answer checked. It corresponds to a monopole on S^3 (with the usual metric) in much the same way as the self-dual Taub-NUT metric corresponds to a monopole on $\mathbb{R}^3$. M is complete and diffeomorphic to a ball, with the deformed sphere at infinity.

There were two posters concerned with HH-spaces or "hyper-heavens". These are complex 4-manifolds with metric having one Weyl spinor, say $\tilde{\psi}_{A'B'C'D'}$ algebraically special but with no restriction on the other, ψ_{ABCD} and with vanishing Ricci tensor. They were defined by Plebanski and Robinson (1977) who succeeded in reducing the field equations to a single-non-linear p.d.e. on a single potential. *J.D. Finley* in his abstract *Algebraically degenerate HH-spaces with real Petrov type* considered the problem of finding the complexification of real Lorentz-signature algebraically special space-times among the HH-spaces by further restricting the potential. He reduces the problem to a system of p.d.e.'s.

K. Rozga in his abstract *Right-conformally equivalent HH-spaces* considers the question of when the relation

$$g_1 = g_2 + h$$

can hold between metrics g_1, g_2 of HH-spaces M_1 and M_2, when h also has a particular form generalising the Kerr-Schild form. He also considers the equivalence relation where M_1 and M_2 have the same anti-self-dual Weyl tensor.

Finally there were two related posters *World spinors for gravity and superstrings* by *Dj. Sijacki* and *Anholonomic affine spinors and holonomic world spinors* by *Y. Ne'eman.* These were concerned with constructing spinors for SL(n, $\mathbb{R}$) and exploiting them in affine and Einsteinian gravity, superstrings and the classification of hadrons.

V.I. Arnol'd (1978) Appendix 2 of *Mathematical Methods of Classical Mechanics* New York: Springer-Verlag

J. Avan, H.J. de Vega and J.M. Maillet (1986) Phys. Lett. B171 255

J. Avan and H.J. de Vega (1986) *Conformally covariant approach to the Integrability of SDYM*, preprint, Laboratoire de Physique Theorique et Hautes Energies, Paris

R.J. Baston and L.J. Mason (1985) Twistor Newsletter 20 47

A.A. Belavin and V.S. Zakharov (1978) Phys. Lett B73 53

T. Futamase and B. Schutz (1983) Phys. Rev. D20 10

N.J. Hitchin (1975) J. Diff. Geom. 9 435

N.J. Hitchin (1984) in *Global Riemannian Geometry* ed. T.J. Willmore and N. Hitchin, Chichester: Ellis Horwood Ltd.

S.A. Huggett and K.P. Tod (1985) *An Introduction to Twistor Theory* LMS Student Text, Cambridge; C.U.P.

J. Isenberg, P.B. Yasskin and P.S. Green (1978) Phys. Lett. 78B 462

B.P. Jeffryes (1986) Class. Quant. Grav. 3 841

C. Kozameh and E.T. Newman (1985) Found. Phys. 15 487

C.R. Lebrun (1980) D.Phil. thesis, Oxford

(1982) Proc. Roy. Soc. Lond. A380 171

(1983) Trans. Am. Math. Soc. 278 290

(1985) Class. Quant. Grav. 2 555

L.J. Mason (1985) Twistor Newsletter 21 49

E.T. Newman (1980) Phys. Rev. D22 3023

D. Page (1978) Phys. Lett. B80 55

H. Pedersen (1986) Math. Ann. 274 35

R. Penrose (1985) in *Quantum Gravity: An Oxford Symposium* ed. C.J. Isham, R. Penrose and D.W. Sciama, Oxford; O.U.P.

R. Penrose (1976) Gen. Rel. Grav. 7 31

R. Penrose (1982) Proc. Roy. Soc. Lond. A381 35

R. Penrose and W. Rindler (1985) *Spinors and Space-time: Volume 1* CUP

(1985) *Spinors and Space-time: Volume 2* CUP

J.F. Plebanski and I. Robinson (1977) in *Symposium on Asymptotic Structure of Space-time* ed. F.P. Esposito and L. Witten,

New York: Plenum Press

R.S. Ward (1977) Phys. Lett. A61 81

(1985) Phil. Trans. R. Soc. Lond. A315 451

E. Witten (1978) Phys. Lett. 77B 394

S.T. Yau (1977) Proc. Nat. Acad. Sci. USA 74 1798